New trends in the utilization of educational technology for science education

New trends in the utilization of educational technology for science education

The Unesco Press Paris 1974

Published by
The Unesco Press
Place de Fontenoy, 75700 Paris
Printed by Snoeck-Ducaju & Zoon, Gand

ISBN 92-3-101143-X

© Unesco 1974

Printed in Belgium

Preface

This volume explores current trends in the utilization of educational technology to improve the teaching and learning of science at all levels. In preparation for the work, a survey was carried out of recent literature in the field, so as to determine the most significant developments. A series of state-of-the-art papers was then prepared, on topics and by authors chosen on the basis of the survey. Such overlapping as occurs between the eight areas selected for these papers has the advantage of providing an opportunity to compare different points of view. The purpose of the papers was to present a comprehensive and balanced analysis of the area in question: activities under way, results obtained, major problems, future needs, etc.

A meeting of specialists in the field was subsequently convened jointly by Unesco and the Committee on Science Teaching of ICSU (International Council of Scientific Unions) from 13 to 16 September 1972 when the eight papers were discussed by their authors together with another nine specialists, as well as representatives of ICSU and Unesco.

In the light of these discussions, the authors undertook to revise their papers and prepare a final version for the present volume. It was also unanimously agreed that two more contributions should be added: a paper focusing specifically on the needs of the developing countries, and an article which would convey to the reader, in a non-specialized language, the highlights of the meeting and the spirit in which the discussions took place.

The book is directed to all those persons who have an interest in or direct responsibility for improving science education at any level through the utilization of more effective methods of teaching and learning. Its purpose is to assist them in their work, by giving them access to information and ideas regarding ways in which educational technology is being used for better education in science.

Unesco wishes to express its appreciation to all those who have helped in the publication of this book, especially the authors of the papers and the other participants in the preliminary meeting, as well as the Committee on Science Teaching of ICSU, co-sponsors of the above-mentioned meeting.

The views expressed in the various papers are the responsibility of their authors, and do not necessarily reflect those of Unesco.

Contents

When the media are not the message
(a layman's impressions of the meeting), Daniel Behrman
11

Computer-based science education, Donald L. Bitzer, Bruce Arne Sherwood
and Paul Tenczar

Summary 18
Introduction 18
Lesson examples 19
Problems 25
Solutions 26
Suggested reading 29
References 30
Appendixes 31

Programmed learning in science education, George O. M. Leith

Summary 34
Introduction 34
Models of programmed instruction 35
Leading concepts in programmed instruction 37
Group-paced programmed instruction 41
Co-operative learning 42
Learning and individual differences 43
Programmed learning and science education in developing countries 45
Generalizations 46
References 48

The use of television in science teaching, J. Valérien

Summary 54
Introduction 54
Current trends in science teaching 55
Television and in-school education 56
Television and further education 60
Conclusion: Promoting the use of television in science teaching 63
Bibliography 64

The use of radio in science education, John C. H. Ball

Summary   70
Introduction   70
Radio itself   71
Radio reception   72
Radio production techniques   73
Radio science education for general listening   75
Radio science education for the classroom   76
Radio and printed support material—teachers   79
Radio and printed support material—pupils   81
Radiovision for science education   83
Radio and the tape-recorder   84
Radio and mathematics   86
Radio and science correspondence courses   87
Some conclusions   88
References   89
Appendixes   90

Learning media: theory, selection and utilization in science education, Arthur I. Berman

Summary   102
Media classification schemes   102
Media subsystems   105
Media selection criteria   117
Basic learning systems   126
Practical utilization of learning media   130
Conclusions   136
Appendix   138

Integrated multi-media systems for science education which achieve a wide territorial coverage, A. R. Kaye and M. J. Pentz

Summary   144
Introduction   144
Trends in the development of distant study systems   148
The influence of educational technology on the design of a multi-media distant study system—the Open University   154
The future potential of multi-media distant study systems   172
Bibliography   175
Appendixes: Open University materials   177

Integrated multi-media systems for science education (excluding television and radio broadcasts), S. N. Postlethwait and Frank V. Mercer

Summary   186
Media—what are they?   186
Media—a classification scheme   188
Some examples of multi-media systems   192
The educational potential and implications of multi-media   207
Current trends in the use of multi-media   208
Achieving the potential   212
Bibliography   214

Educational technology in the professional training of science teachers, A. Perlberg

Summary   222
Introduction   222
Educational technology in teacher education—rationale and purposes   223
Educational technology in teacher education—selected trends and practices   224
Conclusions   235
Bibliography   236

Educational technology applied to the learning of science in developing countries, Isaias Raw

Summary   242

Appendix

Participants: Meeting of experts on the utilization of educational technology in the improvement of science education (Unesco, Paris, 13-16 September 1972)   247

When the media are not the message

a layman's impressions of the meeting

by Daniel Behrman

Just about the time that these words were being written, the sprightly *New Scientist* published a report about the videocassette revolution in the United States that now seeks to make taped television as readily available as frozen orange juice to the American consumer in his own home. 'One can now go into the *Time-Life* Building in New York at any time between 9 a.m. and 5 p.m. and practice putting on a 12-foot green which even returns one's ball automatically. The idea is to publicize the golf lessons with Jack Nicklaus which *Time-Life* Video offer on cassette. Weary of stooping to pick up the golf ball, the New Yorker can sit to watch a speed reading course which is demonstrated free every hour. Provided with a stop watch (which can be bought for $ 9) and drill books, he watches the videocassette unroll the text of *King Solomon's Ring* at increasing speed, while Dick Cavett tells him how to keep pace. If by this stage, his back is playing up after all that putting, he may be interested in one of the medical counselling series, which includes a backache programme. . . .'

This account might convey the impression that here we have the answer to all the questions raised when 18 experts from 11 countries met from 13 to 16 September 1972 at Unesco headquarters in Paris to talk about the 'Utilization of educational technology in the improvement of science education'. But that would be jumping to a conclusion. The videocassette is just another medium—along with teaching films, programmed textbooks, overhead projectors, computer tutors, educational television, radio courses—being used to supplement the efforts of teachers trying to reach more and more pupils of all ages inside and outside school systems. The experts had their differences of opinion but they agreed that in education, unlike entertainment, the media are not the message. To put it another way, in educational technology the education is more important than the technology.

The meeting was convened jointly by Unesco and the Committee on Science Teaching of ICSU, the International Council of Scientific Unions. Through discussion of a series of papers prepared by some participants, the experts tried to draw a picture of the state of the art in educational technology as applied to the teaching and learning of science.

The papers, however, represent only one aspect of the meeting. There was also the cutting edge of comments from certain participants who sought to counsel and to express caution based on their own experience. Through these comments, the papers took on new dimensions which, it is hoped, will become apparent in this introduction by a self-confessed layman who was able to be present.

What might be termed an inventory of the technology on hand was given by the author of the first paper presented, Professor Arthur I. Berman of the Institute for Studies in Higher Education at the University of Copenhagen. He brought out the peculiar demands placed on media by science education, a theme that was to be reiterated during the meeting. Teaching wave mechanics is not the same thing as learning how to putt or cure a backache, even if media manufacturers try to convince the customers that it is. In his paper, Berman states: 'Educational research fails to support the view that courses taught exclusively by television are better than live instruction as measured by the performance of the learner'.

One reason he gave is student reaction against computers, taped television or packaged courses. The lecturer is much decried but, at least, he can vary his delivery on the basis of the responses and reactions of the students. Worst of all, Berman added, the taped television course and other 'high technology' media cannot easily be changed. 'That is the danger, there is never a final word in any subject and one cannot be sure, in any case, that the students really understand the message and are learning from it.'

Berman seemed to favour 'low technology' where teachers can make changes in immediate response to student feedback. One example he gave is the overhead projector which enables notes and diagrams in any form to be flashed onto a screen or a light-hued wall. It humanizes the educational process because the projected image, like the blackboard, can readily display the teacher's own drawings and lettering. All in all, Berman thought that science education would do better to concentrate on reaching pupils at an early age—as early as from two to six years old—instead of developing ever more elaborate hardware. More attention should be paid to software or, as Professor Yves Le Corre of the Faculty of Sciences in Paris called it, teachware.

Comment from the floor was not long in coming. Professor Jerrold Zacharias of the Physics Department of Massachusetts Institute of Technology (MIT), and one of the world's pioneers in the reform of science education, tried to probe more deeply. More than feedback is needed, he said; the student must be involved. Pressure should be put on him so that he thinks about what is going on. 'In making films, I've always tried to make small mistakes so that the student will be alert and something will come out of him.'

Berman agreed, remarking that some media do nothing but show a picture of someone speaking while the student is expected to understand.

This view was challenged by Professor M. J. Pentz, Dean of Studies in Science at the Open University in the United Kingdom which reaches some 40,000 enrolled students (not just an 'estimated audience' of spectators) with a combination of television and radio programmes, home study, and short intensive periods at summer schools. 'We have not subordinated the role of the teacher', Pentz said, 'we have transformed it.' On open-circuit television, where the public can tune in, teachers are put on their mettle. Their attitude is different from what it would be on closed-circuit television or in a lecture theatre. On television, friends, relatives and academic colleagues can see the lecturer. It is this 'shop-window effect' which makes teaching by television worth while.

Next the voice of the developing world made itself heard. Stephen Awokoya, Director of Unesco's Department of Scientific and Technological Research and Higher Education, brought the meeting back to basics. 'In developed countries, there is an important scientific infrastructure that overflows into the school system. In the United States or Europe, you pick up a telephone. But what if I have to shout to communicate in my village? What if educational technology is just something to be read? How do we evolve a strategy to feed innovations to developing countries where there is no infrastructure of science and technology? We must isolate principles that can be applied to developing countries. They do not have a gigantic infrastructure, but they can use intermediate technology as a nucleus that should be made to flower and grow.'

The developing world had another spokesman in the person of Dr Isaias Raw, who is currently at the Educational Research Center of MIT. He was one of the experts sent by the ICSU Committee on Science Teaching and his comments were so numerous and to the point that he was asked to submit a paper at the end of the meeting. He had few illusions about the sales resistance of developing countries. 'It is easy to sell in developing countries when an item is expensive . . . because anything big and shiny then leads to a big inauguration.' He recalled attending a meeting at Tufts University in the United States a few years ago where two floors were taken up with an exhibition of obsolete teaching aids. 'Let's not persuade developing countries to spend money on this junk', he pleaded.

One simple cheap technique is represented by what the British call 'steam radio'. Just audio coming out of the box, no visual. Nothing very complicated, but quite effective as John C. H. Ball of the Voice of Kenya in Nairobi showed in his paper on 'The use of radio in science education'. As he observes, radio is much cheaper than television, at both ends, transmitting and receiving. Consequently, over 3,500 primary schools in Kenya are equipped with radio for educational purposes, but not a single one has television.

An unexpected spin-off from the educational broadcast in Kenya has been a science textbook for primary school children. The British Council offered money to print pamphlets to accompany programmes and these are now selling at the rate of 60,000 a year with half the children in Kenya's primary schools learning

When the media are not the message (a layman's impressions of the meeting)

science. Such material is important, Ball noted, because children must have something to look at while the radio is on, so that eyes and ears can work together. This has led to a technique known as 'radio vision' in which a film strip is accompanied by a radio programme, a sort of poor man's educational television.

Raw wanted to go further than film-strips or pamphlets. He felt that inexpensive kits to teach nutrition, health or the uses of electricity should be developed as a form of educational technology accessible to countries that cannot afford an educational television network.

Zacharias saw another use for radio—an opportunity for the pupil to reply. It already exists on talk shows in several countries where people can phone in. Cheap radio transmitters could provide the same facilities, where telephones are scarce. He was concerned over the way in which television gobbles up channels that radio could use: a thousand radio channels for one television channel. 'We must find mechanisms to handle two-way communications. At least one per cent of listeners should be able to talk back upstream to the people dishing out the programmes.'

Pentz was sceptical about the possibility of replacing educational radio by the 'odd teacher doing his own recording'. Whether on tape or over the radio, expertise is required. The tape cassette has the advantage of controlling playback time since it can be scheduled and repeated as desired, but Pentz reminded the meeting that it costs much less to broadcast than to send a cassette. Professor Arye Perlberg, head of the teacher training department at the Israel Institute of Technology in Haifa, saw a future with radios so cheap that anyone with a radio in his pocket could have access to lifelong education. 'The trouble is that many developing countries want to get out of radio and into satellite communication and television as status symbols. The money put into television programmes could give radio to all instead of educational television just for certain groups.'

Nevertheless, open-circuit educational television has some undeniable achievements to its credit. These were reviewed by J. Valérien, Head of the Department of School Radio and Television Broadcasting in France, in his paper on the use of television in the teaching of science. Closed-circuit television is an interesting solution but cost rules out its general use even in rich countries where it is restricted to certain institutions, such as teacher-training schools. Valérien showed how educational television broadcast over public channels was used to meet an emergency situation created in Italy when pre-vocational training became mandatory for children between 11 and 14 years old and another in France when the school-leaving age was raised from fourteen to sixteen without an immediate increase in facilities. He was particularly sanguine about the prospects for open-circuit television teaching in adult education. Teachers in traditional education systems, however, tend to reject it and one remedy for this would be to use television in centres where teachers are trained or given refresher courses. They would then be more inclined to use it in their own work.

Raw observed that 'the problem in some countries is that schools are very rigid. One must hope that television has not replaced the rigidity of schools by broadcasting another form of rigidity. In several countries, television is used to formalize bad education. Then high school education through television becomes just a panacea providing high school diplomas.'

One of the success stories cited by Valérien was the Open University in the United Kingdom. Pentz and a colleague, A. R. Kaye, contributed a paper on 'Integrated multi-media systems for science education which achieve a wide territorial coverage'. When he commented on their paper, Pentz pointed out that the Open University has proved particularly attractive to teachers who make up some 25 per cent of its 40,000 students. 'There is a built-in bonus. They earn more money if they get a degree from the Open University. After a year with us, they are more prone to use new methods in their own classrooms. There is certainly a pay-off if you can entice teachers into the system.' He thought there was another pay-off in the home of the Open University student whose whole family is influenced by his example. The broadcasts also have an effect on casual viewers who begin to wonder why they should not take advantage of the opportunity.

Pentz nonetheless felt that science education by open-circuit television had a limited future. 'We are only on the second channel and the time span is restricted. If we were to have ten times as many courses as we now have, we would need 300 to 400 half-hour transmissions a week. No, we will be forced to find a way to get an audio-visual signal across for less than £20 per student. We need an instrument that can take synchronized sound and pictures into the home. Then we could forget open-circuit television except for the shop-window effect. There would be an enormous market for such a gadget. Manufacturers charge ten times as much.'

Kaye, Pentz's associate from the Open University, pointed out that today, as compared to ten years ago, there is much less fixation upon a single medium such as television, radio, or the teaching machine. The Open University uses everything, even live teachers at times. It is also putting out kits so that students can make scientific experiments at home.

It was during the discussion of a paper on 'Programmed instruction in science education', presented by Professor G. O. M. Leith of the Department of Research and Development in Education at the University of Utrecht in the Netherlands, that the learning process rather than the educational medium came under scrutiny from the meeting. Leith himself related how much of the early dogma of programmed instruction has vanished. 'At first, a student had to be given only fifteen words in a frame. The reason was simply that the aperture in Skinner's first teaching machine was high enough only for fifteen words. Everyone copied him; the daring extended it to twenty-five words.

'We have discarded most things that made the product look like an advertising image. At first, programmers thought they could teach almost anything without the aid of a live human teacher in the classroom. The dedication of early programmers seems crazy when we look back on it. One even put a research assistant behind a panel and told the pupils to pretend he wasn't there.'

Another shibboleth that has gone by the boards is the claim that programmed learning is necessarily better because it allows individual self-paced instruction. According to Leith, this does not happen all the time. There are situations where, even with programmed techniques, group-paced learning gives better results.

Individual differences in students must not be overlooked. Leith stated that the extrovert seems to learn more effectively through discovery in an environment where he is not directed and he can work things out himself. 'Throughout our educational system, we pretend we are open but it turns out that the pretence is highly structured and very directive.' Kaye, of the Open University, went even further: 'We can't draw lines between introverts and extroverts; when we have 40,000 students, we have 40,000 kinds of problems.'

On one specific problem mentioned by Awokoya at the start of the week, Leith thought programmed instruction could help children in developing countries to relate abstract images, such as a drawing of a molecular model, to three-dimensional objects. Zacharias agreed that it is always hard to go to three dimensions: showing pupils how to arrange six pencils into four triangles is easier than discussing how to make a tetrahedron.

What matters is not the medium, but what one is trying to teach. 'The subject must be a vehicle, not just an end. You're after skill and knowledge. We cannot overlook the *ins* of education: insight and intuition, initiative and ingenuity. Education is a real social problem. We're not trying to provide automatons who will vote the way we want them to vote, we're trying to achieve social ends.

'It is easy in education to get people to do hard intense work so there is no daydreaming. But daydreaming is very important. There are many varieties of daydreams—I daydream gadgets—and we must make room for them.

'We learn by successive approximations, that's what happens in life. Even with television or the computer, it is easy to put in hints so the student can learn on his own. We must emphasize the topic, not just the joys of pushing things around with computerology.'

That remark by Zacharias was a challenge to Professor Donald L. Bitzer who introduced the paper on 'Computer-based science education' that he had written with Bruce Arne Sherwood and Paul Tenczar at the Computer-based Education Research Laboratory at the University of Illinois in Urbana, Illinois. Bitzer accepted it when he not only described his university's PLATO IV computer system but demonstrated it as well. The Platonic dialogue by telephone across the Atlantic, between Bitzer in Paris and PLATO IV in Urbana, was easily the high point of the meeting.

The system works with 2,000 terminals that enable students to converse with PLATO IV in ordinary English, typing out their questions on a keyboard and getting back their answers in letters or images on a flat glass screen. At the University of Illinois, PLATO IV helps teach thirty-five courses that run from population planning to organic chemistry.

Bitzer said that PLATO IV has brought the cost of computer instruction down to about 70 cents an hour per terminal as compared to typical previous costs of five to ten dollars per student per hour, 'adequate to hire a good private tutor'. A 4,000-terminal system is being built at Illinois and, by 1974, it is hoped to reduce the price to 40 cents an hour per student and, by 1980, to 20 cents an hour. Even today, the computer more than pays for itself by carrying out other tasks. Bitzer remarked: 'With what we charge scientists, we'd have to pay students ten cents an hour to be non-profit.'

He had brought a terminal with him and, from a basement office in Unesco House, he was able to question the computer by telephone, 'talking' to it as a student would. In the first demonstration, the student had to aim a rocket at a mountain. The trajectory fell short on the glass screen and PLATO IV commented: 'No mountain here'.

Then Bitzer got down to business, running an experiment in genetics with a wide choice of fruit flies offered on his glass screen. He tried to fool the computer by labelling a male fly 'Mary'. PLATO IV replied: 'That fly is the wrong sex. If you don't know the difference, press "Help".'

But Bitzer did make his point. The computer could simulate genetic experiments classically made with fruit flies to show how hereditary characteristics are transmitted. Then he switched to a course in population dynamics that Illinois offers each year to 300 planners from foreign countries. As an example, PLATO IV compared population growth in the United States and Mexico, to show at what point in the twenty-first century the latter will catch up, depending on the values assigned to various parameters.

In a final experiment, arrows on the terminal keyboard were used to direct a little man bouncing balls on the screen. This course is used to teach computer programming to children in their second year of primary school. Bitzer predicted that 'in ten years, computer literacy in the United States will be as common as the ability to read'.

His demonstration was not complete. He explained that, in ordinary use, each terminal also has a stock of 250 images that can be flashed onto the glass screen within two-tenths of a second when a course requires a high-quality colour picture, as in medicine. This must have been what led the sceptical Raw to observe that he thought students should still come into contact with reality at times instead of sticking to computer simulations, no matter how perfect. 'Would you want to be operated on by a computer?' he asked.

Bitzer took the criticism in his stride. He maintained that the Illinois system does teach by successive approximations and that PLATO IV can speak verbally to the student. It is harder for the student to reply to the computer because it understands words only in a certain context. In a chemistry course, for example, it may ask, 'Is it blue?' and the student will reply, 'No, it is red'. But, in case the student happens to be looking out of the window and anwers 'No, it is cloudy', one might foresee this and programme the computer to reply: 'Sorry, cloudy is not a colour'.

'Computers aren't perfect', said Bitzer, 'but neither are humans. The computer must incorporate an understanding of subject matter, not just simple answers. We are a facility, the user decides what we do. It is like using the telephone. The phone company cannot tell you what to say.'

Educational technology can be a computer. It can also be a gifted teacher using a cord stretched across a room to demonstrate wave phenomena in physics, as Dr Albert Baez, a member of the ICSU Committee on Science Teaching, did for his colleagues at the meeting. Baez' talk preceded a paper on 'Integrated multi-media systems for science education' that had been prepared for the meeting by Professor S. N. Postlethwait of the department of biological sciences at Purdue University in Lafayette, Indiana, and Professor Frank V. Mercer of Macquarie University at North Ryde in Australia. Postlethwait presented it in Paris along with a film that he prepared especially for the meeting and in which he shows in action the multi-media system that he himself uses at Purdue.

His motion picture was worth quite a few thousand words. Postlethwait had set out to give individualized instruction to a class of 600 students. In the film, participants at the meeting could see how he did it. A student comes into a learning centre where thirty-two booths have been set up. He goes there whenever he is ready. He picks up his assignments, enters a booth, dons earphones and listens to his 'tutor' on tape. If he does not understand, he plays the lesson back. Each booth is equipped with an 8-mm projector for a teaching film. As part of the learning centre, there is a laboratory where an experiment has been set up. The student goes to the lab, collects data and prepares an individual analysis. When he has gone through his assignment, he leaves the booth for the next student.

This independent study session is the heart of the system which includes a weekly general assembly for several hundred students and small assembly sessions where eight students meet with an instructor, again once a week, mainly for short quizzes. There are many opportunities for informal sessions, beginning with the coffee room next to the learning centre. Postlethwait himself makes it a point to invite 500 students to his own house for coffee on a Sunday and they can telephone him at home two evenings a week if they are in trouble. He admitted that they do not always come around to his home. ...

That led Zacharias to bring up the difficulties that professional educators face when they try to communicate with students. 'No, the students don't always come to our door. We must be able to talk to them. It's like sitting next to a girl in an airplane. You don't wait for a wing to fall off before you start a conversation. But you need jacks or better to open. We must provide the subject matter so that students can talk about something besides sex and drugs.

'In fact, we leave out all but what our colleagues consider serious business. Are we un-persons? Are we to go on letting education do to students what it has been doing? Up until now, education has been nothing but a major failure, trips to the moon notwithstanding.'

Raw agreed with him. 'You can go to the Unesco newsstand upstairs and buy a horoscope. In the *New York Times,* you can find ads trying to get you to buy a home computer so that you can work out your horoscope. This shows the tremendous impact of science education.'

Zacharias went on with some suggested remedies. 'We must cut the number of required courses so that students can live half-slave and half-free. They should do just enough to get requisite knowledge. Technology is not all-important, we can use Gutenberg, Xerox and larynx.'

'It is examinations that are important. Students learn their courses the last week before an examination. Didn't you? They can't waste time spent loafing for an exam next February. This means that self-paced courses must have examinations ready on time and graded fast. Perhaps Unesco could provide a resource-book of examination questions or a big compendium of examinations.'

He wanted to see more research in education. 'In the United States, about 0.2 per cent of the money spent on education goes into research. Modern industries spend from 5 to 10 per cent of their gross revenues on research and development. In the United States, education and the railroads put 0.2 per cent into research—and the railroads are dying.'

The final paper presented at the meeting dealt with 'Educational technology in the professional training of science teachers'. Its author, Professor Arye Perlberg, head of the Department of Teacher Training at the Israel Institute of Technology - Technion in Haifa, had a tendency to lock horns with some of his predecessors. 'Most education is focused on learning, but we must look at how to forget and how to unlearn in our technological societies that are ever changing. We say that everyone who knows a subject matter knows how to teach—and that is why higher education systems reward research and not teaching. All higher education systems are built on this religion.' Perlberg was especially interested in teaching how to teach. In his paper, he defines educational technology as a systematic way of designing, carrying out and evaluating the process of learning and teaching and of ways and means to bring about more effective instruction. He goes on to describe some of the systems used to improve instruction such as micro-teaching, simulation techniques and observation methods. He concludes by stressing the need for a more systematic approach to education such as the concept of educational technology embodies.

In its final session, the meeting grew less and less formal. Irony of ironies, the educational technologists insisted on eliminating the tape recorder that had been taking their words. They complained that they could not interrupt each other if they had to push a microphone button before starting to speak. Raw asked: 'Is technology ready to be used and recommended? Let's not oversell what is not yet ready to be sold.' Zacharias was among those who thought too much attention had been paid to papers: 'I'm not accustomed to be regimented by papers'. And Pentz pleaded for short summaries of the papers which could be understood by 'ordinary people teaching ordinary kids'. He was getting back to that same basic problem of communication that Zacharias had mentioned before. Zacharias hit it again: 'When we try to reach children, we must get rid of our professional "iffy-tude". So many of us try to talk so that colleagues won't pick some little nit.'

The group reached a consensus that there could be no consensus. Recommendations made by individuals would stand and be published; there would be no attempt to seek group recommendations watered down to the lowest common denominator. The prospective user would have to judge for himself, then make up his mind on the basis of his own requirements. It was on this note that the Meeting of Experts on the Utilization of Educational Technology in the Improvement of Science Education ended and the publication of its results was begun.

Computer-based science education

by Donald L. Bitzer, Bruce Arne Sherwood, and Paul Tenczar
Computer-based Education Research Laboratory,
University of Illinois, Urbana, Illinois 61801, U.S.A.

Summary

Students use computers in science education in two ways: they can *write* computer programmes in order to study complex systems and to learn numerical techniques; and they can *interact* with educational computer programmes written by teachers. The first type of use is widespread, while the second has been severely hampered by the lack of suitable authoring and delivery systems. This paper concentrates on the latter form of computer-based education and gives examples of materials written for their students by biologists, chemists, mathematicians, and physicists. These materials reflect diverse teaching styles and strategies, including tutoring, simulation or modelling, and drills. By the variety and complexity of these examples we hope to dispel the misconception that the role of the computer is limited to 'programmed instruction' or to the presentation of simple multiple-choice questions.

The problems of computer-based education include: (1) The need for an adequate terminal for student use. The common teletype is not adequate in science education—a *graphical* display terminal is required, a device which can rapidly display line drawings, graphs, and pictures. (2) The need for adequate computing power. A weak computer may only retrieve stored questions and recognize stereotyped responses. To go beyond this simple 'teaching machine' function requires enough computing power to *generate* displays and problems and to recognize open-ended responses. (3) The need for good teachers to author materials without requiring the services of expert computer programmers. This implies the need for a suitable authoring language and system. (4) The need to make the cost of computer-based education far lower than it has been. Typical costs have been several dollars per hour per student, which cannot compete with the cost of a human tutor. It has been necessary to invent a new technology in order to make progress toward economically viable computer-based education.

One solution to the problems of computer-based education is the PLATO IV system now beginning operation at the University of Illinois. (1) The heart of the student terminal is the plasma display panel, a flat sheet of glass upon which the computer can light up or turn off any of a quarter-million dots (in a 512 × 512 grid) to display text, graphs, and line drawings. The computer can select colour photographs to be projected on the back of the transparent panel. For technical reasons discussed in the paper, this display device represents a major advance over previous technology, including the cathode-ray tube. (2) The PLATO system is controlled by a large scientific computer with adequate power and speed to permit the presentation of complex material. The system responds to student input within a fraction of a second. (3) Authors write their own materials in the TUTOR language, which is powerful yet easy to use. Computing power in the PLATO system is used to aid authors in their creative process. (4) When fully implemented it is estimated that capital and operating costs will be $ 0.50 per student hour at a terminal. Part of the cost reduction is due to a radical restructuring of the way in which the computer itself is operated: in particular, fast electronic memory replaces slow mechanical memory for many important functions, which leads to greatly improved computer utilization.

Appendices to the paper discuss the contrast between large and small computer-based education and give an example of the use of the TUTOR language.

Introduction

Computers are increasingly being used in science education, both in direct instruction and as calculational tools. Science teachers can prepare computer-based educational materials with which students interact at their own rate, giving the individual student a patient and intelligent tutor which can simulate complex phenomena, provide drill on basic concepts, and diagnose and treat weaknesses in preparation or comprehension. Students may write their own computer programmes and treat problems that transcend the limitations of traditional analytic approaches.

Although computers are used to supplement science education in hundreds of institutions around the world, in only a few schools and colleges have whole courses involved the computer in a major way. Of those few projects which have given birth to complete computer-based courses, fewer still have focused on the engineering issues of making mass utilization practicable; most projects have rather been directed at exploration and small-scale testing. The PLATO IV system [1][1] of the University of Illinois is the culmination of a major research and development effort begun in 1959 and leading to a viable computer-based education system. More than 1,500 hours of computer-based educational materials have been prepared on the PLATO system, including over 20 one-semester courses in diverse fields. This large curriculum base includes lessons representing all the major types of computer utilization and pedagogical styles, and we will use PLATO lesson examples to illustrate the role of the computer in science education.

The lesson examples which follow will help define what we mean by computer-based education: in particular, we hope to dispel the misconception that the role of the computer is limited to the presentation of simple multiple-choice questions. The lesson examples are followed by a discussion of the fundamental problems associated with making computer-based education viable.

The figures which accompany the lesson examples are photographed from the 21 × 21 cm (8.5 × 8.5 in) plasma display panel of a student's individual PLATO terminal. The plasma display panel shows orange text and graphics on a black background; reduced prints are given here for ease of reproduction. It is important to bear in mind that each student has his own individual terminal (with display screen and typewriter keyboard) and that it is highly unlikely that two students would have the same picture on their screens simultaneously. Indeed, it is unlikely that two students would experience identical presentations of the lesson, since the computer interacts with each student on an individual basis.

Lesson examples

Biology

Simulation techniques are used in a genetics course to teach students the laws of inheritance. This computer laboratory allows students to conduct a standard series of fruit fly matings. Each student is presented with stocks of parent flies to examine. Besides flies with normal characteristics, mutant flies can possess such features as white or pink eyes; vestigial or veinless wings; and black, ebony, or striped bodies. The flies are not pre-stored pictures but consist of assembled parts: head, eyes, thorax, wings, and abdomen. Each fly is efficiently coded by a single computer word which specifies the exact type of each body part. This is similar to the biological coding of information in genes located on chromosomes. The computer can construct flies with any combination of normal and mutant characteristics. When the student requests that a mating be made, wtihin seconds the offspring are displayed. Since most mutant characteristics are recessive, they do not appear in the first generation offspring. The student can choose some of the first generation flies as parents for yet another generation (Fig. 1). Students

FIG. 1. Fruit fly genetics. Some of the offspring have white eyes and/or vestigial wings not seen in the parents. The student records the observed characteristics in his notebook.

maintain a scientific logbook of all these experiments so that later they can make statistical tests of hypotheses and hand in the results in a formal laboratory report.

All of the offspring are generated by using random numbers and probabilities based on the Mendelian laws of inheritance. Thus, this computer analogue of the

1. For references, see page 30.

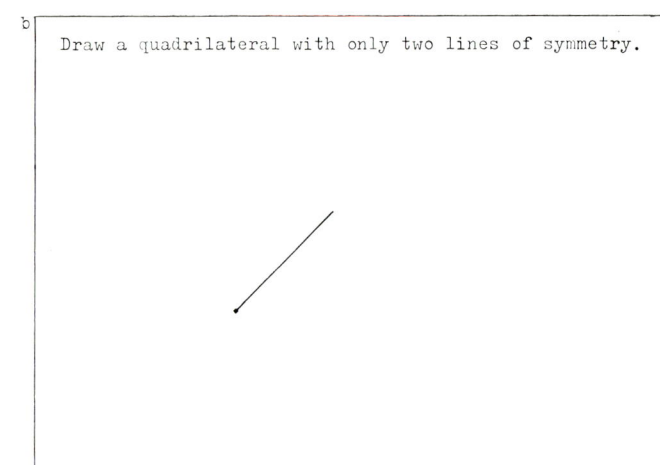
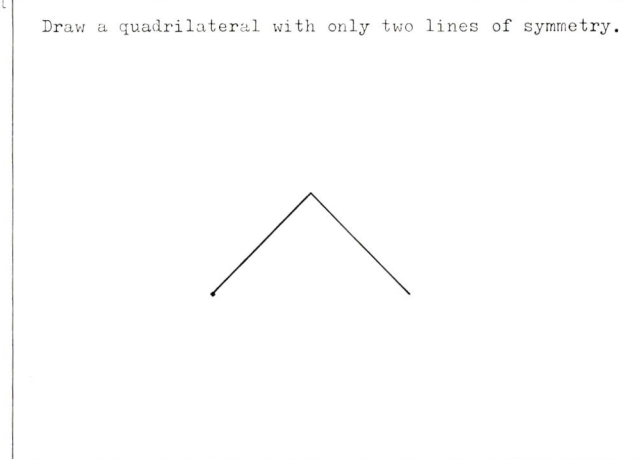
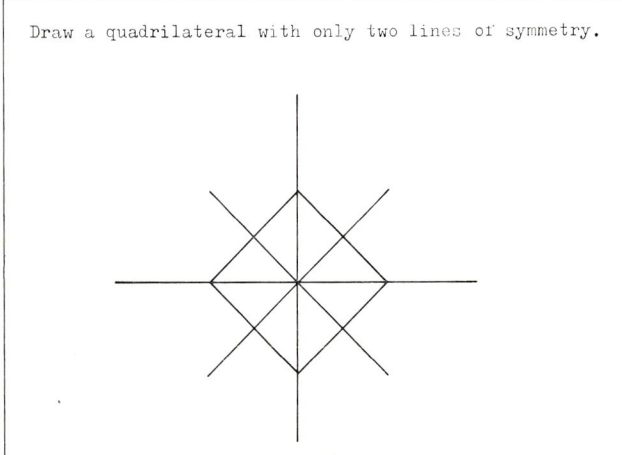

Fig. 2. Drawing a geometrical figure. The student moves a cursor and marks vertices to construct the figure. The computer shows the student that his figure is incorrect by drawing the four symmetry lines.

real biological system produces thousands of possible outcomes and gives each student his own experiment. A series of conventional fruit fly experiments takes several months. Culture medium must be prepared, bottles sterilized, flies examined at odd hours, etc. Using the computer, a student can perform the basic experiments of Mendelian genetics in three to four hours' work. This time compression of the experiments makes the logical flow of the multi-step process much more comprehensible. Additionally, in a conventional course this experiment is almost always performed 'cookbook' style since not enough time or help is available for the student to go his own way. With the computer, the student can explore various experimental strategies, for it takes only minutes to start over, and help is always available.

The biology lesson just described is one of 35 lessons designed to introduce beginning college students to genetics and evolution [2]. The students spend four hours a week with the computer, where all lesson material is presented, followed by a two-hour discussion period with a human teacher.

Geometry

Pattern recognition is a basic feature of a series of 15 lessons designed to teach informal geometry to junior high school students [3]. This course is designed to give students experience with the *facts* of plane geometry (symmetry properties, definitions, etc.) before formal proofs are attempted. The students are asked by the computer to construct specific geometric figures by using a set of eight keys to move a cursor around the screen (Fig. 2). When ready, the student can request that the computer 'judge' his work. Figure 3 demon-

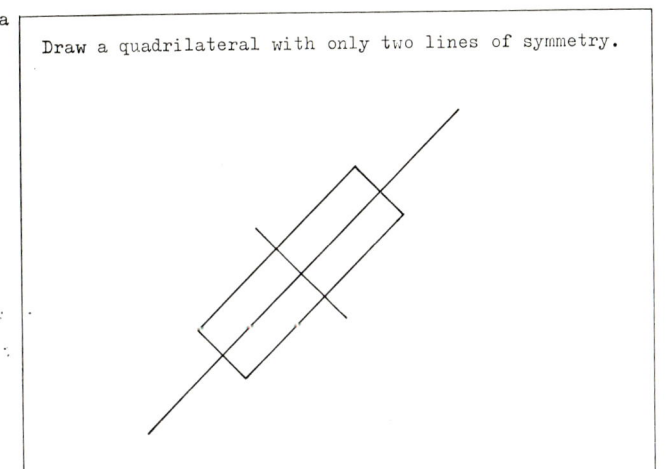

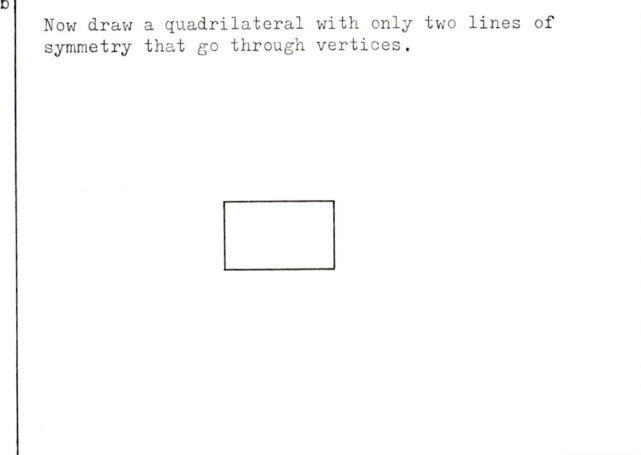

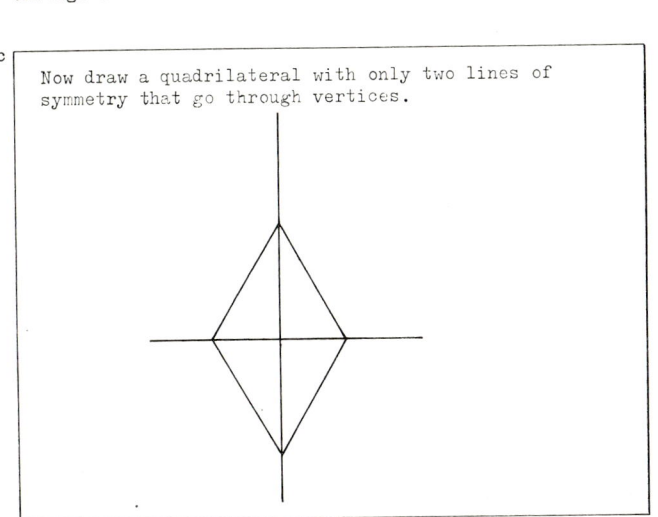

FIG. 3. Constructing prescribed figures. After succeeding in the first problem the computer chooses an appropriate second problem. The student then incorrectly draws a figure of the same type before finally constructing the desired rhombus. Note that the computer is able to classify the figures independent of size, shape, or orientation.

strates a typical sequence of the student-computer interaction. It should be evident that sophisticated 'judging' algorithms must be used rather than stored answers since the size, orientation, and even specific type of figure can be arbitrary. Furthermore, when the student has an improper figure, the computer must do more than just tell the student his figure is wrong or the student will become frustrated. The student's attention must be directed to specific features of his construction that are inadequate.

Chemistry

Dialogue with the computer supplies the student with information needed to identify chemical unknowns in a qualitative analysis lesson [4]. A student-computer exchange, in the student's natural language, tends to maximize flexibility of approach for the student [5]. Here is an example of a college student's conversation with the computer. (Remember that this dialogue actually appears on a display screen, not a teletypewriter. In particular, the spectra are shown on the screen in colour.)

Student: What colour is it?
PLATO: The solid is white and the liquid is colourless.
Student: What is its melting point?
PLATO: The melting point is 19-21°.
Student: What does the nmr look like?
PLATO: This is the 60 MHz proton nmr spectrum. For a table of chemical shifts, press DATA. (The spectrum is shown on the screen.)
Student: ir
PLATO: This is the ir spectrum. (The infrared spectrum is displayed on the screen.)
Student: What is the unknown soluble in?
PLATO: Please provide more information.
Student: Is it soluble in H_2SO_4?
PLATO: The unknown is soluble in cold conc. H_2SO_4.
Student: Elemental Analysis.
PLATO: We can give you results from a sodium fusion. To try it press LAB.
Student: Picric Acid Derivative
PLATO: Please forgive.... The following words are NOT in my vocabulary... picric.
Student: dnp
PLATO: The unknown reacts with 2,4-dinitrophenylhydrazine to form a new substance with a melting point of 249-251°.
Student: Where does it boil?
PLATO: The boiling point is 195-200° at 760.
Student: Permanganate Test
PLATO: The unknown reacts slowly with $KMnO_4$.
Student: Tin and HCl
PLATO: Please forgive.... The following words are NOT in my vocabulary... Tin.
Student: Tollens
PLATO: There is no reaction with Tollen's reagent.
The unknown is
Student: Acetphenone
PLATO: SP
Student: Acetophenone
PLATO: OK
For a new unknown press NEXT.

After the student has obtained enough information about the unknown, he can attempt an identification (Fig. 4). Shown is a student's misspelled but proper identification. The computer recognizes that the student's answer is correct but that the spelling is inadequate. Spelling algorithms such as this must be a basic part of any educational computer system for, as in this case, if the student were merely told his response was erroneous he could easily spend a great deal of time on the wrong track.

The score shown in Fig. 4 refers to an aspect of the lesson in which the student tries to determine the unknown with the least 'cost' in chemical and physical tests. Some tests cost more than others: for example,

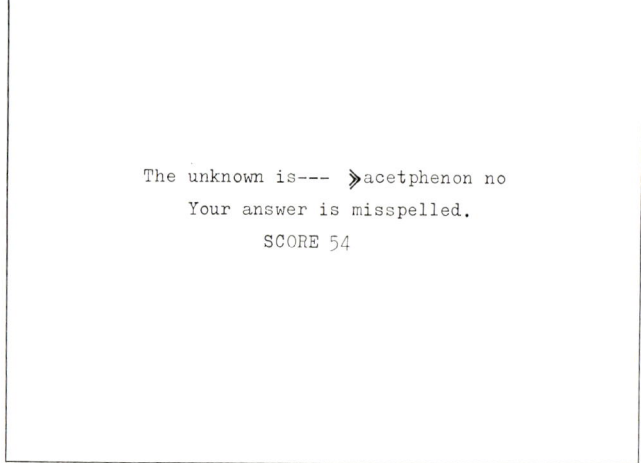

FIG. 4. Qualitative organic analysis. The student has correctly identified the compound but has made a spelling error. The score measures the student's efficiency by charging for chemical tests performed on a scale commensurate with the expense of the test in the laboratory.

determining the infrared spectrum costs ten points while the melting point costs only one point. The chemist who wrote the lesson wanted to encourage students to make simple tests before using expensive apparatus such as an infrared spectrometer.

This lesson is not meant to replace the organic chemistry laboratory. Rather it is meant to sharpen the intellectual process of formulating questions and interpreting results before the student enters the laboratory. Thus, in a matter of hours, a student can logically identify 5 or 6 unknowns—often more than he would identify in a whole semester's work in the laboratory. This is just one of the many chemistry lessons totalling

a

```
If the acceleration is constant, the average
velocity v̄ can be written as a simple function
of the initial velocity v_i and the final velocity
v_f. Write an expression involving v_i and v_f:

v̄ = 》 (v_f-v_i)/2  no

Your expression gives the wrong result. Press
-next- to see why.
```

b

```
Consider a car that speeds up (with constant
acceleration) from 60 to 80 fps to pass a truck.
What would you say is the average speed v̄
during this passing maneuver...

v̄ = 》 70  ok            fps

Right, but your formula gives

      (v_f-v_i)/2 =   10.0

So you must rewrite your expression.
```

c

```
If the acceleration is constant, the average
velocity v̄ can be written as a simple function
of the initial velocity v_i and the final velocity
v_f. Write an expression involving v_i and v_f:

v̄ = 》 (v_f^2-v_i^2)/2(v_f-v_i)  ok

Fine. A simpler form is  (v_i+v_f)/2
```

Fig. 5. Kinematics formulas. The student is shown by example why his formula is invalid. He then gives a valid expression and the computer points out a simpler form. These judgements are made by algorithm, not by searching lists of possible answers.

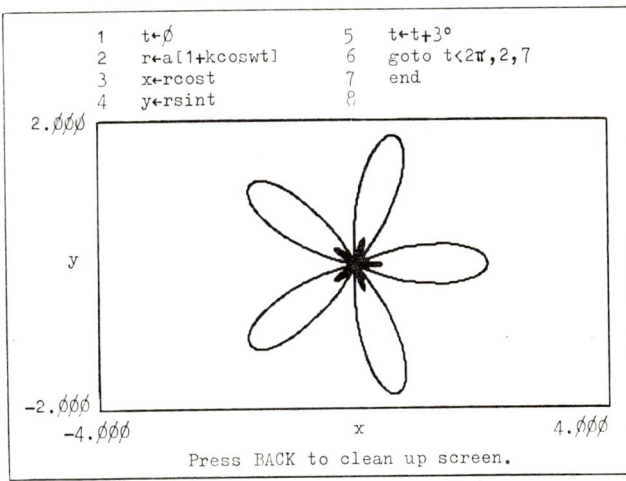

Fig. 6. Student programming. A mathematics student writes a programme to plot a polar function. The angle is 't'. On a separate display page the student specified the plotting variables and their bounds and initialized $a = 0.75$, $k = 1.5$, and $w = 5$.

30 hours which are taught by computer at the University of Illinois [6].

Physics

In an introductory mechanics course, students are asked to participate actively in the derivation of the basic kinematics equations. In Fig. 5a the student has given an algebraically incorrect expression in one step of the derivation: a correct answer is $(v_i + v_f)/2$. In Fig. 5b a simple numerical example shows the student the inconsistency of his formula. This procedure provides the student with an important method for checking the validity of an algebraic expression, and this numerical substitution method permits the computer to handle appropriately all possible algebraic responses, independent of form. In Fig. 5c the student has given a complicated but algebraically correct response and the computer has noted that the response is correct but not in the simplest form. This distinction is made on the basis of the number of arithmetic operations encountered in the numerical evaluation of the student's expression [7]. As in the previous examples, it can be seen that judging student responses by algorithm rather than by comparison with a list of stored answers gives the student great freedom and contributes to heightened interaction.

This example is drawn from a one-semester computer-based mechanics course [8] in which students spend two hours at the beginning of each week studying at a computer terminal. The computer introduces basic concepts, treats applications, simulates phenomena, and tests comprehension. Classroom and laboratory work later in the week build on this solid preparation.

Mathematics

An example of the computer as tool is shown in Fig. 6. The student has written a short programme to evaluate and plot a parametric function, with an angle 't' as the varying parameter. An attempt has been made to keep the computing language [9] as close to standard algebra as possible to avoid inconsistencies with the natural language of direct instruction.

(The most extensive and successful integration into education of the computer as tool has taken place at Dartmouth College, Hanover, New Hampshire where nearly all students, including non-science students, write computer programmes as an integral part of their studies and recreation [10].)

Programming by children

Young children can be taught the basic elements of programming. First, a series of games teaches the child a set of operations which can be carried out by a little man on the screen. In Fig. 7a the child has walked the man, one step at a time, through a maze: in Fig. 7b the child learns how to pick up a ball, carry it, and put it down. (A set of eight keys on the keyboard move the man one step in the eight basic compass directions. The 'plus' key picks up a ball and the 'minus' key puts it down.) After learning the basic operations, the child can write a list of operations for the man to carry out, as in Fig. 8, and watch the man follow instructions [11]. An important aspect of this exercise is that the child can

FIG. 7. Games leading to programming by children. The child walked the man through the maze starting from the lower left corner of the maze. Next the child learns to pick up a ball and carry it into the house.

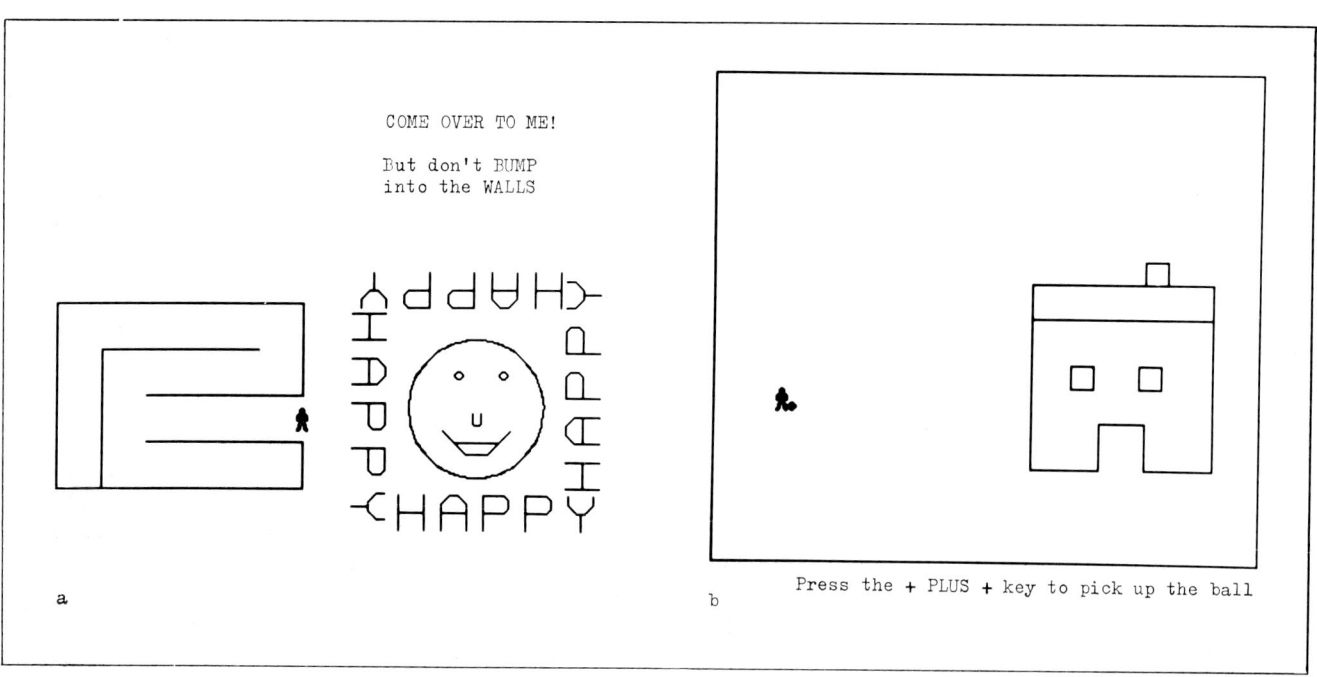

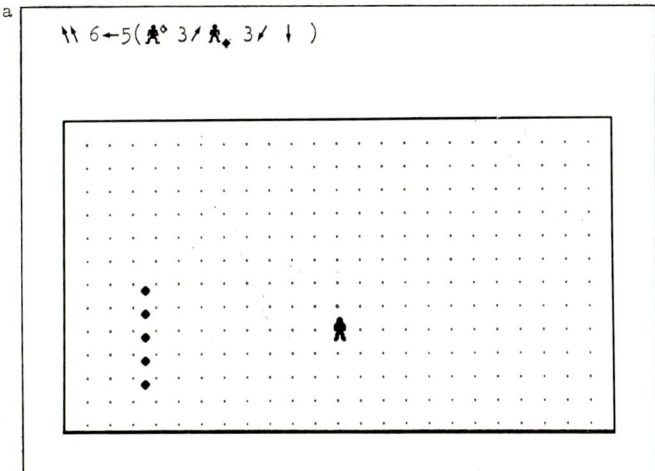

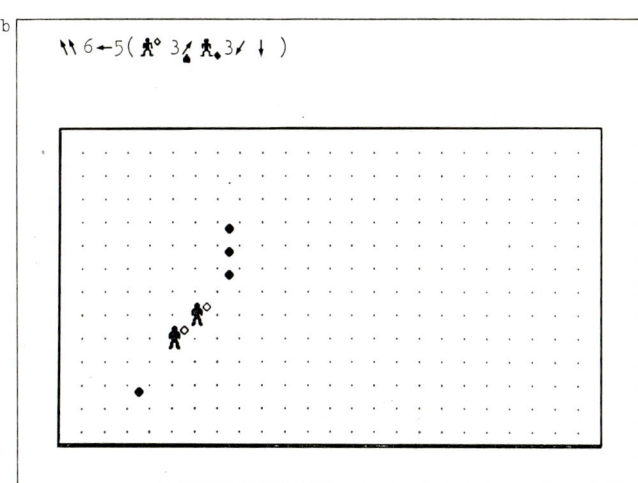

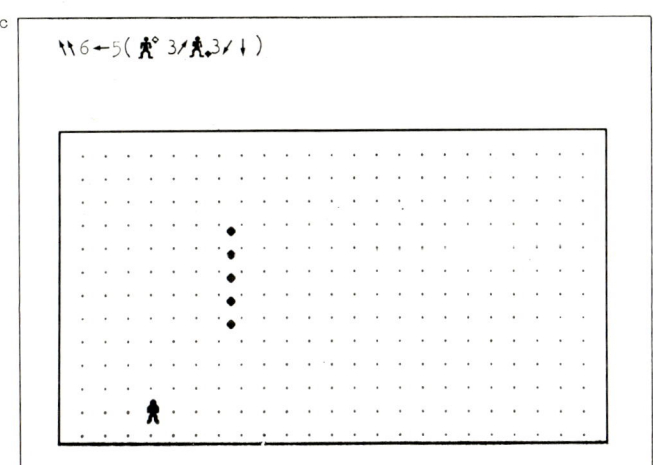

Fig. 8. Programming by children. The programme is built using operations learned in previous games. The child watches as the man carries out the list of instructions.

write an inconsistent programme and receive an error message such as 'There is no ball here to pick up!' The child enjoys giving directions to the man, and he sees the important aspects of a computer and a programme: step-by-step processing, repetitive loops, the concept of an operation, etc.

Problems

We have defined through examples what we mean by computer-based education. Now it is appropriate to ask what are the basic problems in the field of computer-based education.

Many difficulties have restricted the application of computers in education. Educational computer systems have been too expensive for wide-scale use. The limitations of many existing systems have in some educational circles caused 'computer-assisted instruction' to be identified with mere multiple-choice testing or simple drill. (We use the term 'computer-based education' in the hope of avoiding this identification.) The authoring of computer-based lessons has often been very difficult, requiring the services of computer programming experts; teachers find themselves shut out of participation in structuring their courses. We will discuss these and other problems facing computer-based education, then report on progress being made in solving them.

Display

The physical form of communication between computer and student, or between computer and teacher, is fundamental to all other questions. In most cases, the student's communications device ('terminal') has been some kind of alphanumeric display, usually a teletype-writer. While adequate for some special purposes, its slowness, limited character set, noise, and near inability to draw graphs and diagrams make it a poor medium for the full educational message. The expanded capabilities of a fast graphical display device are almost mandatory for most educational purposes, especially in science education. Such devices have in the past been very expensive, but new technology is rapidly changing this situation. Ideally, the student's terminal should permit photographic image projection, two-way voice communication, pointer inputs, etc., in order to fully engage as many senses as possible—visual, auditory, tactile. At some point, economic considerations force compromise with this ideal, but the basic principle should be kept in mind: education is an extraordinarily difficult human enterprise, and it requires a flexible and powerful medium.

Computing power

Too often there has been a narrow conception of the role of the computer itself in 'computer-assisted instruction'. The computer is thought of as a minor control element, choosing and relaying essentially static information to the student and distinguishing among a few standardized replies from him. For such purposes a small or weak computer is sufficient. But for more general purposes, especially in science education, a powerful computer is required to generate (rather than merely retrieve) material for the student and to process open-ended student replies and questions. How is such power to be paid for? Evidently it must be shared among a large number of users; new ways of organizing such large systems have now made this feasible and economical. There is a hint here of the advantages of a large system over a small system. We return to this point in detail in Appendix A.

Authoring

For a computer-based educational system to be viable and to be accepted by the educational community, it must be relatively easy for good teachers to create computer-based lesson materials. Having a powerful rather than a weak computer at the heart of the system makes it possible for the system to help rather than hinder lesson authors in their creative work. Quality improves and costs drop by eliminating programmers and by placing the author in direct control of his medium. In order to achieve this close coupling to authors, it is necessary to create a suitable author language and authoring environment which strongly minimize the need for special computer knowledge. The ideal is to make the system transparent and responsive to the author as well as to the student.

Cost

Another critical issue is cost. The use of computers in direct instruction will be possible on a wide scale only if this is economically feasible, no matter how great may be the supposed benefits. Typical costs of educational computer systems have been about 5 to 10 dollars per student contact hour, which is enough to hire a good private tutor. A cost reduction of a factor of ten is required to make it feasible to use computers in education. It is crucial that overall computer system costs be driven as low as possible, while yet implementing enough power and flexibility to be useful. These conflicting requirements have forced the invention and development of completely new technologies. Contrary to much lay and professional belief, the computer technology of the 1960's was incapable of widespread educational application; the costs were too high, even for rather primitive systems. The new technologies include a radically different display device for the terminal, unique telecommunications, and a drastic restructuring of the computer's software to reflect the interactive educational environment.

Solutions

We have discussed some of the basic problems which have inhibited widespread application of computers in education. We will now discuss the ways in which these problems have been faced in a particular case—the PLATO IV computer-based education system of the University of Illinois.

Display

We have already given examples of computer-based educational materials. They were produced on the

PLATO IV system, and the figures are photographic prints from the student's display screen, which is a flat plasma display panel [12], not a television tube. Some discussion of this device is necessary to explain the nature of the student terminal. The plasma panel is a crucial element in making feasible a sufficiently flexible communications medium.

Until recently, the cathode ray tube was normally used for displaying computer-generated graphical and pictorial information. Because a cathode ray tube must be refreshed thirty times per second to maintain the image without objectionable flicker, an expensive external memory device is required in addition to the television apparatus itself. (In the case of home television the image is sustained by the broad-band video channel transmitted continuously from the television station. For *individualized* use the memory device must be near the display unit, since assigning a video communications channel to each user is prohibitively expensive.) The 'storage' cathode ray tube is a television tube with built-in memory due to the special electrostatic properties of its faceplate. This device is better suited to computer-based education, for the computer need transmit the graphical information only once and no refreshing is required. However, the storage television tube does have drawbacks. One major problem is the difficulty of performing a selective erase of a small portion of the display without disturbing the rest of the picture. Selective erase is necessary for many aspects of computer-based education, including erasing and retyping part of an answer, and in animated sequences performed by drawing a figure, pausing, erasing the figure, and redrawing it elsewhere on the screen to give the impression of motion. Other problems include the long period required to erase the entire screen, the need for frequent maintenance adjustments, and the impracticality of superimposing photographic information on the screen.

The plasma display panel was invented at the University of Illinois to solve these problems. Its memory is at the display unit, inherent to the panel. It permits the selective erase of even a single dot without disturbing the rest of the picture. The display is bright, with high contrast, and free of flicker or fading. The panel consists of two sheets of glass on which are deposited 512 horizontal and 512 vertical conductors (the conductors are transparent). Neon gas between the horizontal and vertical conductors can be made to glow as bright dots at the intersections of the 512×512 grid of conductors. (The resolution is 2.4 dots/mm.) The simple structure lends itself to low-cost mass production. The organization into a 512×512 grid of dots is ideally suited to addressing by a digital computer. The plasma panel makes possible, at low cost, graphical display capabilities that formerly were available only at prohibitive expense. Moreover, the simplicity of the device makes possible additional economies in the design and operation of the telecommunications and of the computer software.

Because it is flat and transparent, the plasma panel can support a rear projection screen for colour slides or movies, selected and driven under computer control, with computer-generated text and graphics superimposed on the plasma panel. This unique combination adds an important dimension to computer-based education. For example, the computer can select a full-colour slide of the human heart for a medical student, then superimpose pointers or animated flow markers on the plasma panel to illustrate dynamically the complex action of the organ. Note that *transmitting* colour photographs from the computer would make communications extremely expensive.

With explanation of the display device, it is important to note that the usual optical distortions of television are completely absent—the flat display panel with its evenly spaced grid gives a display free of distortion or jitter. The resolution is so fine that a viewer is unaware that the text and graphics are actually composed of individual dots.

The fruit fly picture illustrates another important aspect related to the symbols needed for education. In addition to the standard upper- and lower-case letters, numbers, punctuation marks, etc., lessons in some subject areas require a rather large set of additional symbols. For example, when teaching Russian, the Cyrillic character set is needed. When teaching physics much of the Greek alphabet plus mathematical symbols may be needed. The fruit flies are drawn as appropriately positioned symbols—right wings, left wings, eyes, etc. This display mode is many times faster than drawing the flies one point at a time. At the beginning of the student session the computer transmits the required special symbol patterns to the terminal—Russian, Greek, or fruit fly parts—and thereafter the computer need only specify which symbols to plot at what screen positions. The PLATO terminal writes 180 symbols per second, each symbol consisting of an 8×16 grid of dots. Similarly, the PLATO IV terminal has enough intelligence to draw the many dots comprising the lines in the geometry lesson simply from endpoint specifications sent by the computer, at the rate of 60 connected lines per second [13].

There is much more of a technical nature that could be said concerning the nature of a student terminal useful in education, but hopefully the heredity and geometry examples illustrate the basic point: for educational purposes a sophisticated terminal is required. As an exercise, imagine transferring the pedagogical approach of these two examples to a system with typewriter terminals. It would be impossible to preserve the essential aspects of these educational materials, proof that the nature of the student terminal largely determines the possible pedagogical approaches.

The effect of the type of terminal on the range of educational possibilities has been too often underestimated. We have seen interesting pedagogy created following the introduction of each new terminal capability. Other devices under development that have already generated unusual lesson material are a random-access audio device and a touch-panel that permits the computer to recognize where the student is pointing at the display screen. In recognition of our present ignorance of what may prove to be valuable in the future, the PLATO IV terminal has extra input and output connectors for easy attachment of new devices.

Computing power

The need for computing power is well illustrated by the introductory examples. The fruit flies are generated randomly, following the statistical laws of inheritance. No two students will experience the same results, except in the statistical sense. It is the biological algorithm of Mendelian genetics that is programmed. The algorithms of the geometry lesson involve much computation to achieve the accurate pattern recognition of the student's open-ended geometrical response. Understanding the chemistry student's free-form questions requires organized searches of a rather large data base of vocabulary words and basic concepts. All of these aspects of computer-based education require a powerful computer, as opposed to the meagre computing requirements for simple multiple-choice materials. Because the memory banks and other non-computational parts of a computer system comprise a major portion of the total system and are similar in cost whether the computational unit is powerful or weak, a weak system can easily be more expensive than a powerful one. The weak system may be capable only of simple programmed instruction or multiple-choice testing which can be done much more cheaply with books and other media. Only a powerful system can, through its enhanced capabilities, justify its costs.

Authoring

The fruit fly lesson was written by a biologist, the geometry lesson by a mathematician, and the lesson on qualitative organic chemistry by a chemist. These authors were able to create the sophisticated materials on their own, without the aid of programmers. This relates directly to the need for strong computing power in the system to lift much of the programming burden from the lesson author, yet place him in direct control of the medium.

One of the major tasks in building the fruit fly lesson was the generation of the special characters used to assemble a picture of a fruit fly. The biologist drew the characters directly on the screen, then used them in this lesson. To create the dialogue lesson, the chemist constructed a list of the relevant vocabulary words, stated word synonymy, listed the basic concepts and the corresponding responses. The system took care of transforming the wide range of student responses into forms which would match the basic concepts and yield an appropriate response. The mathematician's task was facilitated by powerful calculational capabilities easily accessible in the system for performing his pattern recognition task. All three authors benefited greatly from the system's responsiveness, for they could switch in a few seconds from authoring the lesson to testing it as a 'student', then back to writing and correcting it. This speed of transition is enormously useful in lesson creation.

All of the PLATO materials are written in the TUTOR language which is specially designed to facilitate the creation of computer-based lessons utilizing graphical display terminals. We give an example of TUTOR programming in Appendix B.

Cost

We have already discussed two important factors which influence cost: the plasma panel makes possible an inexpensive graphical display, and an appropriate authoring procedure enables authors to create their own materials. Another major cost area is the computer itself, and it is appropriate to discuss briefly the novel computer utilization in the PLATO system [14].

A 'time-sharing' computer, which seems to service many users simultaneously, actually serves only one user at a time. The computer services a user for a few thousandths of a second. If the computer manages to finish its work for all the users within a fraction of a second, each user has the illusion of complete control of the machine. In going from one user to another, the computer must save the first user's programme and status and load the second user's programme and status. This procedure is called 'swapping'. It takes place between the computer's high-speed memory banks and a mechanical, rotating disc or drum of magnetic recording material. Unfortunately, the mechanical speed of these devices is extremely slow compared with the electronic speed of the computer, so that the swapping procedure involves a heavy overhead. The computer is frequently either waiting for a programme to work on or involved in the difficult decision of whether to swap or what to swap in order to maximize its overall efficiency. As a result such systems tend to have high computer costs because the computer is doing useful work only a fraction of the time. To put it another way, the computer can handle only small numbers of simultaneous users and the cost per user is proportionately high. Moreover, computer-based educational materials administered by such a system tend to be of a simplistic frame-presentation nature, because the constraints of a slow swapping procedure require that the material be organized in a linear sequence of very short segments. This is a severe limitation: richness of cross-connexions is needed to provide quality materials.

One obvious solution would be to keep the students' lessons and individual status information in the computer's memory and avoid swapping. This has almost never been done because even auxiliary bulk computer memory is far more expensive than disc or drum memory. PLATO started from the premise that this scheme should nevertheless be used to improve quality and to improve computer utilization. It is overall performance that matters, and increased memory costs are offset by the elimination of the high swapping overhead, with drastic improvement in quality. While a student on the PLATO system studies his lesson, no swapping to disc or drum occurs: the swapping is to a special auxiliary computer memory of extremely high speed (the Control Data Corporation 'Extended Core Storage'). To maximize the usefulness of this memory, lessons are shared, with only one copy of a lesson in the memory no matter how many students are studying it. (In disc-swapping systems, students usually have to have their own copy of the lesson, as well as their individual status in that lesson.)

Because the entire lesson is available, corresponding to one or two hours of student study, PLATO lessons are usually quite complex in the interconnexions of their parts and rarely resemble the frame-by-frame question-and-answer format so prevalent in the field of computer-assisted instruction. Again we see that, as with the type of terminal, *the system design has an important bearing on the possible styles of pedagogy*. This point has been systematically ignored by too many researchers who have thought that questions of system design were minor compared to pedagogical questions, not realizing that the limitations of their systems were distorting their research results. Only if the system is sufficiently powerful to pose few constraints on possible educational approaches do details of the system cease to matter.

The result of this restructuring of the computer utilization in the PLATO system is that the computer ceases to be the most expensive part of the computer-based educational system, because a large computer can now run hundreds rather than tens of terminals. This order-of-magnitude improvement is due mainly to the elimination of swapping, but is partly due to the simplicity of the plasma panel terminals and associated telecommunications equipment. For a discussion of overall costs, see the articles under reference 1. Total costs including capital and operating costs are estimated at about $ 0.50 per student hour at a terminal.

Suggested reading

There recently appeared a two-part article by science reporter Allen Hammond on the present state of computer-based education in the United States [15]. It discusses the range of uses of computers in education and the current large-scale projects funded by the National Science Foundation.

Dartmouth College has been active in the computational use of computers in education and is the nucleus of a large network of schools and colleges engaged in these activities and utilizing the Dartmouth computer system. (See reference 10.)

An important centre in Europe, directed by Yves Le Corre, is the 'Ordinateur pour Etudiants' [16] of the University of Paris, where work has been done in physics and in biology.

The Physics Curriculum Development Project [17] directed by Alfred Bork at the University of California, Irvine, has produced a considerable body of material in physics. Both direct instruction and computing have been introduced into physics courses at Irvine.

A group led by Wallace Feurzeig at Bolt Beranek and Newman, Cambridge, Massachusetts, has created the LOGO language for computational applications that need not be of a numerical nature. This and several other groups, including one led by Seymour Pappert at the Massachusetts Institute of Technology, have had students of various ages, including young children and college students, write LOGO programmes to study mathematics and problem-solving [18].

There exists a voluminous literature on computers in education, but the field changes so rapidly that publications earlier than 1969 tend to be of little use now. The utilization of computers in education is almost as widespread as computers themselves, so we have cited only some representative projects whose size and commitment have permitted the creation of significant quantities of curriculum materials.

REFERENCES

1. BITZER, D. L.; BLOMME, R. W.; SHERWOOD, B. A.; TENCZAR, P. The PLATO system and science education. In: *Proceedings of a Conference on Computers in Undergraduate Science Education.* IIT and Commission on College Physics, 1970. (Available from American Institute of Physics, 335 East 45th Street, N.Y., N.Y. 10017.)

    ALPERT, D.; BITZER, D. L. Advances in computer-based education, *Science,* vol. 167, p. 1582 (1970).

    BITZER, D. L.; JOHNSON R. L. PLATO: A computer-based system used in the engineering of education. *Proceedings of the Institute of Electrical and Electronics Engineers,* vol. 59, p. 960 (1971).

    LYMAN, E. A summary of PLATO curriculum and research materials. *Computer-based Education Research Laboratory (CERL) report X-23.*

2. HYATT, G. W.; EADES, D. E.; TENCZAR, P. Computer-based education in biology. *BioScience,* vol. 22, p. 401 (1972).

3. DENNIS, J. R. Teaching selected geometry topics via a computer system. *CERL report,* X-3a (1969).

    DENNIS, J. R. Identification of pictorial responses in computer-based geometry instruction. *CERL report, X-3b* (1971).

4. SMITH, S. G. The use of computers in the teaching of organic chemistry. *Journal of chemical education,* vol. 47, p. 608 (1970).

5. TENCZAR, P.; GOLDEN, W. M. Spelling, word and concept recognition. *CERL report,* X-35 (1972).

6. SMITH, S. G.; GHESQUIERE, J. Computer-based teaching of organic chemistry. In: vol. IV of *computers in chemistry and instrumentation,* ed. by J. S. Mattson, H. C. MacDonald, and H. B. Mark, Jr. New York, Marcel and Dekker, Inc., in press.

7. SHERWOOD, B. A. Judging algebraic expressions and equations. *American journal of physics,* vol. 40, p. 1042 (1972).

8. SHERWOOD, B. A.; BENNETT, C.; MITCHELL, J.; TENCZAR, C. Experience with a PLATO mechanics course. In: *Proceedings of a Conference on Computers in the Undergraduate Curriculum.* Dartmouth College (1971).

    SHERWOOD, B. A. Free-body diagrams (a PLATO lesson). *American journal of physics,* vol. 39, p. 1199 (1971).

9. This 'GRAFIT' computing language, created by B.A. Sherwood, is itself written in the TUTOR language, as are all the lessons discussed in this article.

10. KEMENY, J. G.; KURTZ, T. E. Dartmouth time-sharing. *Science,* vol. 162, p. 223 (1968).

    NEVISON, J. N. The computer as a pupil: the Dartmouth Secondary School Project. Kiewit Computation Center Report (1970), Dartmouth College. *Biennial Report 1969-1971,* Kiewit Computation Center Report.

    LUEHRMANN, A. Should the computer teach the student, or vice versa? *AFIPS Conference Proceedings,* vol. 40, p. 407 (1972).

    Also see a sequence of five papers in *AFIPS Conference Proceedings,* vol. 34, p. 649-689 (1969).

11. This pictographic programming language was created by T. Tenczar. It is written in TUTOR.

12. JOHNSON, R. L.; BITZER, D. L.; SLOTTOW, H. G. The device characteristics of the plasma display element. *IEEE transactions on electron devices,* vol. 18, p. 642 (1971).

13. STIFLE, J. The PLATO IV Student Terminal. *Proceedings of the Society for Information Display,* vol. 13, p. 35 (1972).

14. TENCZAR, P.; BLOMME, R. W.; PARRY, J. H.; SHERWOOD, B. A. *PLATO IV system software* (to be published).

    STIFLE, J. PLATO IV architecture. *CERL report,* X-20 (1972).

    STIFLE, J.; BITZER, D. L.; JOHNSON, M. Digital data transmission via CATV. *CERL report,* X-26 (1972).

    BITZER, D. L.; JOHNSON, R. L.; SKAPERDAS, D. A digitally addressable random-access image selector and random-access audio system. *CERL report,* X-13 (1970).

15. HAMMOND, A. L. Computer-assisted instruction: many efforts, mixed results. *Science,* vol. 176, p. 1005 (1972).

    HAMMOND, A. L. Computer-assisted instruction: two major demonstrations. *Science,* vol. 176, p. 1110 (1972).

16. *Rapport d'activité O.P.E.*, Laboratoire de l'O.P.E. (Paris-VII), 84 p. (1972).
    FISZER, J. The use of a computer in teaching biology. Current trends at the O.P.E. laboratory. *Symposium über programmierte Instruktion und Lehrmaschinen.* G.P.I. (Hochschuldidaktik), Berlin 5-8 April 1972.
17. BORK, A. M. The computer in a responsible learning environment—Let a thousand flowers bloom; and M. Monroe, Physics computer development project—computer assistance in student problem assignments; both in: *Proceedings of a Conference on Computers in the Undergraduate Curriculum*. Dartmouth College (1971). The project has produced many internal reports in addition to publications.
18. FEURZEIG, W.; LUKAS, G.; FAFLICK, P.; GRANT R.; LUKAS, J. D.; MORGAN, C. R.; WEINER, W. B.; WEXELBLAT, P. M. Programming-languages as a conceptual framework for teaching mathematics. In four volumes, Report No. 2165, Bolt Beranek and Newman Inc., Cambridge, Massachusetts (1971).

Appendixes

A. LARGE SYSTEM VERSUS SMALL SYSTEM

There has been much discussion of the merits of large versus small computer-based education systems. As proponents of the large-system concept, it may be helpful to discuss our reasoning on this matter.

It should first be made clear that we are not talking about the question of 'centralization' versus 'decentralization', which is essentially a different issue. If a student or author has the full power of a large system available at any terminal, whether near or far from a large computer, that system is decentralized as far as the user is concerned. Conversely, if a part of the authoring process for a small system must be carried out in a different place, on a special authoring computer system, then a critical part of the operation of the small system is inconveniently centralized.

Years of detailed data collection on the PLATO III system show that *average* processing and information transfer requirements for a student are remarkably independent of what subject he is studying, method of presentation, age, etc. For example, an elementary-school student working on a simple drill goes through material rapidly but this material requires little processing or display for each interaction. On the other hand, a college student studying complex scientific material thinks a long time between interactions, but this material requires a great deal of processing and display generation for each interaction. The product of interaction rate and computer processing or display requirements per interaction turn out to be approximately the same in both cases. To be specific, averages of approximately 1,000 computer processing operations per second and about 15 displayed characters per second (approximately 150 words per minute) characterize our findings. Since the PLATO III system is characterized by processing and display rates of over twenty times these average rates, the observed average requirements presumably reflect physiological constraints.

In the design of a viable system these averages are not the whole story: the peak requirements are just as important. The science student, on the average, thinks for a long time between interactions, but the system must respond instantly so that the student can continue his line of reasoning without interruption. It would be disastrous to force the student to wait a long time for the reply. There is therefore an enormous difference between the average and peak rates. Without going into the details of 'queueing theory', it should be clear that only a large system has the necessary reserve power to work rapidly through the huge peak requirements represented by the science student's interactions. Additionally, the larger the system, the less damage is caused to system responsiveness by statistical fluctuations in the number of students simultaneously requiring service. Roughly speaking, if N interactions per second are anticipated, the number observed will be $N \pm \sqrt{N} = N\left(1 \pm \dfrac{1}{\sqrt{N}}\right)$ due to Poisson statistics; the probability of overload conditions scales like $1/\sqrt{N}$. These factors favour the large system.

An advantage of a large system that is difficult to quantify is that one large computer can perform much more complex tasks than can a group of small computers of comparable aggregate power. Free-form dialogue, complicated display generation, rapid extensive calculations, powerful authoring procedures—all of these are essentially out of the reach of the small computer. The reason for this is rather subtle. A time-sharing computer services only one user at a time. During the fraction of a second that the computer is working for an individual user, *all* of its basic resources are devoted to him: fast memory, processing unit, data transfer channels, etc. The more powerful these resources, the more sophisticated will be the service. For example, a large fast memory with high-speed transfer from bulk memory permits operations on a large data base of vocabulary for natural-language dialogues Unlike the swapping medium (bulk computer memory or disc memory) whose total cost is proportional to the number of users (each of whom needs some average bulk memory allocation), the basic computer resources used during the actual fractional-second processing *are not duplicated for each user*. The larger the number of users of the system, the more can be paid for basic computer resources to permit more and more sophisticated processing. With a small number of users, a weak processor with small amounts of fast memory and inadequate transfer capabilities is all that can be paid for, at the same cost per user as will buy much more capability in a large system. (One might object that the cost of the central processor is proportional to the number of users and their required number of operations per second. However, the more expensive processors have added capabilities as well as increased speed, so processing requirements do not scale linearly. Additionally, the processor usually represents only about 10 per cent of the total system cost.)

There are therefore two related but different reasons why many PLATO capabilities could not be duplicated in systems designed to serve a small number of terminals. One is that the peak demand by a student may take an unacceptably long time to process (and cause queuing problems for other students), and the other is that the basic resources may be inadequate to perform some tasks at all (insufficient memory to manipulate a data base, etc.).

Within a factor of two or three, the management and administration of a large system is comparable in cost to that of a small system. In both cases, there must be a director, an assistant director, some computer operators, etc. This makes the management of multiple small systems expensive.

There is no way of making many small computers temporarily look like a large computer in order to carry out heavy computational tasks, such as sophisticated analysis of educational data gathered by the system. A large system can handle both the student interactions and standard computing jobs as well. The processing of standard administrative and research computing jobs helps pay for the system, whereas a small system incapable of this performance is purely an add-on expense. A related point is that the management and distribution of a large data base of curriculum materials is best handled by centralizing the storage of these materials. This permits teachers to monitor students performance at a distance and ensures that lesson material can be updated for all students, everywhere in the network.

B. THE TUTOR LANGUAGE

All PLATO materials are written in the TUTOR author language, which was originated by Paul Tenczar in 1967. Figure 9 gives a simple example of TUTOR programming and its use by a student. Note that the lesson author did not list 'tringle' as a possible misspelling of 'triangle': the misspelling was detected by the algorithms of the TUTOR 'answer' command. In addition to the simple display and judging commands illustrated here, TUTOR has a large repertoire of display, judging, calculational, and branching capabilities which makes possible the complex lesson examples described earlier in this article. Because TUTOR is a full language, not a format for administering standardized items, authors are not restricted to a particular pedagogical strategy or presentation mode. (In fact, some authors have even constructed TUTOR lessons which administer standardized items drawn from a structured data base, so this capability is also available.)

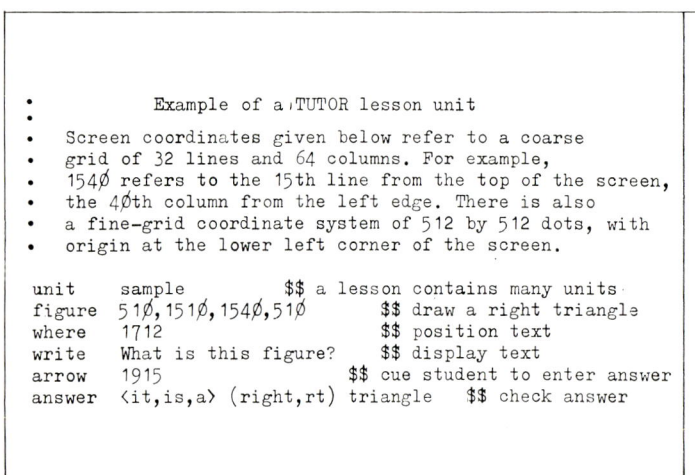
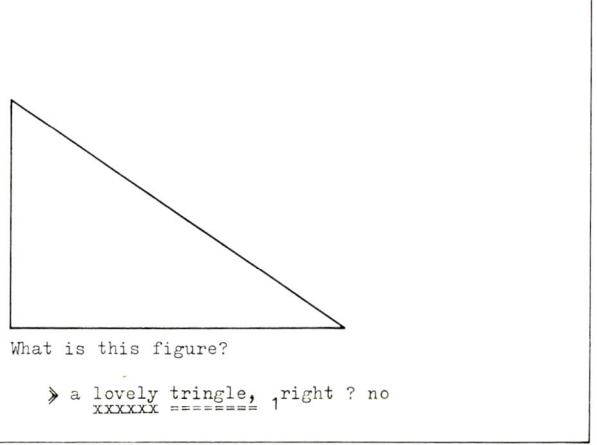

Fig. 9. The TUTOR author language: a lesson unit and how it looks to the student. In the 'answer' command the words 'it', 'is', and 'a' are specified to be unimportant, and 'right' and 'rt' are to be considered synonyms. The reply to the student is much more than a simple 'no'—the word 'lovely' is crossed out, the misspelling is underlined, and the word 'right' belongs to the left.

Programmed learning in science education

by George O. M. Leith
Department of Research and Development in Education,
Rijksuniversiteit te Utrecht, Maliebaan 5, Utrecht, Netherlands

Summary

Programmed instruction is a set of techniques and principles for designing effective learning situations. A programme is therefore a means of ensuring that students achieve intended objectives. This is accomplished by carrying out analyses of objectives, as well as of the learning tasks involved in their achievement, of learning processes which students must employ, and conditions (including media) which will facilitate learning. In addition, the instructional materials and their environment must be set up and tested, that is, the programmer must try out his programme and obtain feedback to discover what modifications are needed for students to master the tasks.

During its evolution a number of relatively restrictive models were developed which were initially regarded as rivals. It is now recognized that the earlier forms of programme limit development too much and a generalized systems approach is what characterizes programmed learning.

In particular, the idea that programmed learning is self-paced, individual learning has been abandoned. For example, it has become clear that group-paced learning is often as good as or better than self-paced learning. This has opened the way for the employment of audio-visual media and much greater effectiveness is found when, e.g., television lessons are programmed. At the same time, small-scale methods such as roller boards or programmed lectures, with classroom feedback from students (who display a coloured card or light to indicate their responses) have given good results and found easy acceptance where there is resistance to greater changes.

It has also been found that social interaction may form part of a programmed environment so that co-operative learning and discussion will sometimes be indicated to achieve particular objectives. One aim of programmed learning—to individualize instruction—is nearer to fulfilment. It has been found that students may differ in learning style so that what is helpful for one kind of individual may hinder another and vice versa. In general, discovery seems to facilitate learning and transfer of extroverted students of secondary and tertiary level while introverts are favoured by more highly prompted strategies.

However, there are many difficulties in mapping out this area. Thus primary school children seem to differ in success with discovery according to their degree of anxiety rather than extraversion. Anxiety also changes role with age in helping or hindering younger and older students learning from programmed instruction whatever the approach.

The success of programmed learning of science in developing countries seems to vary according to the flexibility of approach in adapting to specific characteristics of students and to the style of schooling which prevails—as well as to the needs for instruction. Programmed learning is time-consuming to prepare but some progress has been made in adapting existing materials to other cultures and languages, though some serious problems of perceptual and intellectual development differences are apparent.

Introduction

Programmed learning has contributed to education in a number of important ways. On the one hand, it has focused attention on several critical aspects of the learning situation (need for student activity, learner-centred emphasis, importance of objectives and feedback). On the other, it has helped to promote the acceptance of systems thinking in education (system-objectives, analysis of substructures and their relationships, internal evaluation and modification and system-evaluation).

Its impact has been not only at micro-levels but also at the wider macro-levels of innovation, teacher-training and planning. This penetration of concepts into wider and wider fields of educational thinking has to a large extent confounded and diffused the 'image' which programmed learning fostered initially. In addition, changes in techniques of preparation and presentation have made programmed learning situations difficult to recognize.

Paradoxically the success of programmed learning is to be measured, in part, by the degree to which it loses identity as one among many educational methods and media and becomes absorbed into the general framework of educational polices and practices. Such assimilation and influence is inevitable since—unlike television and audio-visual hardware, the project method and so on—programmed learning is a set of meta-principles for selecting, organizing, developing, modifying and evaluating more specific approaches, techniques, media and methods.

The fact that programmed learning may be thought of as a method of instruction, a medium, a set of techniques, a systems methodology and a philosophy can lead to ambiguity when its introduction is considered or its effectiveness in use is evaluated. There seem indeed to be at least nine identifiable motives for introducing and using programmed learning:
1. To enhance the efficiency (and effectiveness) of learning.
2. To overcome a shortage of teachers.
3. To improve pre- (and in-) service teacher training.
4. To individualize instruction.
5. To introduce a specific curriculum reform.
6. To provide models of teaching methods.
7. To help in the reorganization of teaching (e.g., team-teaching, destreaming).
8. To act as a research tool in educational investigation.
9. To initiate a systems approach.

While some of these reasons for adopting programmed learning are not incompatible with each other, several of them would be difficult to make consistent in a project. For example, if a project with limited resources aims to introduce a new curriculum (e.g., modern mathematics) a substantial effort will be needed to select and train a group of programmers, to prepare curriculum objectives, methods and materials and to prepare mathematics teachers to use the materials in a useful manner. To give widespread training in programme construction at the same time would be a dilution of effort. Yet it appears that planning for optimal success of programmed learning does not always appreciate that innovational strategies have to be selected according to the objectives of the innovation.

The introduction of training in programming methods in a teachers' course might aim to improve their classroom teaching performance or to help them to prepare programmed texts [29].[1] In the former case the teachers may never construct or use a programme in school—but the goal of improving performance might be attained. In the latter, teachers may become materials producers in a teaching team and never 'perform' in the classroom. What the training aims to accomplish seems rarely to be made explicit or what it achieves to be measured.

Models of programmed instruction

Programmed learning initially evolved four models for designing learning situations. The earliest, 'adjunct', method was in effect an insistence on frequent diagnostic tests with immediate knowledge of results and corrective feedback to accompany units of instruction. 'Intrinsic' or 'branching' programmed instruction went further in providing additional remedial sequences following correction, skip gates for learners not requiring the section, and an elaborate set of rules for constructing programmes [40]. Linear (extrinsic) programmes were based on a particular account of learning and included many prescriptions for eliciting, shaping and reinforcing responses [44]. The fourth model 'Mathetics' shared with linear programming a behaviouristic analysis of learning but developed a radically different approach to constructing learning systems [13].

Some years ago programmed learning was defined in terms of a set of essential principles or characteristics.
1. Construction of programmed learning materials is steered by an explicit analysis of objectives, formulated in operational terms ('students will be able to...').
2. Programmed learning involves active responding (frequently phrased to mean 'overt responding') by the learner.
3. The programme consists of a progressive (logically ordered) sequence of small steps.
4. Students receive immediate knowledge of results about the correctness of responses.
5. The programme is modified until students make few errors as they respond (5 per cent linear, 15-20 per cent branching).
6. The learner proceeds at his own individual pace.

Almost all of these 'rules' have now been demonstrated to be invalid and many programmers show a sceptical disregard of the 'essential' rules. Indeed those which now appear to be critical are:
1. Formulation of objectives and criteria of successful achievement.
2. Analysis of tasks and design of learning situations.

1. For references, see page 48.

3. Validation of learning situations and evaluation instruments.

Research over the last ten to fifteen years has gradually stripped away many of the features which seemed especially characteristic of programmed learning.

Thus small steps, overt responding, self-pacing, feedback prompts, low error-rate were shown to be only sometimes helpful, and often unnecessary or even handicapping. The successful use of group-paced audio-visual programmed instruction as well as studies showing the effects of forced-pace learning invalidated individual self-pacing as a defining rule [24].

Systematic variation in size of step has revealed that this factor must be related to subject-matter, learner-maturity, etc. [24]. Again, while the effectiveness of overt responding has been upheld for specific situations, others have been defined in which *covert* responding is reliably better [24]. These and other rules of thumb which once defined programmed learning have been shown to be dubious, sometimes but not always applicable, and often misleading or false.

Mathetics has a different set of rules, not all of which are established, but which, since they are a departure from other 'paradigms', can be briefly summarized for the sake of contrast. The unit of learning is the operant (S-R connexion). Operants are said to differ in 'span', that is (oversimplified for conciseness) the remoteness or degree of complexity of a response which is reliably elicited by a stimulus. There are allowed to be covert connexions between S and R (S-r-s-R) which may extend the 'span'. Mathetics programmers like to estimate an operant span which is a bit larger than the target population can stretch to rather than one which is too small—in contrast with the 15-word frames of early linear programmes. These S-R connexions are organized into *chains* in which a previous response is a stimulus for a following response. Also distinguished are 'multiple discriminations', where the learner might confuse the particular S and R relationships, and 'generalizations' or concepts.

Facilitating methods are described which emphasize focusing attention, preventing interference, etc.

The most unusual aspect of Mathetics is that some things are organized so that they are learned from the final stage (remove the hypodermic syringe, rub the site of the injection and clear up) back through to the beginning (collect equipment, assemble the syringe... etc. [1]). Chains tend to be set up backwards in order to maximize the motivation of completing the job, and to ensure that interference (forgetting) in the middle part of the chain does not occur. (Knowledge of results prompts are usually omitted—the 'reinforcement' being thought to lie in successful completion of the task).

Additional features are: recognition of the effects of retro- and pro-active interference and means of preventing them; the use of simulation (apparatus, models, paper and pencil representation) and advocacy for no more theoretical material than is required to achieve the objectives. There are also somewhat complex rules for determining the sequence for teaching units of instruction.

Mathetics, which has been used mainly in industrial training, approaches task analysis less from the angle of subject matter than from analysis of what a 'master' does. The aim is to give learners mastery of the job (complex task) whatever this may be.

The approach has been considered suitable particularly for learning procedures, skills and other types of tasks which are relatively low in the taxonomy of educational objectives but it can in fact make contributions at many levels. Examples of Mathetics programming include 'Interpretation of infra-red spectra' and 'Transfer of RNA' which involve learning conceptual principles and, in the former, interpretative judgement.

In Table 1 (pp. 38, 39) main characteristics of five approaches to programmed instruction are listed. The fifth represents the present phase of evolution.

Though training in linear and branching methods of programming is still practised there is a growing tendency to ignore the prescriptions laid down earlier and to choose techniques from an ever widening range. For example, programmes may include simulation and gaming elements, employ tape-slide and loop-film components, can be presented by television or in multi-media form as well as by means of print. Teaching machines, which limited the range of techniques available and added little or nothing to learning, have become redundant.

From the critical features of programmed learning a number of concepts have emerged which have attained importance in their own right.

---

1. The operant span may include collect ... assemble ... check ... fill or each of these may be a span or each may be broken into smaller units depending on the learner's repertoire.

Leading concepts in programmed instruction

Mastery learning. Mathetics is explicit about the goal of mastery. Other methods have emphasized the attainment of high levels of performance (the 90-90 criterion: 90 per cent of learners will achieve not less than 90 per cent of total test score). Use of criterion frames, unit tests, remedial branching and validation procedures attempted to ensure mastery. When programming is released from particular constraints it emerges in one form as mastery learning in which no student proceeds to the next instructional unit until he has mastered the one before [6, 45].

Criterion-referenced evaluation. While Schonell's carefully designed Diagnostic and Remedial Tests foreshadow this notion (the pupil's difficulties and error patterns are elicited so as to design individual remedial teaching for him) the present distinction of criterion- and norm-referenced testing has emerged from programmed learning. A norm referenced test ranks pupils in order of merit—a criterion referenced test indicates that a student has or has not achieved mastery. Hence the emphasis changes towards diagnosis (analysis of subordinate tasks) and insistence on 'readiness' training [37].

Learning hierarchies. Mastery learning is evaluated by criterion-referenced tests. These notions in turn imply (though not in a strict sense) that reaching a terminal-objective requires a succession of subordinate tasks to be mastered in a particular order. This point of view, elaborated by Gagné, has received limited support from research and seems particularly fitting in science education where knowledge can be arranged hierarchially. For example, the Principle of Conservation of Momentum states a relationship between Velocity and Mass and thus subsumes them. Velocity is a concept which unpacks into speed and direction and so on [12].

Taxonomy of learning. The notion of hierarchy involves not only subsumption (vertical sequence) but also a horizontal lay-out of levels, each level being characterized by a position in a learning taxonomy. For example, a principle integrates concepts. Concepts are abstractions which may be formulated as rules about the classification of members and inclusion or exclusion of instances (e.g., cats are mammals with fully retractile claws). These are two levels of a taxonomy of learning such as that given in Table 2 (p. 40).

Task analysis. An educational task analysis differs from a subject-matter analysis (such as programmed learning used to involve) in making explicit attempts to link the terminal objectives with the structure of knowledge and capabilities involved and to define these in terms of learning activities. For example: to give the capacity to generalize a scientific principle and apply it to new cases, a task analysis will specify a particular type of activity, e.g., personal eduction of the principle from problem situations rather than meaningful reception (understanding the meaning of a principle and learning to formulate it). Thus the taxonomy of learning processes is a guide to constructing the first version of a learning programme by specifying the kinds of knowledge and capacity needed to acquire the content of a curriculum outline and achieve its objectives. It also helps to determine the sequence of learning activities and, because it has a kind of network analysis function, to point up missing parts of the sequence.

Feedback and objectives. Programmed instruction has distinguished two kinds of feedback—to the learner and to the programmer or instructor. Feedback to the learner has the function both of informing him that he is proceeding correctly and of giving him reinforcement. That to the programmer/instructor comes from the performance of students. If too many students are unsuccessful at predetermined nodal points of a learning programme then it must be modified and improved. Investigation of students' errors is carried out to diagnose faults in the sequence of learning activities. This appraisal applies both to progress checks throughout the programme and to the total effectiveness of the programme in achieving the stated objectives.

It should be noted that many inadequacies in learning come about through interference of one part of a learning sequence on others—leading to forgetting. Much can be done to identify potential sources of conflict and competition and introduce facilitating means to avoid them. For example the use of mediating material—mnemonics, perspective-giving rules, pointing up differences—has already been referred to. It helps also to provide summary reviews following each unit of learning [27].

TABLE 1. Characteristics of five programming methods

Programming method	Background concepts	Presentation mode	Response mode	Immediate knowledge of results	Size of steps	Use of visuals, audio-visual media performance tasks, etc.
Adjunct	Thorndike's connexionism	Print, lectures	Multiple-choice tests	Yes. Student continues selecting until correct answer reached	'Lesson'	Little implied
Branching	Eclectic	Teaching machines 'scrambled' texts	Multiple-choice-selection	Yes. Remedial instruction follows false choices	Several hundred words or less	Diagrams and pictures (random-access tape too expensive)
Linear	Operant conditioning	Teaching machines, print	Overt, constructed responses (occasionally multiple-choice)	Yes. Small error rate	Fifteen to twenty-five words (exceptionally more)	Diagrams and pictures, Nowadays tape/slide, closed circuit television, etc.
Mathetics	Operant conditioning	Print, audio/visual tactile presentation, models, simulators, etc.	Multiple-choice (matching, ranking, etc.) constructed responses, covert responses	Infrequently used	As large as possible	Extensive use of simulators (sometimes ingenious paper and pencil models), check-lists, algorithms, performance tasks, etc.
Learning systems design	Eclectic	Single or multi-media (as Mathetics)	As determined by task analysis	No rule	Determined empirically	As Mathetics, in addition, group tasks

Programming method	Rules for sequencing	Typical teaching method	Accommodation of individual differences	Integration with other methods	Typical programmer	Utility and base of preparation
Adjunct	None	'Normal' teaching	None	Feedback classroom	Teacher	Cheap, quickly set up, effective in 'robust' teaching situations
Branching	Logical subject matter analysis	Didactic exposition, sometimes inductive	Remedial branches skipgates (usually not effective)	Sometimes integrated 'live' teaching	Subject matter expert sometimes with programmer or programming team	Branching of questionable value, cannot deal with some tasks, moderately easy to prepare
Linear	As branching	Ruleg (statement of rule, applications) Egrule (examples, rule educed)	Difference in pace of learning (group-pacing as good)	As branching	As branching	Useful for some tasks and immature learners; danger of boring academically able, moderately easy
Mathetics	Symbolic flow chart, system of weights to estimate order of units and topics	Demonstration imitation, application (but no method barred)	Matheticist would prepare more than one programme	Teacher monitor sometimes programmed-in	Mathetics programmer or team aided by 'master'	Wide range of application, difficult to prepare in complex areas
Learning systems design	Task analysis, network analysis	Methods chosen to match objectives (including 'drill')	Recognition of differences in learning style, preparation of different methods and modes	Teacher integration, peer-learning, project-learning, team-teaching	Team approach — objectives/task analyst, programmer, subject-matter consultants	Few constraints except validation, requires many special skills

TABLE 2. Taxonomy of learning

Learning process	Example	Methods to facilitate learning
Stimulus discrimination	Distinguishing cytological specimens	Presentation of extreme differences, high-lighting, labelling, giving distinguishable names, knowledge of results, gradual removal of prompts (fading or vanishing—compare Skinner's technique)
Response learning	Learning to whistle	Practice in making the response, hints to help to achieve it, successive approximations monitored by knowledge of results (compare Skinner's shaping of responses)
Hook-up (association)	Correctly naming anatomical parts	Various forms of drill practice unless mediating associations exist when mnemonics or meaningful procedures are helpful. A mixture of repetition and completion (response prompting and constructed responding) may be best. (Gilbert's Multiple Discrimination is a complex case.)
Response integration (Serial learning, chaining)	Responses are individually known but must be assembled in a particular order or a new order, e.g., a list; a procedure; a new, polysyllabic term	Long strings of response can be learned by Gilbert's backward chaining. Shorter ones benefit from mediating links (e.g., mnemonics)
Learning how to learn (Learning set formation)	'Knacks', intuitive judgements. Essentially one learns to eliminate faulty responses and select relevant ones	Explicit rules do not enter the learning because the learner is at prior stage of development or understanding (e.g., the rule is complex), because the experience required to give meaning has not yet been achieved, or there is no clearly formulable rule. Learning is achieved by mastering the solution of problems which must range widely in variety
Learning concepts	Concepts are 'defined' by a rule though the learner need not be able to formulate it. Mammal, disordinal interaction, reference group	If the learner has related concepts and experience the new concept can be learned by giving the rule. Alternatively it must be derived from instances and non-instances
Concept integration (Principle learning)	Principle of moments	Concepts related together by logical, mathematical, probability or other links can be learned by pointing up the concepts and having the learner discover the relationship or by formulating it directly
Hypothetico-deductive problem solving	Surprising event C occurs: if X were the case (a hypothesis) C would be a matter of course (C. S. Peirce)	Practice in formulating hypotheses, devising test situations and gaining evidence needed. Range of practice, guidance and obtaining of feedback important
Learning schemata	Structural framework of thinking, e.g., two-value logic	Learning of schemata may need 'dissonance' with existing framework and perspective-giving rules
Self-evaluation	Learning to make independent judgements and appraise one's own work	This probably needs group interaction in which criteria are developed and applied with mutual support before independence is reached

Another source of conflict, however, is failure to recognize that multiple objectives need special analysis to work out how they can all be accomplished at the same time. For example, learning rules for verbatim recall and for application in problem solving may require different methods of instruction. Hence, to employ one or another method will lead to achievement of one but not both objectives. Many courses, for example, require learning of practical capabilities as well as theoretical. Discovery of optimal methods for reaching all objectives is a difficult task.

Group-paced programmed instruction

Programmed Instruction is frequently thought of as self-paced instruction. Apart from the fact that the organizational problems of self-pacing are not always welcomed by teachers this idea shuts out the possibility of mass-media presentation and group-paced learning. From the point of view of developing countries, especially, it should be known that there is a lot of evidence for the success of group-paced programmed instruction.

Principles have been developed for the audio-visual presentation of programmes which have been compared with self-paced learning of the same materials (either audio-visual or some other format). For example, in testing a variety of methods for teaching the new decimal currency in Britain, adults who learned from a group-presented tape (with printed materials and coins to manipulate) learned better than from a self-paced version of the programme (as well as an adjunct programme). What is more, the difference between adults of above and below average intelligence was significant in the two self-paced conditions but not the group programme. The same was found with junior school children. On the other hand, fifteen-year-old schoolchildren were poorer than adults with the tape methods though they did as well with the self-paced methods. The reasons are probably to do with study skills and degree of attentiveness [28].

In another case young children were taught elementary electricity (by a discovery method). They learned better in the group-paced condition than by self-paced study [48].

Attention can also be drawn to studies of learning by programmed closed-circuit television (the approach including use of simple apparatus while viewing the monitor). Group pacing was again as good or better than self-paced study of the same materials which were on mathematics and physics [32].

These experiments involved children with a fairly wide range of intelligence. The group-pacing was preceded by trials to assess the median response time for each 'frame' of the programmes. There is a high degree of robustness in this method which seems able to accommodate a spread of as much as two standard deviations along an intelligence test scale.

Since these findings have been repeated many times (with roller-board, overhead-projector, audio-tape, slides and television) it would seem safe to advocate that television instruction and other forms of audio-visual teaching should be given the reinforcement of programming [36, 17].

Group-paced programmed instruction can of course be extremely economical in materials, learning time and skilled teachers. In situations of shortage, where the complex requirements of a completely designed new learning system are beyond reach, the improvement in learning to be gained by relatively inexpensive and easily administered methods cannot be refused merely because more sophisticated methods would accomplish more complex objectives. In this context, mention may be made of the several methods of ensuring active responding, obtaining feedback about how students are learning and giving them knowledge of results, designated as the 'feedback-classroom'. They incorporate elements of programmed instruction without necessarily involving a fully-fledged analysis, design and validation —though the validation can be carried over from lesson to lesson or class to class.

In its most sophisticated forms, expensive electronic equipment is required to equip each student with response buttons. Multiple-choice questions are answered by selecting a button. A teacher's console sums and displays the number of correct and incorrect answers on a display panel and students themselves are informed by means of coloured lights at their places [10]. Such systems have been miniaturized and made into portable equipment which can be wheeled from class to class [50].

In another case, a battery-operated box is used to display coloured lights to the teacher, while a yet more simple device is the 'Cosford Cube' which is a small cube bearing a different-coloured patch on each face. The learner can hold it cupped under his chin to show his answer to the teacher and no-one else [47]. Whereas many researches show how little attention is paid in lectures, the effectiveness of heightening attention by questioning rather than presenting information in statement form has received confirmation [43].

The introduction of systems methods may well be facilitated by this relatively simple adaptation of class-teaching techniques. The teacher (or course designer) is obliged to plan and structure the lesson more carefully, to choose occasions for student responding and feedback and, since overhead projection or other erasable displays are used, the lesson itself can be modified in the light of revealed defects—unlike integrated-media lessons using films or slides which are relatively inflexible and resistant to modification [18].

Co-operative learning

Much of the work on co-operative programmed instruction has been vitiated because the learning situations employed were not adapted to group interaction. For example, joint reading of printed frames merely irritates many learners. Where, however, genuine co-operation can be evoked and where social interaction is believed to have an educational role in addition to individual instruction there have been significant pay-offs.

Working in mixed ability groups of four, 8-year-old children, ranging in IQ from 75 to 130, learned about magnetism using a kit of apparatus for discovering information and inferring rules. These groups achieved learning gains which were equal for both less and more intelligent children. Moreover they gained more than even bright children working individually. Similar results were found with co-operating pairs of 10-year-olds who were learning principles of moments from a self-correcting model of the conceptual system. Low-ability children working together were not a successful combination, but paired with an able learner both kinds of pupils were superior in learning and transfer even to co-operators who were both bright [1].

At secondary school the pattern becomes quite complex [2]. For example, the sociometric effect of knowing the partner often appears to detract from learning (random assignment significantly better than own choice partner). The major effect, however, is the influence of differences in temperament. An interaction analysis was made of the dialogue between partners who had been systematically paired in all combinations of high and low ability, extroversion and anxiety [3]. The results of the analysis, in which four interaction categories had significant weighting in a multiple regression analysis, are shown in Table 3.

TABLE 3. Comparison of pairs having the same or different anxiety levels (heterogeneous ability pairs in brackets, homogeneous without brackets)

'Opposite anxiety pairs'	Achieved	(32 %) 74 %	more on the post-test than 'same anxiety pairs'
'Opposite anxiety pairs'	Achieved	(113 %) 98 %	more on the transfer-test than 'same anxiety pairs'
'Opposite anxiety pairs'	Spent	(36 %) 59 %	more time in showing solidarity, raising other's status, giving help and rewarding than 'same anxiety pairs'
'Opposite anxiety pairs'	Spent	(132 %) 121 %	more time asking for orientation, information, confirmation, than 'same anxiety pairs'
'Opposite anxiety pairs'	Spent	(21 %) 19 %	less time in disagreeing, passively rejecting, withholding help than 'same anxiety pairs'
'Opposite anxiety pairs'	Spent	(25 %) 49 %	less time in showing antagonism, deflating other, asserting self than 'same anxiety pairs'

The interaction analysis was made of a sample of children in a large-scale experiment the full results of which show that pairing children of opposite anxiety level who are both extroverts or both introverts gives the best results. Thus not only is the outcome clear but the processes of social reinforcement which bring it about are apparent [3].

Whether these results would obtain in other cultures must of course be determined by experiment. The point is that this series of experiments gives reason to believe that the very same children can either help or hinder each other's learning depending on the mixture of temperaments. While other studies do not disconfirm the evidence given above, there appears to be no other investigation in which all possible combinations of four temperamental qualities with intelligence has been attempted. Failure to do so, however, confounds the competing effects of same and opposite extroversion and anxiety levels and the effects are washed out.

One study of computer-assisted instruction may be cited in connexion with these findings. Some of the students worked alone and some co-operated in pairs. It was found that high sociability, low test anxiety students learned better in pairs while low sociability, high test anxiety students learned better as individuals. Submissive students working with dominant ones were favourable to computer-assisted instruction as were dominant students working individually [46].

Learning and individual differences

For several years there has been serious discussion and research on the question of aptitude/treatment interactions. The suggestion is that different methods of instruction, which are so often found to give equivalent learning outcomes, are appropriate for different kinds of students. Because relevant categories of students are not usually distinguished, the analysis of differences between methods shows wide error variances and fails to show an advantage for any method. It is suggested, however, that method A might be favourable for X-type students and method B for Y-type students.

There are two sets of dimensions which have revealed consistent interactions:

1. Degree of structure of learning tasks ranging from heavy guidance and prevention of errors to learning situations which require the student to derive structure and, in doing so, commit errors which are eliminated by prompting him to think again.
2. Personality traits, particularly the orthogonal dimensions of introversion/extroversion and emotionality/stability (general anxiety, neuroticism).

Extroverts are characterized not only by sociability, and 'outgoingness'. They are relatively more distractable than introverts, less given to persistence at boring tasks (i.e., tasks involving repetition of similar operations, fine detail etc.) and have a greater tolerance for ambiguity and uncertainty.

The last group of characteristics especially has profound effects on learning styles. Reviews of research on factors influencing academic achievement seem to agree in finding that introverted students (in school and university) are relatively more succesful than extroverts [11]. A reason given is that the more sociable nature of extroverts distracts them from study. Just as important seems to be the formality of teaching systems.

It appears necessary to find ways of enhancing the academic achievements of extroverts who are at present, apparently, dismissed as not being good students.

Evidence for personality/treatments interaction is summarized below to elucidate the kinds of instruction which helps or handicaps extroverts.

An early experiment attempted to define a dimension of 'discovery-direct instruction' by setting up problem-solving tasks which were organized in sets obeying a particular principle. These were arranged progressively, becoming gradually more difficult and more complex. One point along the dimension was as described above. Another point added a statement of each principle and a third point had in addition to that an indication of the correct solution. These were termed guided discovery 1 and 2 and complete guidance. 'Discovery' was the least structured set of tasks in which the problems were posed not in sets, but in a random order (except that the first few were easy, warm-up problems). Immediate feedback of the correctness or incorrectness of a solution was provided.

The four treatments did not represent a consistent dimension, however. For example, discovery 1 caused interference. Consequently the unstructured and most highly structured and prompted methods were compared (there were no between-treatments differences).

Extroverts were more succesful than introverts in further problem-solving tests following the 'discovery' method while introverts were superior to extroverts if they used the complete guidance method [23].

A further experiment was devised in which Mendelian genetics was programmed in a direct instruction manner and also as a 'discovery' approach. In this case students attempted to find a principle to explain the inheritance of characteristics from a three-generation family tree. Their attempts were corrected by pointing out factors which remained unexplained, further attempts to derive principles were made and so on. All students were tested by means of problems requiring application, extension of principles and new organization of principles.

Extroverts were superior to introverts by the discovery method, introverts better than extroverts by the direct instruction method. Students were tested again without warning after a month. Discovery group extroverts now had even higher scores and introverts much lower ones.

The direct method showed the same pattern as before but the difference was reduced. No overall difference between methods appeared on either test [23].

At secondary school level different approaches to teaching addition of vectors were set up to test the interaction of tolerance for structure with the degree of extroversion. The test was made more rigorous by employing methods in which children worked out practice examples either preceded by or followed by the rule or principle to be applied [31].

In another case they were asked to work out and write down the rule and some were then given a statement of the rule as a check and some were given no check. Immediate knowledge of results within the programmed learning sequences containing particular examples was given to all pupils [25].

It was argued that rules before examples would support introverts and constrain extroverts while rules after would permit extroverts to engage in testing and structuring in their own terms. Again the absence of check rules was expected to disturb introverts but not extroverts, and it was also anticipated that having to formulate and write down their own rule and match with an official rule would interfere with extroverts' learning. (N.B. the difference between 'rules after examples' and 'formulate rule, write it and check' is a difference in amount of constraining structure).

Extroverts were significantly better in learning and transfer (substraction of vectors) if they received rules *after* examples, while introverts were slightly better if they had rules *before* examples. Above average ability (but not below) there was an interaction of check and extroversion on the transfer test, extroverts being better without checks, introverts with. Below the average IQ, checks gave better results presumably because children were unable to work out the rules successfully. Another test of the hypothesis was made by teaching elementary meteorology to secondary school children by four programmed instruction methods: (1) small-step linear programme; (2) large-step text material followed by test questions; (3) linear programme—each section followed by a large step as in (2); (4) skip-branching —each large step followed by linear programme section to be skipped if test items were succesfully completed. Above- and below-average-ability children showed the same trends. Extroverts were more successful with (2) and (3), introverts with (1) and (4).

Cutting across these interactions of methods and extroversion is a relatively strong effect relating general anxiety (in the trait sense) and learning achievements. At secondary school level and beyond in the research described above, it seems to have no interaction with methods of instruction. For example, in the genetics teaching experiment anxious learners were much poorer in test scores than unanxious. However, the pattern is complicated by the fact that (a) the correlation with achievement is non-linear; (b) the direction of the influence changes. Thus younger secondary children, if they are in the upper and middle thirds of the distribution of anxiety, have tended to do better than the non-anxious remainder. At a somewhat older age, the optimal amount of anxiety is the middle range, while with college students the most anxious third appears to be handicapped compared with the remainder [30].

Furthermore, there is evidence [1] that anxiety rather than extroversion interacts with the methods of instruction among primary school children (structuredness helping anxious children) [23].

While other research programmes have also discovered interactions of this kind (e.g., inductive/deductive teaching by levels of anxiety) it is as yet difficult to make a coherent pattern. Often the individual difference measures used in different inquiries are difficult to relate to each other. Some anxiety tests, for

---

1. Confirmation has been recently forthcoming in a study of junior school mathematics learning now at the analysis of data stage (by E.A. Trown).

example, may also measure introversion with the result that one investigator's neurotic introverts might correspond to another's anxious subjects, and non-anxious to stable extroverts.

Programmed learning and science education in developing countries

Unesco has been responsible for a number of curriculum development projects in science education which have involved the use of programmed instruction materials.

Many important lessons about programming and about introducing programmed learning methods can be derived from experiences in Africa, Egypt, India and countries of south-eastern Asia, in particular. They present between them a conspectus of the problems which exist, and a variety of approaches.

In a few cases an attempt was made to organize training, production, experimentation and information services in a research and development centre [22, 14]. In others, the attempt to use programming methods has had to be piecemeal and intermittent, and was interrupted by many different crises. Sometimes curriculum reform and programming have gone together [15]. At other times, conflicts between curriculum change and serving understaffed schools have been evident [41].

One promising venture in central Africa was not maintained in its development [15]. However, there are a large number of important results from work in central and west Africa which appear to reinforce each other. Roebuck's [39, 40] project found major difficulties in accommodating programmed learning methodology to the existing patterns of teaching science and mathematics and study habits. In particular the change of role for teachers from that of focal authority and disciplinarian to manager of learning resources was hard to implement—especially since the turnover of teachers was very high. At the same time pupils were unused to checking their own responses. A very important factor in these African countries, moreover, is the fact that a second language is the medium of instruction though Okunrotifa [35] found it possible to make changes in the wording and arrangement of frames so as to adapt them to his subjects' level and style of reading.

As well as problems of changing traditional classroom methods, attitudes of pupils and study habits, level of reading and form of presentation in frames (e.g., the construction of sentences, the spacing of response blanks and the comprehensibility of answer prompts) there are even more fundamental problems for science teaching. These are related to spatial aptitudes and the perception and interpretation of drawings, diagrams and models.

They are also related to the development of scientific concepts in non-Western cultures. McFie's finding of improvements in spatial aptitudes following two years of science instruction [33] may indicate that projects of longer duration should be carried out. It seems necessary to pay attention to the differences not only in vocabulary but in conceptualization which form the background of readiness and capability upon which to build up scientific knowledge [7, 9].

In India, where there is a well-established Association for Programmed Learning, and where many university centres have set up training and research programmes, there is a different perspective. While many tests have paralleled work done elsewhere there have been serious attempts to adapt methods and invent techniques to match local conditions. For example, elements of programming were experimentally introduced into television teaching in 1966 [21]. To overcome shortage of materials and to fit a formal teaching climate, programmes have been presented on roller-blackboards or even orally to a whole class simultaneously and found to be as good as individual programme use. Programmes made up of sequences of pictures have been constructed to teach illiterate adults in the health field. In such work, group interaction has played an important part. Group study has also been introduced by requiring groups to discuss and discover solutions and answers to simultaneously presented frames. As well as this, role-play and desensitization processes have formed part of successful programmes for training family planning field workers (and for husband-wife co-operative learning in attitude and behaviour change) [4].

Among established successes in science and mathematics teaching has been a growing development of out-of-school and off-campus (correspondence) programmed learning in which high-school and university students have been found to learn mathematics and statistics without classroom supervision [5]. In brief, Indian effort and enterprise have been realistically directed towards giving greater strength to the classroom teaching of inadequately trained teachers as well as to investigating Mathetics, practical, discovery and group interaction types of programmed instruction and making use of integrated systems. Japan, which has experimented with programmed learning for over twelve years, has developed sophisticated teaching/learning systems for science education which integrate classroom feedback devices, teaching machines, audio-visual and television presentation and programmed texts. On the other hand, Egypt, beginning in 1966, has produced only a handful of programmes and experiments (in biology, mathematics and psychology) [16].

Patterns of development and stages in the acceptance of programmed learning principles are widely different. In Japan, university research, innovating teachers and commercial enterprise have stimulated interest and distributed knowledge quite widely. In Singapore, government interest has promoted a large number of introductory courses for teachers. In the Philippines, a 'cascade' principle has been adopted and differentiated training (for those who teach programming, prepare programmes and use programmed learning situations) caters for different needs for knowledge [34].

One thing to be emphasized about programmed learning is that, in spite of its development from particular viewpoints, it is not tied to any theory or approach. This means that it can be adapted to any background system. Unfortunately, in a large number of cases one type of programmed learning (e.g., linear self-instruction texts) has been introduced into school organizations which, indeed, are in need of reinforcement but, because of existing weaknesses, cannot adapt to a change in the teaching system. If the requirement is to support inadequate teaching strength, it seems likely that those proved benefits of programmed instruction or 'quasi-programmes' which involve a minimum of change in teaching style can be of great service. Such means as simultaneous presentation of programmed material with immediate feedback from the teacher or monitor, involve a minimum of teaching method change yet would bring a big improvement in effectiveness of learning. To make radical modifications (e.g., small-group, project-centred programmed instruction or individualized learning prescriptions), however, requires extensive retraining of teachers.

In other words, the programmer must take account of environmental constraints, be ready to offer alternative solutions in highly resistant as well as welcoming climates and sometimes evaluate his intervention in terms of: qualified teachers freed for other work; schools enabled to provide instruction which would otherwise have been omitted; and students taught who would have been out of reach of school; rather than the optimistic achievement of almost perfect mastery by almost all students. On the other hand, he must sometimes take as his objective the restructuring of the teaching-learning system and take measures to retrain and reorientate teachers for their changed roles.

Generalizations

A number of general points can be made about programmed instruction in order to give perspective. The first is that programming is not an approach which can be appraised as an alternative to other methods. It is a general methodology for clarifying goals, constructing learning situations and sequences and conducting continuous and overall evaluation. Depending on the constraints which obtain, programming is capable of offering different solutions to problems of educational systems design.

A second is that, because of its pragmatic nature, it is constantly changing and improving its techniques. It is nourished by many sources of ideas and research—in recent years, for example, network analysis, logical decision trees and many research findings from the psychology of instruction have elaborated and strengthened its framework.

A third point is that, as in other fields, there is a tendency for certain orthodoxies to arise and become fixated. Just as in microteaching, it is automatically assumed that students of teaching cannot serve effectively as pupils in microclasses or that microlessons last for five minutes, certain features of early programmes and methods seem to have become accepted as foundation stones. These orthodoxies can, because of their persuasive coherence, prevent a wider view, more flexible techniques and readiness to modify practices.

Just as evolutions in technology are apt to preserve non-functional forms from a prior stage (railway *carriages,* disposable *spoons* for stirring), features of non-programmed instruction still adhere. For example, the influence of the classroom textbook and retention of verbal-learning methods instead of providing objects and apparatus for personal and group discovery experiences, is still marked. Similarly, the early formulae of programmed instruction tend to persist, e.g., the dogma that it is individual, self-paced instruction—a principle which highlighted the need for every pupil to be involved in the learning process through active participation—but which no longer needs to emphasize 'individual' or 'self-paced'.

At the risk of subscribing to another dogma there seems to be a place for the invention and validation of physical (or paper and pencil) models of concepts which can be employed to give initial manipulatory experience and self-corrective feedback. Such devices as the apparatus for principles of moments, the double-balance and a model for teaching vectors illustrate the point. By operating on them pupils learn the set of relations in a concrete manner which supports later stages of abstraction. Such 'structured apparatus' appears to have a point from early stages of number learning to university levels of science. Thus a succesful method of teaching about RNA employed cardboard models in a Mathetics programme. Students gained insight into the notions of 'messenger' and 'template' by manipulating patterns of coloured cardboard shapes. Even so it is important to carry out research to determine if and when the appealing ideas of Bruner, Piaget and Talyzina can be applied and whether they imply particular strategies of instruction.

Similar remarks apply to other aspects of educational technology which have inherited their own creeds. Thus it was found difficult to convince a television producer that drawings were required for a particular sequence, not a videotape of three-dimensional reality. Spilling over from one medium or mode into another can be difficult because each one seems to acquire a wish to do everything in a self-contained way. In programming television 'lessons' it was necessary to include pupil use of apparatus for discovering and verifying principles—but there is a tendency to wish to employ the medium in its own terms. Again, programmed instruction has not always taken opportunities to go outside the constraints of text or machine, for example, in controlling the access to later units by means of human monitors or by using fellow-pupils to provide feedback.

When programmed instruction is regarded as a dynamic system in which goals and targets have been analysed and set, learning situations have been planned and developed and preliminary trials of the units of the system as a whole begun, for the purpose of testing the need for revision, only completely spontaneous aspects of instruction fall outside. Even in this case the role of spontaneous intervention can be foreseen and in some degree controlled.

The question therefore arises whether any and all learning can be programmed. This question need not be metaphysical if certain limitations of systems design are acknowledged. Some systems, for example, involve probability rather than determination. Hence, if, as may be the case, some things are learned best by encountering random events rather than systematically ordered events, the fact that unpredictability is essential to learning can be used in developing a programme. It has been held that certain high-level processes are not within the scope of programming. To this it must be said that if instruction to achieve such an objective is repeatable, it is programmable (and perhaps capable of improvement). Teaching students to be scientifically critical, capable of generation hypotheses and of establishing personal criteria for self-evaluation were objectives approached successively by programming techniques. It involved an overlapping succession of classes of students which undertook activities based on those of the one before and elaborated in testable ways. The approach involved establishing dissonance (e.g., through conflicting outcomes of research). Having determined how to reach this aim to a point where students were motivated to engage in inquiry, groups were formed which developed testable ideas. Each group was maintained in a relationship of increasing independence from the instructor. At an interim stage, students collaboratively developed criteria for evaluating the work being done and finally applied the criteria to: (a) other groups' work; (b) others' contributions within groups; and (c) their own personal contribution [26].

While this does not correspond to the image conveyed by a programmed textbook, it nevertheless fulfils the criteria laid down earlier (testable objectives, task analysis, evaluation and modification). On the other hand, though the building up of courses in modular units (sometimes with alternate modes), each unit fitting into an overall plan to reach objectives and having evaluation tests for student self-checks and unit improvement, may appear closer to the earlier patterns of programmes, such courses may also depart widely from prescriptions in using a variety of media and group activities [45].

REFERENCES

1. AMARIA, R. P.; BIRAN, L. A.; LEITH, G. O. M. (1969). Individual versus co-operative learning I. *Educ. res.*, vol. 11, p. 93-103.
   Describes experiments with primary and secondary school children who learned elementary physics, in co-operative groups or pairs or individually, from programmes using apparatus. The evidence favoured co-operative learning and supported heterogeneous ability co-operation.
2. AMARIA, R. P.; LEITH, G. O. M. (1969). Individual versus co-operative learning II. *Educ. res.*, vol. 11, p. 193-199.
   Further evidence is given of the effects of co-operation. In this study the personality of the co-operators was systematically varied. Interim evidence indicates marked influences on learning and transfer as a result of compatibility or incompatibility of temperaments.
3. AMARIA, R. P.; LEITH, G. O. M. (unpub.). Individual versus co-operative learning III.
   A sample of children from this and the former study were tape-recorded and their dialogues submitted to interaction analysis. The best learning and transfer occurred when anxious and non-anxious children worked together. Other pairs engaged in negatively reinforcing interactions. A completely systematic arrangement of anxious/non-anxious, introverts and extroverts in over six schools confirmed this and indicated that the pairs should be composed either of extraverts or introverts who are anxious and non-anxious.
4. BASU, C. K. (ed.) (1969). *Programmed instruction in industries, defence, health and education,* Indian Association for Programmed Learning.
   A collection of articles many of which describe Indian experiments and adaptations of programmed learning to meet educational needs.
5. BASU, C. K., The use of programmed instruction in correspondence education. In: Patel, I. (n.d.).
6. BLOCK, J. H. (1971). *Mastery learning.* New York, Holt Rinehart and Winston.
   A compendium of chapters and abstracts describing methods of mastery learning and the results of experimental trials.
7. BUNYARD, J. K. (1969). The use of programmed learning in the Northern States of Nigeria. In: Dunn, W. and Holroyd, C. (eds). *Aspects of educational technology II.* London, Methuen.
   Programmes on science have been developed and used at different levels of schooling. Emphasis is placed on the development of verbal associations to enrich concept meanings, on the need for practical work in science programmes and reference is made to successful learning in pairs with programmes.
8. BUNYARD, J. K. (1972). A comparison of the learning achieved by Nigerian and English children from programmed material. *Programmed learning,* vol. 9, p. 7-17.
   A programme on general science prepared in Nigeria was given to 1st-year secondary pupils and also to English 1st-year comprehensive school children. African children were less favourable towards getting immediate knowledge of results answers (other might cheat) and towards experiments (dislike of manual activities) than English children. Correlations between mental tests and criterion scores suggest that a programme writer should be aware of differences in, e.g., spatial ability and level of verbal comprehension between cultures.
9. CLARKE, D. W. (1968). *The use of programmed learning to accelerate the acquisition of the concept of numbers.* Unpub. M.Ed. thesis, Univ. of Birmingham.
   An empirical test of a method of accelerating the acquisition of number concepts which is based on the notions of algorithms and learning set formation procedures.
10. DUNN, W. R. (1969). Feedback services in university lectures, *New university,* vol. 3, April, p. 21-22.
    Describes the use of coloured cards to provide feedback by students during lectures in Glasgow University Medical School. The Department of Physiology employs an electrical feedback system with over 100 positions and a console for registering the proportion of correct responses. Students are also given individual knowledge of results by means of coloured lights.
11. EYSENCK, H. J. (1972). Personality and attainment: an application of psychological principles to educational objectives. *Higher education,* vol. 1, p. 39-52.
12. GAGNÉ, R. M. *Conditions of learning.* New York, Holt, Rinehart and Winston.
    Application of psychology of learning principles to the problems of improving learning. Evidence for a hierarchical analysis of scientific knowledge in order to promote learning.
13. GILBERT, T. F. (1962). Mathetics: the technology of education. *J. mathetics,* vol. 1, p. 7-73.
    The first of two consecutive 'articles' explaining principles of Mathetics and procedures for designing instruction.
14. HAWKRIDGE, D. G. (1966). First results of programmed learning research in Central Africa. *Programmed learning,* vol. 3, p. 17-21.
    Describes the validation of a programme on 'contours'.
15. HAWKRIDGE, D. G. (1969). Programmed learning and problems of acculturation in Africa. In: Dunn, W. and Holroyd, C. *Aspects of educational technology II.* London, Methuen.
    The paper identifies two specific areas for research: pictural perception and scientific thinking. On the former Hawkridge describes his experience in validating 'contours' in which he found special difficulties with African children in relating two and three dimensional representations—difficulties not experienced with European children in validation trials. In comparing stages of intellectual development between African and European children some acceleration might be achieved through the use of programmed instruction designed for this purpose.
16. ISKANDER, K. Y. (1971). Some research activity in the field of programmed instruction in the United Arab Republic. *Visual education,* October, p. 77-78.
    Reports of three experiments in programmed learning. One compared learning about flies from a short programme in an elementary school with performance of children who did not learn. The second on fire extinguishing

compared programme alone, programme with class text and class text only. Programmed learning was found better than text alone. The third used a programme on psychology and compared with lectures. The programme was more effective and took less time (see also Iskander, K. Y., in: *Visual education,* June 1971).

17. Issing, L. J.; Schellenberg, C. (1969). The application of the principles of programmed instruction to the design of ETV broadcasting—a comparative study. *Programmiertes Lernen und Programmierter Unterricht,* vol. 1, p. 57-65.
   Comparison between programmed text, programmed television and conventional television versions of a broadcast on magnetism was made using 6th-grade students. Both programmed versions were better than non-programmed television but not different from each other.

18. Kenshole, G. E. (1969). An experiment in undergraduate teaching using audiovisual aids. *Physics education,* vol. 4, p. 157-160.
   Describes an integrated media system in engineering teaching and draws attention to the fact that it is difficult to modify after its creation.

19. Komoski, K.; Green, E. J. (1964). *Programmed instruction in West Africa and the Arab States, A report on two training workshops.* Paris, Unesco, Educational Studies and Documents, no. 52.

20. Krishnamurthy, G. B. (1969). Programmed learning in health programs. In: Basu, C. K. (1969).
   Describes programming techniques which use pictograms for illiterates, role play in group programmes and a method of guiding group discussions by programming a structured series of problems. An extension of this is used to overcome communication barriers in interaction between individuals (e.g., husband and wife).

21. Kulkarni, S. S. A review of studies conducted in India. In: Patel, I. (n.d.).
   This article gives a good impression of the range and variety of work on programmed learning in India. It includes studies on: How well does it teach? (very well); Who can it teach? (5th grade can learn 8th grade mathematics though they take longer); What can it teach? (family planning, group problem solving as well as the expected things). Experiments on programmed television, correspondence education, group administration and adaptation of foreign programmed texts are reviewed.

22. Lawler, C. J. (1969). Programmed learning in the developing countries of Africa. *Programmed Learning,* vol. 6, p. 189-196.
   Reviews use of programmed learning in Africa and suggests that single technique programmes are unsuitable but variety of techniques including Mathetics and group interaction methods should be used. Programmes from elsewhere need revalidation. Special attention must be paid to African problems such as the use of a second language, difficulties of knowledge of results (no K.R. works according to the writer), problems of interpretation of visual representation, there is need for prolonged, co-ordinated research and for permanent centre.

23. Leith, G. O. M. (1969). Learning and personality. In: Dunn, W. and Holroyd, C. (eds.), *Aspects of educational technology.* London, Methuen.
   Summarizes a series of experiments on the interaction of method of instruction ('discovery' versus direct teaching) and personality. Using programmed materials in science and mathematics it was found that extroverts learn better than introverts from relatively unstructured situations, while introverts are better than extroverts when the situation is highly structured. This was the case in secondary and tertiary education. Primary children had a similar interaction but of anxiety and method.

24. Leith, G. O. M. (1969). *Second thoughts on programmed learning.* London, Nat. Council for Educ. Technology, Occ. Paper no. 1.
   Reviews the evidence about methods of programmed learning and claims made by different factions. Many of the fundamental tenets are shown to be unsupported or false but the conclusion is that programmed learning has been freed from many of its constraints.

25. Leith, G. O. M. (1970). The acquisition of knowledge and mental development of students, *J. educational technology,* vol. 1, p. 116-128.
   Presents results of the author's researches on how students learn and the conditions which facilitate particular kinds of learning.

26. Leith, G. O. M. (1970). *Non-conventional methods of education.* Paper read to the International Education Year Congress, Buenos Aires.
   Describes alternatives to traditional 'direct frontal teaching' which include development of courses in which students formulated their own objectives and criteria of evaluation and carried out evaluation of each other's work as groups and individuals' contributions to groups. The approach involves gradual withdrawal by the teacher from his presenting and directing roles towards guidance and organization of learning resources.

27. Leith, G. O. M. (1971). Conflict and interference: studies of the facilitating effects of reviews in learning sequences, *Programmed learning,* vol. 8, p. 41-50.
   Summarizes a series of five experiments, most on programmed instruction in science and mathematics. All are on how to overcome the effects of retro- and pro-active interference in learning. In two of them it is shown that material which causes conflict of ideas can be learned by using mediators, which give an overall perspective if they are given *between* conflicting material. Also helpful is to present the unfamiliar or difficult material before readily grasped material.
   The next three experiments generalize these findings to cases of consecutively arranged sets of concepts and principles. Provision of a summary after each sub-section was consistently found better than giving advance summaries (junior, secondary, and tertiary level experiments).

28. Leith, G., et al. (1970). An evaluation of training systems for teaching conversion to decimal currency and exchange in mixed coinage. *Programmed learning,* vol. 7, p. 140-151.
   Experimental tests of methods of programmed instruction to teach the new decimal currency before its introduction in the United Kingdom. Three methods: modified linear, audio-visual and adjunct were compared using adult workers, technical school children and primary children. Each method was successful—but for different

groups, depending on age, intelligence and personality. Among adults the group audio method was the optimal and was successful with both above and below average intelligence learners but non-anxious extroverts were good with the adjunct method. Group audio reduced intelligence level differences in primary school but was less successful in secondary.

Study skills and attention focusing are reasons for these differences. Departures from assumptions about prescriptions for programmes are noted and a group competitive game as part of a programme is mentioned.

29. LEITH, G. O. M.; BRITTON, R. K. (1973). The influence of learning techniques of programmed instruction on teaching performance in school. *Educ. res.* (to appear).

In a second year cohort of over 200 student teachers, one half was given instruction in principles of Mathetics and preparation of programmes and the other half took other courses (e.g., audio-visual education). All levels of teaching were represented from infant school to upper secondary. Mathetics-trained students of all levels were significantly better on teaching practice performance than control group students.

30. LEITH, G. O. M.; DAVIS, T. N. (1972). Age changes in the relationships between neuroticism and achievement. *Res. in educ.* (to appear).

Over 300 children aged 12 and 13 years from several parts of England were taught five topics of the curriculum including mathematics, science and geography by means of programmed instruction. The whole sample was divided into upper, middle and lower thirds on a measure of neuroticism (emotional stability or anxiety). The 12-year-olds had higher attainment if they were anxious or moderate than if they were non-neurotic. At 13 the middle group had scores higher than those of high and low neuroticism.

Older children got higher scores than younger ones but the high anxiety groups of both obtained the same average achievements. A further experiment with 19-year-old students had a pattern in which non-anxious and middle range students were better than highly anxious ones. Other evidence too points to a swing round in the effects of anxiety on achievement with increases in age.

31. LEITH, G. O. M.; TROWN, E. A. (1970). The influence of personality and task conditions on learning and transfer. *Programmed learning,* vol. 7, p. 181-188.

Secondary school children learned about addition of vectors from programmes prepared so that one-half had a statement of rule followed by examples and the other had examples followed by rules. Introverts were slightly better if they received rules first, extroverts were significantly better if they had rules following examples.

32. LEITH, G. O. M.; AMARIA, R. P.; WILLIAMS, H. (1969). Applications of the principles of programmed learning to the preparation of television lessons in elementary science and mathematics. *Programmed learning,* vol. 6, p. 209-230.

A report of a study in which economical methods of preparing and validating television lessons were developed by using overhead projector trials to verify the effectiveness of presentation, the types of response and the duration of response intervals before provision of knowledge of results. As the same time, television, slide-tape, overhead projector-tape versions (all group-paced) were compared with programmed texts (individual, self-paced). All versions contained practical experiments, sometimes involving group work in the audio-visual modes. Results showed that self-pacing has no advantage over forced-pace learning—and may be significantly worse. Criteria for choice between audio-visual media are economic considerations and hence size of audience. An overhead projector/tape is easiest and quickest to prepare and is a step towards validation of more costly presentations.

33. MCFIE, J. (1961). The effect of education on African performance on a group of intellectual tests. *Br. J. educ. psychology,* vol. 31, p. 232-240.

34. MUNARREZ, N. (1972). *Report to a Unesco seminar on programmed instruction.* Tokyo, May 1972.

35. OKUNROTIFA, P. O. (1968). A comparison of the responses of Nigerian pupils to two sets of programmed instruction materials in geography. *Programmed learning,* vol. 5, p. 283-293.

An American programme was compared with a version adapted to Nigerian culture, by lowering language level, changing grammatical structure, shortening frames, modifying answer prompts, etc. Pupils using this version got higher scores, took the same time and liked the task better than those using the original version.

36. PATEL, I. J., et al. (n.d.). *A handbook of programmed learning.* Baroda, Indian Association for Programmed Learning.

A collection of articles on various topics (including Mathetics) some of which summarize researches on programming techniques in science and mathematics.

37. POPHAM, W. J. (1971). *Criterion referenced measurement.* New York, Educational Technology Pubs. Inc.

A collection of writings on the concepts and problems of criterion-referenced testing.

38. PRICE WILLIAMS, D. R. (1969). *Cross-cultural studies.* London, Penguin.

Collects together many articles on cultural differences in development and performance and perceptual differences.

39. ROEBUCK, M. (1968). *Research project in programmed learning.* Institute of Education, University of Ibadan, 10, 2, 1968.

An interim report which contains a list of the programmes for trial in Nigerian schools and spells out some difficulties of adapting programmed learning to the educational context.

40. ROEBUCK, M. (1969). Using programmes in a developing country: an interim report of the Ministry of Overseas Development Research Project in Western Nigeria. In: Dunn, W. and Holroyd, C., *Aspects of educational technology II.* London, Methuen.

See Roebuck (1970).

41. ROEBUCK, M. (1970). *Factors influencing the success of programmed materials in under-equipped classrooms and inadequately staffed schools.* Report on Ministry of Overseas Development Research Project in Western Nigeria 1967-1968 (mimeo).

The project lasted nearly two years and intended to train programmers and produce programmes in science and mathematics. Administrative and other difficulties reduced the training time but seven programmes were prepared and sixteen published programmes were tried in nineteen schools (1,500 pupils). Experiments with four programmes attempted to establish some points about African use. Findings draw attention to high correlations between reading ability and success, the large differences between schools and the lack of difference between practical work and non-practical versions of programmes.

Problems were the structure of schooling which was inimical to individual self-instruction, the quality and mobility of teaching staff, absences of pupils, use of English as second language as medium, restricted language code (Bernstein), study skills and attitudes which made knowledge of results prompts into 'cheating', heterogeneity of classes.

Attention is drawn to a conflict between curriculum reform and providing for inadequate teaching staff. One conclusion is that in the circumstances group-paced materials might help.

42. SHARMA, M. M. (1969). Application of programmed learning technique under group conditions. In: Basu, C. K. (1969).

An experiment in mathematics teaching which compared individual programmed learning with the same programme on a roll-up board found that the group administered programme was as good as the text even after one month retention interval.

43. SIME, M.; BOYCE, G. (1969). Overt responses, knowledge of results and learning. *Programmed learning*, vol. 6, p. 12-19.

A comparison of lectures using transparencies either giving statements or asking questions. Questions were more effective than statements.

44. SKINNER, B. F. (1954). The science of learning and the art of teaching. *Harv. Educ. Rev.*, vol. 24, p. 87-97.

45. SULLIVAN, A. M. *Psychology 100 at Memorial University of Newfoundland 1968-9, 1969-70*. Interim Report, Institute for Research on Human Abilities, Memorial University.

A mastery learning system was developed to take account of increasing numbers of students. The course is divided into weekly units. Students take pretest, learning materials and criterion-referenced test. If they have failed to master a unit they take another mode (e.g., television; personal tutorial) and when mastery is achieved go on to the next unit. Fast students help slower ones. Comparison with conventional lecture course shows that a very much larger number of students achieved high grades in the mastery course and that lower grades were almost unoccupied. One of several unique features is the possibility to start the two semester course in September or January (by holding a summer semester). In this way validation can be carried out from one semester to the next.

46. SUTTER, E. G.; REID, J. B. (1969). Learner variables and interpersonal conditions in computer-assisted instruction. *J. educ. psychology*, vol. 60, p. 153-157.

Investigation of the interaction of sociability and test anxiety as factors in CAI. One group of students learned alone, a second in pairs and the controls had no learning. High sociability low test anxiety subjects were better in pairs. Low sociability, high test anxiety subjects were better alone. Best attitude to CAI was from submissives working with dominants or dominants working alone.

47. TAPLIN, G. (1969). The Cosford Cube: a simplified form of student feedback. *Industrial training international*, vol. 4, p. 218-219.

Feedback classrooms usually involve sophisticated electronic equipment or simpler electrical devices such as a battery-operated light system. The Cosford Cube (from the name of an RAF station) is a cube having coloured circles on each face. Students show a face to correspond with their selected response. This is used in conjunction with an overhead projector.

48. TOBIN, M. J.; BIRAN, L. A.; WALLER, J. (1968). *A study of audiovisual programmed learning in primary schools*. Birmingham, Nat. Centre for Programmed learning, Res. Reps. no. 29.

Comparisons were made of text and audiovisual versions of a programme on science. Audiovisual was found better both for poorer and better readers and group-pacing did not detract from learning.

49. WALTHER, R. W.; CROWDER, N. *A guide to preparing intrinsically programmed instructional materials*. Springfield, Va. Clearinghouse for Federal Scientific and Technical Information.

Tells how to write, validate, edit and prepare branching programmes for text or machine presentation.

50. WEBB, C. (1970). Feedback teaching systems. *Industrial training international*, vol. 5, p. 243-245.

Describes an electronic classroom feedback device using semi-conductors. The current need is low, cables are of small size and the whole system is easily transported from one room to another. In use an overhead projector presents alternatives for student selection. Use has also been made of circuit boards and practical kits. Administering and scoring tests are facilitated by the device.

The use of television in science teaching

by J. Valérien
Educational Actions Department,
OFRATEME (Office français de Radio-Télévision Scolaire Moderne),
37, rue de la Vanne,
92 Montrouge, France

Summary

If he is to understand the modern world and play his full part in it, man must have a more thorough grounding in science than ever before. And the evolution of techniques is such that this basic knowledge must continually be brought up to date and supplemented by means of lifelong education. In order to meet this growing need, it would seem logical, in a world where technology is omnipresent, to have recourse to the most modern educational technologies, and in particular television techniques.

There has been a profound regeneration in the actual organization of science teaching: courses are based on a small number of more general concepts, and observation and methodology are replacing the traditional acquisition of knowledge.

How, in this setting, can educational television be used? Before answering the question, we must make two important distinctions: education may be in-school or post-school, and television may be closed-circuit or open-circuit.

For in-school instruction, closed-circuit television can be used in a single class, in one establishment or in a group of establishments by means of wire relay systems. Generally speaking, the total cost of this system is high and there are technical problems in the way of its widespread adoption. Its main interest lies in the teamwork done by teachers: they indeed are the ones who benefit most from it.

Open-circuit television, because of its scope and complexity, requires the assistance of professionals, who work in close collaboration with teams of educators to produce series of programmes for broadcasting. These may be stop-gap broadcasts, intended to deal with urgent problems (such as a sudden prolongation of the period of compulsory school attendance) which the traditional educational system cannot cope with; they may be supporting broadcasts, designed to supplement and illustrate the traditional lesson; or they may be broadcasts forming an integral part of an educational system that has been completely redesigned with the idea of using the television medium. (The Open University is at the moment the most striking example of this last type.)

For various reasons (inadequate equipment, teacher resistance, etc.), science instruction through television is marking time in the traditional scholastic system. It may thus be that television will find its true place in a field which has not yet been organized as a 'system', that of further or post-school education.

As we have seen, closed-circuit television can be used for teacher training. This use can be extended to the vocational training of adults in general and in addition to training in certain specialized technological disciplines.

But it is above all in open circuit, and as an integral component of a multi-media system, that post-school television proves effective. The author gives a detailed description of several experiments which have been carried out successfully in different European countries. (Be it RTS/Promotion in France or Project 'Delta' in Sweden, the results are very encouraging.)

It seems then that television develops freely where—as in adult education—no rigidly organized system already exists. In the traditional school system, the attitude of teachers must be changed. If they have been trained by means of television, retrained by the same means, they can no longer refuse to their pupils what has been of benefit to themselves.

Introduction

To fulfil himself in the modern world, to understand an environment in which industrial technology and science play an ever more important role, man has need today of a solid grounding in science. The initial baggage he is given in the course of regular schooling (when regular schooling exists, or is widespread among the population, which is not always the case) is no longer enough. From the point of view both of knowledge and of methods, his education must continue throughout his working life. Certain countries, like France, have taken legislative and administrative steps to provide this lifelong training.

The growing need for scientific training, due essentially to the evolution of the modern world and the rapid spread of education, has been analysed and is now widely recognized. In face of that need, it would seem logical to lean heavily on the most modern educational technologies, and in particular to make a widespread mass use of television techniques. While it is true that many countries have engaged in research with a view to applying such techniques to science teaching, it must be acknowledged that this use of television is far from being general.

Current trends in science teaching

Before reviewing a number of experiments that seem very promising for the future and analysing the problems that have to be solved, we shall make a brief examination of the present-day organization of science education.

In the countries where this education is most widely available, that is, in the industrialized countries, it is spread over several stages: pre-school (ages 2-6), primary (6-11), intermediate (11-15), secondary (15-18) and higher (18 and over). At the pre-school and primary levels, science is taught by a general teacher, but from intermediate school on there are teachers specializing in various disciplines: mathematics, physics, chemistry, geology, biology, medicine, technology, applied sciences. Each discipline has its specific teaching methods; and co-ordination of the work is not always adequate.

Whether the teacher has a general training only or is a specialist in one discipline, instruction is for a group of pupils meeting in a classroom inside a school or a university (which is used only a few hours a day and a few days a year).

Though it has never been scientifically attested, one teacher to twenty-five pupils has gradually come to be accepted as the proper ratio and, under pressure from teachers' organizations, this is tending to become the general rule.

Science instruction takes various forms: lecture courses, individual lectures, lectures followed by class discussion, practicals, directed work, talks on particular problems, demonstrations, exercises; at the higher level there are also seminars. A number of examinations, of various types, round off each stage of study and lead to the award of diplomas.

In the field of further education there are yet other possibilities: correspondence courses, evening courses, university summer schools, etc. But the traditional in-school pattern tends to take over here as well.

At the present time we find that two trends are becoming general in all countries:

First, the accumulation of new knowledge deriving from the progress of industrial technology and the results of scientific research is necessitating a renewal of the content of education, reflected in the formulation of new syntheses and the introduction of *new curricula*. For instance, data on ultra-structures obtained through electronic microscopy are bringing about a new approach to biology and physics and are introduced in teaching earlier and earlier: taken up only in the course of higher studies less than ten years ago, they are now an integral part of secondary-school curricula.

Second, the ineluctable change in the relations between teacher and pupil is bringing about an increasing reliance on *active methods* which give an important place to learner participation. Nowadays it would be unthinkable to develop a training programme for adults without consulting the learners. The same principle is not always supposed to apply in the school-room: and yet the schoolmaster is no longer the sole purveyor of knowledge, since it is estimated that pupils acquire about 80 per cent of their information outside school.

The place of youth in the family unit and therefore in society is not what it was when today's adult generation were young; and the necessary lessons must be drawn from the fact that this is so. We need only recall the recent experiments in space and think how many science teachers had to rearrange their courses and answer their pupils' questions after the Apollo rocket launchings.

These trends necessitate a *thoroughgoing regeneration* extending to all the scientific disciplines.

First, we find that for each discipline the entire syllabus has been reorganized around a few central, more general concepts. For example, in biology, studies of individual subjects or species are being superseded more and more by an ecological, molecular and genetic approach.

The various movements of regeneration have another feature in common: in each discipline, the aim is rather to develop the acquisition of method than to increase the sum of knowledge to be retained.

We can also observe in each discipline the attempt to initiate students into a scientific approach which can be represented by the diagram shown overleaf.

Each discipline should contribute, with its own specific means and field of experiment, to this introduction to

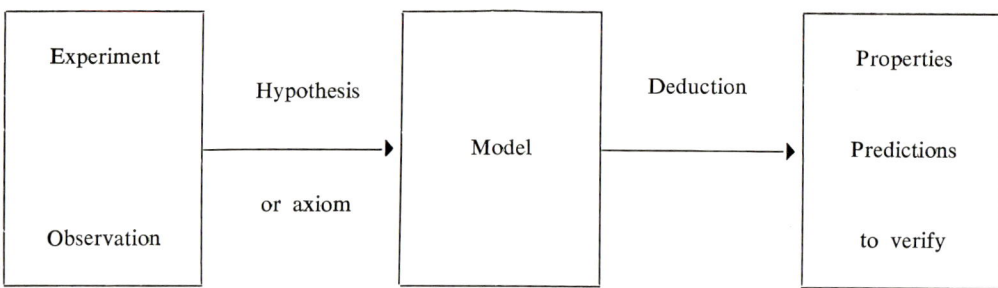

the scientific approach. But the hard fact is that the way in which reforms in science teaching have been gone about varies from one discipline or level to another. Thus in France, where new curricula introducing modern mathematics have been set up at every level, the aim in primary instruction is to develop observation and the forming of hypotheses, whilst that aspect is too often excluded at the secondary level; in secondary education (15 to 18 years of age) the 'applied mathematics' aspect is non-existent—which runs counter to the reform under way in physics, chemistry and biology teaching.

Finally, we must bring out the distinction that exists between the exact sciences, such as mathematics, and the experimental sciences, such as physics or biology. In the experimental sciences one has to start from an experiment, isolate one or more variables and get the pupils to perform certain operations so that they can acquire the method and behaviour that characterize the scientific spirit.

This conspectus of the existing situation necessarily provides our starting-point for discussing the use in science education of the richest and most varied means for representing and interpreting the world through image and sound—that is, television.

We shall consider two levels: the initial education given at school or university and the further education given teachers and other adults.

Television and in-school education

There are two forms in which television can be used for science teaching in schools:

Internal television may be used for a class, for a whole school, or for several establishments linked by wire relay; this is called 'closed circuit'.

Alternatively, omnidirectional television may be used, the messages being broadcast by waves over a whole region or an entire country: this is 'open circuit'.

Closed-circuit television

Closed-circuit television, which requires relatively simple equipment, of industrial standard, has had something of a boom in recent years.

Its use may be restricted to one classroom. The installation usually consists of a camera that can be used in horizontal position, a lens (on which the quality of the whole installation depends), a receiving set and a video tape recorder. With an installation like this the teacher has a *collective magnifying glass* which enables the class to observe photographs, transparencies, maps, graphs, small objects, micro-sections, etc., under good conditions. The camera can easily be adjusted to a microscope or reproduce the screen of a cathode-ray tube.

Dynamic phenomena can in some cases be recorded and reproduced by means of the video tape recorder. The apparatus presents no particular problems as to use and does not demand specialized staff.

Use of the installation supports the instruction given by the teacher but does not alter its nature. Certainly, only the wealthiest schools can provide every science classroom with such equipment. Its general adoption at all levels of teaching would require large outlays of money. Moreover, for it to be used well there must be a competent and dynamic teacher on hand.

Its use may be extended to a whole establishment. This involves a more elaborate installation including a broadcasting centre (studio and control room, usually with two or three cameras, a telecine projector, a mixing console, a special effect generator, one or more video tape recorders, a control console, etc.), to which a certain number of classrooms equipped with receiving sets are linked by wire relay. Such an installation requires the presence of specialized staff (director, maintenance assistant, cameraman, programmer, researcher, etc.). The initial cost, plus amortization and operating expenses, have to be added to the regular expenses of the establishment. That is why it has as a rule only been possible to instal such equipment in schools of an experimental character.

For closed-circuit television to be used well, there must be teamwork among teachers within a given discipline and co-ordination between the various teams. That calls for a certain flexibility on the part of the school administration (arranging timetables so that teachers can work together) and goodwill on the part of the teaching staff, who are not necessarily convinced of the need for teamwork and do not always accept what is a fairly sizeable addition to their work-load.

At the level of a single establishment, it is in this way possible to produce broadcasts which follow a certain progression and are integrated into regular instruction. Obviously the teachers have to agree as to what the progression will be (and undertake to respect it in their own classes), and also decide at what points a broadcast can suitably be introduced. Such broadcasts necessarily represent a compromise between 'industrial' television, which simply shows one unadorned datum (such as the flow of a liquid), and 'entertainment' television, which calls for a certain staging.

In science, the images presented, whether they be static or dynamic, are usually significant (the picture of a flower represents a flower) and authentic. But they have to be explained, analysed, commented on. . . . The image is no longer what it was in the days of the 'magic lantern'; over-used, misused, it loses its potency. It has to be more and more carefully worked up to be effective. And closed-circuit television is a superlative tool by means of which the teacher, or team of teachers assisted by technical staff, can 'work up' the image, so as to make it more impressive (fade-ins or fade-outs, dissolves, superimposing of all kinds, juxtaposition of two pictures on the same screen, masks of all sorts, reversing episcopes, play of coloured or polarizing filters, scan reversal, simple animated cartoons—countless devices are there for the trained teacher to make use of).

The main benefit of the fairly numerous experiments carried out in many countries lies in the teamwork done by teachers. But what with the cost of installation, the administrative flexibility necessary in the establishment and the goodwill required of the teachers, it is usually impossible to bring closed-circuit television into general use: it remains the prerogative of rich or experimental establishments with resources and facilities which most establishments do not have.

Its use may be extended to a group of establishments. In various countries (Canada, the Federal Republic of Germany, the United Kingdom, for example) a number of closed circuits each serve several establishments. The production centre is located in an establishment with a good-sized documentation centre, and programmes are relayed to a greater number of classes.

The conception we have here is rather close to that of open-circuit television. In the county borough of Hull, in England, mathematics programmes are relayed to all the elementary schools of the area by means of a wire relay system.

Two conclusions seem to emerge from the experiments that have been made:

First, the cost is high. For that reason it is impossible even in the rich countries to generalize the installation of closed-circuit television in ordinary schools. Obviously the countries of the third world cannot contemplate so costly an expedient. General adoption of closed-circuit television runs up against technical problems too. There is no compatibility as among different makes of installation, and this limits the possibility of exchanges among schools. Moreover, there are a number of technical innovations—for instance certain cassette procedures—that are operational at laboratory stage but have not yet been placed on the market. In the circumstances, schools and education authorities are reluctant to invest in any one particular type of equipment and are waiting till the processes that have been announced become available.

Second, the principal benefit of such experimentation lies in the teamwork done by teachers. They learn a great deal from this work. Hence the idea has arisen of reserving closed-circuit television for teacher-training centres. Many experiments are conclusive: self-viewing, micro-teaching, analysis of significant teaching situations, analysis of pupil behaviour—all these are highly promising techniques which closed-circuit television makes it possible to exploit for teacher training.

Open-circuit television

Open-circuit television requires, from the outset, specialized equipment and specialized personnel. Programmes are produced on film or video tape by teams of specialists (directors, chief cameramen, cameramen, vision engineers, sound engineers, electricians, designers, cutters, etc.) in close collaboration with teams of educators responsible for determining the aims and content of *series* of broadcasts.

Since the programmes are broadcast by aerial transmission and are therefore available to the general public, they necessarily borrow from the techniques of staging and scene-design. But they are unlike popular-science programmes produced by regular television organizations in that they are really intended only for a particular viewing public which receives, in advance, the accompanying printed material enabling viewers to prepare for the broadcasts and to make proper use of them.

In a general way, these broadcasts may be classed in three groups:
stop-gap programmes;
programmes in support of normal instruction given according to the standard pattern (one teacher, 25 pupils);
programmes forming an integral part of a teaching system that has been entirely redesigned.

Stop-gap programmes. It sometimes happens that government decisions create exceptional situations which the educational system cannot cope with. For instance in Italy, the constitution provided for pre-vocational training for children of 11 to 14 years of age, but shortage of funds and staff made it impossible to open all the establishments provided for by law. Similarly in France, the minimum school-leaving age was raised from 14 to 16, but the necessary number of places were not immediately available. In both these cases the government called on educational television to fill the gap.

In Italy, beginning in 1958, the Italian Radio-Television Service broadcast a series of courses intended for pre-vocational instruction. These courses, worked out with the pedagogic collaboration of the Ministry of Public Education, covered the following subjects: Italian, history and geography, mathematics, scientific observation, technical draughtsmanship and manual work, French, religious instruction, music, domestic science. For each subject, the courses were supplemented by a low-priced 150-page handbook. Some exercises in mathematics for instance, were collected by the co-ordinator and then corrected in Rome by a team of teachers specially employed for the purpose. The experiment was decisive and achieved its object. It came to an end once schools and teachers were in a position to provide pre-vocational education.

In France, the sudden prolongation of compulsory schooling showed up a scarcity of intermediate teachers. Television was asked, for instance, to supplement the teaching of mathematics. For the four years of intermediate schooling, educational broadcasts were accordingly undertaken which took over part of the course. Work with individual pupils and supervision of exercises were in the hands either of trained teachers, who were thus able to see to a greater number of children, or of student-teachers. The broadcasts were received in the classrooms; they were accompanied by printed cards distributed to pupils free of charge. The experiment was decisive and ended when there were enough trained teachers to provide direct instruction.

Thus educational television has given proof of its effectiveness, especially in the field of science. Direct televised instruction with systematic broadcasting of courses for a school-age public is feasible, and in this way a television organization can make a major contribution to the traditional educational structures. Bearing in mind, however, that despite everything they have not assimilated television, it must be admitted that these structures are themselves somewhat rigid.

Supporting programmes. The formula that has proved most acceptable to the educational system has been and still is that of supporting programmes broadcast during class time, designed to give the teacher, while actually teaching, materials and information which it is hard for him to come by directly. These broadcasts serve to supplement conventional lessons. Most school television broadcasts keep strictly to this formula. They are adapted to the different instructional levels and cover a number of scientific disciplines (mainly biology, physics, chemistry, technology and mathematics). They are usually accompanied by printed materials for the teacher, to help him prepare for the broadcast and make use of it subsequently. More rarely (in Belgium, for example) there are accompanying materials for the pupils.

While it is relatively easy to assess the effectiveness of

stop-gap systems, it is much more difficult to measure the impact of television on a child who undergoes the multiple and various influences of school life. The instantaneous and fleeting nature of television transmissions and the number of variables involved limit the possibilities of scientific verification of the results achieved. It is reasonable to suppose, however, that these broadcasts exert a favourable effect. 'It is felt ... that the introduction of television into schools has in no case produced any setback in school-activities. Whenever it has been well used, it has proved a stimulus for the activities of the classroom; it has improved the general atmosphere and been a tonic for the mind. Far from standardizing teaching practices, television seems to encourage diversity and originality in schools.'[1]

This optimistic assessment by Henri Dieuzeide, however, should not lead us to forget that:

It is far from common for classrooms to be equipped with receivers (the total number of receivers in use is in fact increasing faster than the number in use in schools).

Broadcasts are not followed regularly. Except in primary schools and certain intermediate classes, where teachers and pupils are available throughout the school day, it is often hard to arrange a timetable that makes viewing possible.

The teacher is still free to use the broadcasts or not, viewing being left to his discretion. Deliberate choice, difficulties in the way of reception, over-documentation, lack of interest or information, all these are factors which may perhaps explain why the broadcasts have, when all is said and done, a *limited audience*.

Two points deserve attention. Japan is the first country to have made widespread use of video tape equipment, which makes it possible to record programmes and then broadcast them at the request of teachers. This has resulted in raising the utilization factor of school television from 17 per cent in 1968 to about 71 per cent in 1970, after the compulsory introduction of video tape recorders. Most studies show that it is the best, the most active, the most dynamic and the most inquiring of the teachers who use the broadcasts, while the more unenterprising teachers reject them. One cannot help thinking that often the broadcasts are not used by those who stand most in need of benefiting from a teaching aid.

All this probably explains why the 1970's have not fulfilled the promises and hopes of the 1960's as far as the development of televised science education is concerned.

Government authorities are raising questions as to the cost-effectiveness ratio of broadcasts, and recent seminars of the European Broadcasting Union (EBU) clearly show that school television bodies—whether, as in most countries, they form part of broadcasting organizations or, as in France, come directly under the Ministry of Education—are looking for new roads to follow.

Efforts are accordingly being directed to broadcasts designed for a public not comprised within the traditional academic structures, namely pre-school education and adult education. In the scholastic field proper, the emphasis is on primary education, where pupils are more receptive and more often available. For secondary education, multidisciplinary formulas of the 'science magazine' type are under study: programmes are broadcast outside of school hours, and are subsequently drawn on in class by teachers in the different disciplines. There is also some thought of having broadcasts which would provide the principal basis for exchange of ideas among pupils engaged in scientific activities (e.g., in clubs).

Programmes forming an integral part of new structures. Midway between stop-gap and supporting programmes come programmes for a school or post-school public which form an integral part of an educational system that has been completely redesigned with a view to using television. These broadcasts are scholastic in the sense that they aim at providing a course of instruction leading up to an officially recognized examination: in this sense they are rather different from further education or refresher courses.

A number of experiments in teaching isolated pupils who cannot be brought together in schools have already been tried, without success. More significant are certain experiments now under way such as those in Niger and the Ivory Coast, for primary education; the 'TV school', for intermediate and secondary education; the Open University, for higher education.

In some countries of the third world, the traditional educational system has not reached a degree of development sufficient to meet national requirements. Moreover, people in those countries are increasingly aware that the system they have is directly inspired by

1. H. Dieuzeide, *Teaching through television*, p. 26. Paris, Organization for Economic Co-operation and Development, 1960.

the Western model. If they are to go on proliferating indefinitely current educational forms based on the historical models of the West, the third-world countries risk going headlong into economic catastrophe and social bankruptcy. That is why some of them are moving in the direction of new structures which draw heavily on modern educational technologies.

In Niger, an experiment was carried out covering the whole elementary course; in the Ivory Coast, in 1971-72, the first year of primary instruction took the form of 30 weeks' educational television, and there are plans for extending this programme in the near future to cover all levels and all classes in the country.

The television broadcast, in mathematics for instance, is presented by a 'tele-teacher'. It is the compulsory basis for in-school activities under the classroom teacher's guidance. The system may sound inhibiting, but in point of fact pupil participation is very great.

In the Federal Republic of Germany, the Volkswagen Foundation and the State of Bavaria have set up a combined system of television broadcasts, printed materials and group instruction. This programme is for young people who have finished their compulsory schooling and are entering working life. It consists of 468 lessons, 221 of which are in science (mathematics 78, physics 65, chemistry 13, biology 13, technical drawing 26, electrotechnics 13, industrial chemistry 13).

The television broadcast, which the pupil follows at home, is accompanied by printed materials. Every week, usually on Saturday morning, the pupils meet in a place near where they live, under the direction of a teacher; every three weeks the teacher corrects a written composition.

We can gauge the scope and effectiveness of this programme from certain figures: in 1967, of 30,000 persons eligible, nearly 15,000, or 50 per cent, enrolled (at a fee of 25 DM); 5,000 pupils participated regularly in the Saturday morning sessions, and 3,000 candidates took the final examination in 1969. The proportion of failures was 3.5 per cent, and grades were on the average slightly higher than those obtained in conventional schools.

This 'TV-school' began its fourth course of studies in 1971; 2,700 pupils are enrolled. At the level of higher education the most interesting experiment is certainly the one that is going on in the United Kingdom under the name of the Open University.

The system combines radio and television broadcasts, printed materials, homework assignments done by correspondence, tutorial visits, and summer sessions and seminars in universities of the traditional type. Science education has an important place.

Enrolments closed in August 1970; of the 42,000 applications for entrance, only 25,000 could be accepted, according to plan. This new British university began broadcasting its courses in January 1971.

In conclusion, we must emphasize that the development of science teaching through television is marking time in the traditional school system based on the pupil-teacher relation. It can even be said that that system tends to repudiate technological innovations in general and educational television in particular. Although technology is made use of in every field, education alone would remain at the level of craft activity. It accordingly seems appropriate to recall one of the findings reached by the *Unesco Chronicle* in a summing-up of the activities of International Education Year: 'The time has come to ask whether education must remain the only major human activity in which technology may not increase man's potential, and to denounce the strange and pernicious paradox whereby education is required to change the world without any concession that it must itself be transformed for good and all.'[1]

Ideas, models—these are rather to be looked for in the field of further education, which is not yet organized as a 'system'. That is why a relatively large part of this report is devoted to that field.

Television and further education

As we have emphasized, the tendency is for scientific educational television to be aimed at an out-of-school public, by way of adult education.

Closed-circuit television

To begin with, we should recall the definite advantages closed-circuit television has for adult professional and vocational training. It is used more and more for teacher training, whether this involves initial prepara-

1. Volume XVII, no. 2, February 1971, p. 50.

tion at the end of regular studies or further training at study sessions and seminars.

Two techniques are of special interest for teacher training in general. The first is the practice of self-viewing. The teacher is recorded in a real or simulated situation, through closed-circuit television. Then, alone or in a group, he sees the play-back and is able to take a new look at his pedagogic behaviour. Television is thus a 'looking-glass' that makes objective analysis possible. The other is micro-teaching, which makes it possible for a group to study meaningful teaching situations. Ways of directing practical work, teaching this or that notion, etc., are analysed through several recordings. This technique does away with the need for frequent class visits, and the recordings selected are always representative.

Closed-circuit television is of particular value for certain types of specialized technological training. It is an invaluable intermediary between the theoretical training obtained in the school and university environment and the period of production training, at which genuine responsibilities must be exercised.

Usually closed-circuit television is installed in the training centre (teacher-training colleges, universities, pedagogic research centres like the ones at Rabat, Abidjan, etc.). But its sphere of activity can be expanded in many ways:

by organizing, at the centre or elsewhere, courses or sessions based on video tape recordings (e.g. the PERMAMA project for the further training of teachers in Quebec, run by the University of Quebec);

by establishing suitable links between the centres themselves. In France, each of the seven teacher training colleges in the 'Académie' or educational region of Clermont-Ferrand is equipped with a small production centre. The installation at the Regional Centre of Educational Documentation (CRDP), however, is a large one, with both stationary and mobile equipment. The Regional Centre co-ordinates the productions of the seven training colleges and, even more important, arranges fruitful exchanges among them;

by the use (in North America, for example) of cable teledistribution, which makes it possible to reach one very specific public.

Open-circuit television

Our main analysis, however, is necessarily directed to science education programmes based on a multi-media system in which television provides the principal driving force. This is probably the field where the effectiveness of television is most clear-cut. We shall examine a few significant experiments, one by one.

In France, there are a number of specialized establishments for adult education, like the Conservatoire National des Arts et Métiers, which prepares students for diplomas and degrees in engineering and different technical subjects through part-time courses taken over several years.

Courses in certain departments where enrolment is particularly high are televised and broadcast live in various receiving centres. The auditors enrolled at these centres benefit from the presence of an assistant who answers their questions afterwards and sets them the applied exercises which constitute their directed work. In 1971-72 the televised courses were as follows: elements of mathematics, elements of physics, general mathematics, basic electronics, information science.

Certain other television broadcasts are intended either to give a heterogeneous mass audience general information in new scientific, technical or economic fields, or else to give an audience composed of senior technical or managerial staff specialized information on current developments in industrial techniques. In 1971-72 these broadcasts had the following themes: recent developments in manufacturing industry, innovations and economic development, introduction to data processing, introduction to probability calculus, basic electronics.

A wider and more comprehensive service is provided by RTS/Promotion, a major subdivision of the French educational radio and television department, which broadcasts eight and a half hours of television programmes a week for a very large adult audience with an educational level corresponding to primary or intermediate studies. The programmes are mainly broadcast on Saturday and Sunday; some are repeated during the week. They are supplemented by printed materials offering explanations, fuller discussions and exercises with keys for self-correction. The exercises for some series are corrected by computer. Other series are supplemented by homework done by correspondence.

The RTS/Promotion audience can be estimated at 3 million viewers: of these, 30,000 subscribe to the accompanying printed materials and 3,000 are enrolled in the correspondence courses.

The educators who run this service hope that they will be able before very long to institute a system of performance assessment and examinations leading to academic credits, which would spur more viewers on to personal work and effort.

Besides this, a great effort is being made just now to bring television viewers together, either in school establishments outside class hours or at their place of work during working hours. The leaders of the groups that are formed in this way receive a special set of background materials and are remunerated.

The principal scientific disciplines covered by RTS/Promotion are mathematics (two series), electricity, electronics, technical draughtsmanship, automatic processes and statistics. A series on computer science is in preparation.

Since 1970, the German Democratic Republic has been broadcasting a course on data processing, with one half-hour broadcast a week. The course as a whole consists of: 26 thirty-minute broadcasts, three handbooks, one glossary, exercise sheets corrected by IBM computer. There are no group meetings, for lack of trained leaders. Student work is evaluated by the University of Mainz. At the end of the course a diploma is awarded which is not recognized by the State but is by the schools. The handbooks cost 35 DM and in the first year 140,000 copies were sold, while 55,000 persons signed up (at 18 DM) to take the diploma examination.

In Sweden, a new mathematics syllabus was introduced in the autumn of 1970. There were many changes in theoretical content; moreover, the authorities wanted to promote new teaching methods. When they launched the project they realized that the changes could not be made unless the teachers then working in the schools received further training. This meant courses for 40,000 general teachers in the primary system and 4,000 mathematics teachers in the secondary. Thus it was that Project 'Delta' was undertaken—and carried to a successful conclusion. The syllabus was divided into six parts. A 20-minute television broadcast presenting the main concepts in a stimulating way was followed by a 15-minute radio broadcast giving instructions in use of a pamphlet which was sent to all participants, who then studied twenty pages of this pamphlet and did a test which they corrected themselves. This cycle was repeated several times for a given part of the course.

Periodically, the participants filled in a form containing twenty questions each with five possible answers. The completed forms were fed through an optical reader, and the results were stored in a computer. The content of the course was designed in such a way that studies did not take more than 75 hours.

The entire project was carried out by co-operation between the National Education Council, *Sveriges Radio* (Radio Sweden) and Hermods, the biggest correspondence school in the country.

It is interesting to look at the financial aspect: the direct cost of the operation came to 60 Swedish kroner per participant. This is low in comparison with the cost of traditional teaching, for had it been materially possible to organize sessions of studies covering the same syllabus the cost would have been of the order of 2,000 kroner per participant.

A number of educational television services have also developed programmes for further training of serving teachers. In France, for instance, equal weight is attached to questions of theory and classroom practice. In this way television gives teachers a singular opportunity for exchanging information about their different experiences and experiments. It offers the advantage of 'live' exchange, and the message is not transmuted by successive codings. This method is therefore far superior to describing an educational experiment in a specialized journal.

In the science sector, the activities of the French radio and television department have been mostly directed to teachers. Thus, in mathematics and science (especially physical science) educational television has helped to bring the new syllabuses into effect by means of broadcasts with supplementary material (distributed free of charge to all mathematics teachers) which can be used for systematic instruction following each broadcast.

In the physical sciences the contribution of television (broadcasts and accompanying materials) has taken three forms, reflected in three different types of broadcast: keeping teachers abreast of recent advances and discoveries; giving practical help with the particular difficulties involved in teaching an experimental science (laboratory techniques); and contributing to research and the exchange of results (e.g., dissemination of the results of research into the teaching of mechanics revolutionizing both the general pedagogic approach and the type of experiments employed in this subject). This work has been taken up at national level by the Association des Professeurs de Physique; it has led to a series of experiments in some *lycées*, at upper secondary level, and the commission responsible for reorganizing teaching of the physical sciences has taken it into account.

We should also emphasize the scope and originality of the contribution television has made to multi-disciplinarity, with teams or individual specialists from various disciplines working together on a co-ordinated series of broadcasts (for instance, 'Mathematics at the Service of ...').

Conclusion: Promoting the use of television in science teaching

While the use of television for science education progressed rapidly between 1960 and 1970 thanks to the dynamic work of certain pioneers, one has to admit that today it is only marking time—probably because of the resistance put up by the traditional education system.

It seems it can develop freely where there is no highly organized system already in existence—for instance in the field of adult education, or in the countries of the third world.

Elsewhere, given the considerable advantages television offers, it would undoubtedly be desirable to try to overcome the resistance of the teaching profession, by a policy of inducements, not by coercion.

Towards that end, it seems that various measures might be proposed with a view to familiarizing teachers with television education.

It is no doubt a truism that the teacher is still conservative and is very strongly attached to the educational models he himself knew as a pupil. It would accordingly be appropriate to bring closed-circuit television into operation in teacher-training centres for both initial training and advanced or refresher courses.

As far as further training is concerned, television should in fact be the *prime medium*. In this way it is possible to reach without further ado the entire body of serving teachers, for whom television broadcasts afford the best means of exchanging experimental data. Other forms of in-service training are conceivable, but they have the drawback of being more costly and less effective. A system of training courses, sessions and seminars can reach only a very few teachers, while the 'pyramid' technique, where the training imparted in a particular discipline in a single nation-wide course is passed on down through regional courses, then through local ones, does not always give the results that were anticipated. For one thing there is often distortion of the message from level to level; for another the teachers who volunteer for these courses are often the best informed ones, those least in need of persuasion. Only through television is it possible to act quickly and reach every member of the profession (even the most isolated teachers) without any distortion of the message: and the quality of the message can be kept under check at national or regional level by the best specialists.

The fact remains that in using television for the further training of science teachers we cannot get at every member of the profession unless we provide incentives. Compulsory methods are to be avoided, for they do not always work: they can, for example, lead teachers to ask for (and obtain) their release from service, which clearly detracts from whatever economic advantages have been obtained. On the other hand, performance assessment along the lines adopted in Project 'Delta' can be envisaged; and for teachers who get a satisfactory assessment there might well be improved career conditions (speedier promotion, increase in the percentage of salary calculated towards pension, etc.). Such steps would in all likelihood have the desired effect.

Once the teacher is trained, or given further training, with the aid of television, he will no longer refuse for his pupils what has proved instructive for himself. Then, and only then, will educational television be able to make its full contribution to science education. Not only will it be possible for televised broadcasts to be included in school activities, but also the use of modern technologies by teachers should greatly increase their effectiveness.

In the immediate future two trends should develop in school instruction: at primary school level, general teachers should be more and more willing to have their teaching supplemented by supporting materials; at secondary school level, multidisciplinary 'science magazine' broadcasts outside school hours may interest pupils and so lead them to put pressure on their teachers —who will as a result be keener to follow the broadcasts that are specially intended for them.

BIBLIOGRAPHY

APTER, M. J. *The new technology of education.* London, Macmillan, 1968. Bibliog.
  Introductory guide to television, programmed instruction and computer-assisted instruction; applications in teaching of mathematics.

ASSOCIATION FOR SCIENCE EDUCATION. Using broadcasts. *Science for primary schools,* chap. 5. London, John Murray, 1970.
  The aim of this chapter is to help science teachers integrate radio broadcasts into their courses.

BEATON, W. G. Glasgow ETV—a quinquennial report. *Education and training* (London). November 1970, p. 436-437.
  Brief description of the evolution of Glasgow educational television since its creation in August 1965. It is mainly the hardware that is described, but mention is also made of modern mathematics programmes for secondary pupils and biology programmes for teachers produced by the university and broadcast after class hours. The relevant financial figures up to 1970 seem to indicate that this interesting but costly experiment represents a valuable long-term investment.

BOUHOT, G., PAILLE, A. *Circuit fermé de télévision et enseignement supérieur.* Ecole Normale Supérieure de Saint-Cloud. Centre Audio-Visuel, 1967, 21 p. (Centre Audio-Visuel Publication R-15; mimeographed).
  Report on the experimental use of television at the General Physiology Laboratory of the Faculté des Sciences, Orsay, to illustrate a series of practicals.

BOUHOT, G., et al. Les techniques audio-visuelles dans l'enseignement des sciences naturelles: le Thyrses. *Bulletin de l'Association des Professeurs de Biologie et de Géologie* (Paris), no. 1, 1967, p. 88.

BRETZ, R. *Color TV in instruction.* Santa Monica (Calif.), Rand Corporation, June 1970.
  Several studies have failed to establish the superiority of colour over black and white in the field of teaching. This situation may change when the technology of colour television has evolved.

BUREAU D'ÉTUDES TECHNICO-ÉCONOMIQUES RELATIVES À L'ENSEIGNEMENT AUDIO-VISUEL (BETEA). *Etude technique et comparative des différents procédés de réalisation de films ou de séquences filmées destinés à être utilisés dans les programmes de télévision.* Paris, BETEA, 1969. 110 p. (mimeographed).

BETEA. *Inventaire général des procédés techniques de plateau utilisés pour la réalisation électronique d'émissions de télévision.* Paris, BETEA, 1968. 120 p. (mimeographed).

CHITTOCK, J. TV cassettes—friend or foe? *Educational broadcasting international* (Wrexham, Wales), vol. 5, no. 3, September 1971, p. 164-167.
  This study is confined to systems of electronic projection (on the television screen), considered more 'convenient' than optical projection systems. EVR, Teldec video discs, video tapes, telecine, Selectavision (which uses laser beams and holography) and other systems are described. While their cost alone may inhibit the use of cassettes in education, they can be used in existing closed-circuit designs.

CHU, G.; SCHRAMM, W. Learning from television: what the research says. In: S. Tickton (ed.), *To improve learning.* New York, Bowker, 1970.
  The research referred to in the title was carried out in 1967 by the Stanford University Institute for Communication Research at the request of the United States Office of Education. Here are some of the findings: under favourable conditions, pupils learn successfully through television; there is no proof that comics or animated cartoons facilitate learning; insertion of questions does not seem to improve learning, whereas a break does make for more efficient work; whether the broadcast comes at the beginning or the end of a lesson is immaterial; discussion by the teacher after a broadcast is more effective than a re-broadcast of the programme; the size of the audience does not matter; the impossibility of asking questions or discussing is a disadvantage for (especially) the more advanced pupils; note-taking during the broadcast is a hindrance to learning; television can be useful in the developing countries.

Computers: Hungary looks ahead. *Nature* (London), vol. 232, no. 5308, 23 July 1971, p. 213.
  A brief paragraph announces that provision has been made in the fourth Hungarian five-year plan for training about 20,000 data processing experts. A teaching centre equipped with closed-cicruit television is to be constructed.

COUNCIL OF EUROPE, Council for Cultural Co-operation. *L'enseignement direct par la télévision. Rapport du deuxième séminaire, Scheveningen, 1968.* Strasbourg, Council of Europe. Council for Cultural Co-operation, 1969, 50 p.
  Report on the seminar on adult education held at Scheveningen in 1968. Descriptions, *inter alia,* of TELEAC—an experiment in evaluation carried out in collaboration with the University of Amsterdam and the Television Foundation at Hilversum—and of a pilot project for evaluating the results of series of experimental ETV programmes. An economic study is suggested; this ought to be of interest to all the countries taking part.

CROCKER, R. K.; ROWE, F. B. *Some observations on an elementary science education course by ETV.* St. John's (Newfoundland), Memorial University [1971].
  Description of a one-semester introductory course in science teaching for primary teachers. Video copies of the course were produced for teachers working in isolated regions of Newfoundland. Student-teachers were given sound cassettes so that they could record questions they wanted to ask the instructors; but this did not solve the problem of communication. Cases containing laboratory material were also provided.

DIEUZEIDE, J. Teaching through television. Paris, Organization for Economic Co-operation and Development, July 1960. 72 p.

DRIZUN, I. L. *Syllabus and method for the use of educational television in chemistry teaching at secondary school.* Leningrad, Ministerstvo prosvescenija SSSR (Ministry of Education of the USSR). Herzen Pedagogical Institute, 1969.

Educate teachers for TV. *The Guardian* (Manchester), 10 June 1971.
  The BBC and the ITA consider that teacher training colleges are not doing enough to prepare teachers for televised instruction.

EL SHENAWY, W. An education television pilot project in Cairo. *Education television international* (Wrexham, Wales), vol. 4, no. 4, December 1970, p. 301-304.

   Brief report on a project in teaching science and English in secondary schools, meant to serve as model for a nation-wide televised teaching project. Some practical problems arose: (1) Science films belonging to the Ministry of Education Audio-visual Aids Department and one belonging to the Science Museum could not be used, their investment contracts forbidding televised showing; (2) the teachers' notes were not always delivered in time.

Engineering education by 'tele-study'. *Die Deutsche Universitätszeitung* (Bonn), no. 8, April 1971, p. 241.

   Report on the first assembly of the organization responsible for televised teaching of third-year technology and civil engineering at the University of Tübingen. Televised teaching must be used in conjunction with other aids and with new teaching and learning methods. Creation of a central office of information and co-ordination is envisaged.

ERMELING, H.; STÜPER, F. J., Erfahrungen mit dem Schulfernsehen (Experiments in school television). *Mathematische und Naturwissenschaftliche Unterricht* (Bonn), vol. 24, no. 3, 1 May 1971, p. 171-177.

   In the plans for the Max-Planck School at Gelsenkirchen-Bauer (inaugurated Easter 1966) the intention had been to make use of televised instruction (particularly for science). Later the school was asked to produce the science television programmes for all the city's schools. The camera was used freely as a 'magnifying glass' for science teaching. The article describes the advantage of television in this respect; it also gives a detailed script for televised recording of an experimental demonstration of the molecular mass of carbon dioxide. Pupils have found the programme profitable as a supplement to the course, but not as a substitute for the teacher.

FIRNBERG, J. W. An evaluation to closed-circuit television as used in teaching selected courses at Louisiana State University. *Dissertation abstracts international. A.* (Ann Arbor, Mich.), vol. 30, no. 4, 1969, pp. 1330-1331. (Abstract of the thesis.)

FORSMAN, M. E. Graduate engineering education via television. *1968 IEEE International Convention Digest.* New York, Institute of Electrical and Electronics Engineers, 1968.

   Description of the University of Gainesville's closed-circuit television (GENESYS), which has been reproduced with some variations by other universities.

GORDON, G. N. *Classroom television. New frontiers ITV.* New York, Hastings House, 1970.

   A study of school television covering twenty years proves that television is more welcome in the classroom now than in 1950. Video tapes are bound gradually to replace radio broadcasts. The author describes the opportunities for teaching and learning opened up by television (including closed-circuit television). He also examines the process of integrating televised lessons into the classroom curriculum.

HANCOCK, A. A model for ETV. *Educational broadcasting international* (Wrexham, Wales), vol. 5, no. 1, March 1971, p. 46-51.

   The author tries to reconcile the points of view of teachers and of television specialists. A development model shows that educational methodology should progress towards ingrated teaching patterns based on the technology of education and the use of educational television. Multi-media systems will become the rule.

HAWKRIDGE, D. G. *The teaching of science to students at a distance.* Address delivered at the annual congress of the American Association for the Advancement of Science held in December 1970.

   Brief account of the educational system of the British Open University, which exploits a wide range of teaching methods and reaches 25,000 students. In 1973 there should be 10,000 science students. Instruction is for adults, nearly all of whom work full time. Courses are based on units of self-instruction and evaluation but they also include radio and television broadcasts and personal contacts at regional centres and at the summer university.

HOLROYDE, D. J. G. The use of television in university teaching. In: D. Layton (ed.), *University teaching in transition.* Edinburgh, Oliver and Boyd, 1968.

   The author examines the improvements in learning achieved by use of television in higher education.

HUDSON, B. The future of educational television. In: S. Tickton (ed.), *To improve learning.* New York, Bowker, 1970.

   The author examines the opportunities offered by communication satellites in the educational television field. He thinks that their role will be far more important in the developing countries than in the United States.

KELLER, G. Unterricht im Medienverbund (Instruction through integrated teaching media). *Mathematische und Naturwissenschaftliche Unterricht* (Bonn), vol. 24, no. 3, 1 May 1971, p. 180-181.

   Account of the demonstration of a new teaching medium by an educational publishing combine. Schroedel-Dicterweg-Schönough (SDS) at Frankfurt-am-Main in September 1970. The material presented includes a book by Dr. U. Lubeseder, *Quantities-Forms-Relations,* which can be used in conjunction with a televised programme, 'Introduction to Modern Mathematics'; a handbook to go with radio broadcasts on mathematics, provided with a set of 230 plastic bars of different shapes and colours; and a cassette film to be used in association with a chapter of H. R. Chrisent's *Introduction to chemistry.* This group of publishers works in collaboration with a film company, 'Videophon'. They propose to integrate the following educational media for chemistry teaching: pupil handbooks; 60 five-minute filmstrips; 10 twenty-minute televised broadcasts; and experimental material for teachers and pupils. They plan to leave blanks in the filmstrips so that pupils can give written answers to questions posed over the air. The article also reports on the discussion which followed this presentation, and which bore on the chief problems of multi-media teaching.

KHANNA, P. N. Observing surgical operations by closed-circuit television and direct observation—A comparative study. *Educational television international* (Wrexham, Wales), vol. 4, no. 4, December 1970, p. 314-318.

   A comparison carried out with 36 students at the All India Institute of Medical Sciences in New Delhi. The results show that closed-circuit television is more effective and that its superiority increases in proportion to the number of people in the operating theatre.

LEFRANC, R. *L'utilisation de la télévision en circuit fermé dans l'enseignement supérieur aux Etats-Unis.* Paris, Institut Pédagogique National [1963]. 366 p. (Doctoral thesis under the direction of M. Debesse; mimeographed.)

MACLEAN, R. Television in higher education. In: D. Unwin (ed.), *Media and methods,* London, McGraw-Hill, 1969.

The needs of colleges and universities may be classed in several categories, from the simple audio-visual aid requiring a mobile unit to the full-scale broadcast calling for a well-equipped studio and a broadcasting system. The author describes Glasgow educational television. Video tapes are better than films in that they enable the teacher to keep closer to reality.

MEYER, G. R. Science by radio and television in a developing country—Mauritius, a case study. *Australian science teachers journal,* vol. 16, no. 2, August 1970, p. 21-27.

A study of the situation at the end of 1969 shows that the science broadcasts which were being given at six in the evening should be given during the school day, with the television programmes recorded and rebroadcast in the evening. There should be a special educational television organization, linked to the central institute of education, specifically for teacher training; this would also make it possible to provide science instruction for young adults in rural areas.

NATIONAL ACADEMY OF ENGINEERING. Instructional Technology Committee. *Educational technology in higher education.* Washington (D.C.), National Academy of Engineering, September 1969.

This preliminary report examines the advantages and disadvantages of educational television and computer-assisted teaching. It stresses the advantages (more flexible timetables) of the new video tape recorders. The attitude of the teaching profession to educational television and assisted teaching is often negative. That of students is generally less so. Learning situations ought to be varied (for instance by alternating direct instruction and television). Estimates indicate that in American schools assisted teaching costs approximately three dollars per hour per student. On the operating plane, time-sharing, though indispensable, is hard to bring about. It will take several years yet to establish a simple means of communication between computers and lay users. Computer programmes ought to be interchangeable.

NATIONAL ASSOCIATION OF EDUCATIONAL BROADCASTERS. *Toward a significant difference: final report of the national project for the improvement of televised instruction, 1965-1968.* Washington (D.C.), the Association, 1969. [Analysed by J. S. Clayton in *AV communication review* (Washington, D.C.), vol. 18, no. 1, Spring 1970, p. 88-91.]

NATIONAL ASSOCIATION OF EDUCATION BROADCASTERS (Research and Development Office). Television-in-instruction: the state of the art. In: S. Tickton (ed.), *To improve learning.* New York, Bowker, 1970.

Analysis of different possible uses of television for teacher training: scaling-up of laboratory experiments; recordings of student-teacher activities for self-analysis; presentation of films to a wide public; use of television: combined with teach-yourself systems (video-tapes) and correspondence courses; as an electronic blackboard, especially for mathematics; as a medium of communication; as part of a curriculum or a supplement to it; and in conjunction with programmed instruction. The paper stresses that not all applications of television have had an important or lasting impact in the field of American education. Closed-circuit television is bound to become the main transmitting medium in the future: satellites will merely be useful adjuncts.

NATIONAL EDUCATIONAL CLOSED-CIRCUIT TELEVISION ASSOCIATION. *Directory of educational closed-circuit television in Britain.* London, the Association, 1970.

Description of all the closed-circuit television systems in operation in universities, university administrative offices, teacher training colleges, etc.

NISHIMOTO, M. *The development of educational broadcasting in Japan.* Ruteland (Vt.), Charles B. Tuttle Company, 1969. [Analysed by W. B. Emery in *AV communication review* (Washington, D.C.), vol. 18, no. 2, Summer 1970, p. 210-211.]

Description of the growth of Japanese radio and television (acknowledged to be among the best in the world), which are intended 'to meet instructional needs at all levels of learning'. Includes a report on teaching by correspondence in conjunction with radio and television. The author writes: 'The problem facing the teachers of today and tomorrow is how they are to use school broadcasting, programmed learning and textbooks as an effective unit.'

PENTZ, M. J. Science teaching at the Open University. *Physics bulletin* (London), vol. 21, October 1970, p. 452-454.

Brief description of an integrated system of multi-media instruction including radio and television broadcasts, booklets, textbooks, sets of experimental material, self-evaluation tests. The course consists of 34 units spread over 42 weeks.

RADCLIFFE, P. Canadian report. *Educational television international* (Wrexham, Wales), vol. 4, no. 4, December 1970, p. 287-292.

Report on experiments in educational television carried out in four Ottawa schools and at the University of Toronto. The schools experimented with a 'self-service' television system. A dial telephone on the classroom wall gives access to a library of 2,500 films and video tapes, linked to the school by coaxial cable. Such a system would not cost more than a conventional library if it were expanded to cover the city. The Scarborough College Television Centre at the University of Toronto has been termed a 'spectacular flop' because of its huge cost; in any case, several faculties at the university (sciences, engineering, medicine) have their own television systems.

SCHRAMM, W. The future of educational radio and television. *Educational television international* (Wrexham, Wales), vol. 4, no. 4, December 1970, p. 282-286.

At the present moment more than 50 countries have educational television and more than 100 educational radio. The problem now is to improve the programmes and co-ordinate efforts. One solution would be a wired network making it possible to obtain the desired programmes at long distance (e.g., mathematics students in Palo Alto use consoles connected to a Stanford University computer). Educational television can be developed by point-to-point distribution; here satellites will play an important part. There would seem to be a great future for the use of EVR (Electronic Video Recorders). The 'incubation period' for bringing about major changes can be estimated at '25 years in the industrialized countries; 50 years or more in the developing countries'. People are not quite sure how to combine programmed instruction effectively with radio and television. Information centres on educational radio and television would be very helpful. Radio can perform many of the same tasks as television at a fifth of the cost.

SCHRAMM, W., et al. *The new media: memo to educational planners.* Paris, Unesco International Institute for Educational Planning. 1967. 175 p.

Studies of 23 examples in 17 countries.

Science and technology courses at the Open University. *Educational broadcasting international* (Wrexham, Wales), vol. 5. no. 2, June 1971, p. 92-94.

SEOW, P. The 'Discovering Science' series—producer's eye view. *Educational broadcasting international* (Wrexham, Wales), vol. 5, no. 2, June 1971, p. 98-91.

Description of educational television in Singapore, including the use of broadcasts to supplement the teacher's work in class. Wall charts (as well as other audio-visual material) are widely used prior to and following the programmes.

SPAULDING, S. Advanced educational technologies. *Prospects in education* (Paris, Unesco). vol. 1, no. 3, 1970, p. 9-19.

The author emphasizes that a true educational 'technology' includes the 'setting of goals' and the 'continuous renewal of curriculum'. He stresses the fact that efforts are being made to adapt teaching both to individual and to mass requirements. He notes the lessons learned from educational radio and television experiments in Samoa, Colombia, Niger, El Salvador, etc. He also analyses the reason for the interest educators take in programmed instruction.

Systèmes multi-media: compendium international. Munich. Internationales Zentralinstitut für das Jugend- und Bildungsfernsehen, 1970. 244 p. (mimeographed).

Description of eleven projects for combined teaching systems in eight countries: Brazil, Canada, France, Federal Republic of Germany, Japan, Poland, United Kingdom, United States of America. Compilation prepared for the course organized in Munich (29 April-5 May 1970) by the Internationales Zentralinstitut and the Council of Europe on the theme: 'The application of combined teaching systems and the new aspects and functions of education deriving from them—methods of total or partial programming'.

Systems multi-media dans l'éducation des adultes. Munich. Internationales Zentralinstitut für das Jugend- und Bildungsfernsehen [1971]. 260 p.

Description of twelve projects in nine countries: Austria, France, Federal Republic of Germany, Japan, Netherlands, Poland, Sweden, United Kingdom, and United States of America. Revised and enlarged version of the compendium cited above.

TAYLOR, F. J.; BARNES, N. Some reflections on educational research. *Educational television international* (Wrexham, Wales), vol. 4, no. 2, June 1970, p. 92-94.

Some basic questions as to the value of educational broadcasting still remain unasked. Can television be used successfully for teaching facts, new attitudes, problem-solving or decision-making? And under what conditions? The article also mentions the recent attempts in several countries to use radio and television for professional and vocational training. With the advent of the Open University in the United Kingdom, integrated research becomes imperative.

Television provides best medium for continuing education. *Product engineering* (New York), vol. 40, no. 19, 1969, p. 26-27.

Colleges of technology (particularly those at the Universities of Stanford and Michigan) are making ever increased use of closed-circuit television.

TROTTER, B. *Television and technology in university teaching.* Toronto, Committee on University Affairs and Committee of Presidents of Universities of Ontario. 1970. 84 p.

Proposal for establishing in Ontario a new institution rather like the Open University. The study recommends the creation of an educational development centre in order to make a detailed systems analysis, as well as a planning structure for this new autonomous institution. The students would be mostly full-time wage-earners; they would go regularly to a regional centre to hear and see courses, meet with tutors, etc. Staffing would be at the ratio of one teacher to 50 students.

UNESCO INTERNATIONAL INSTITUTE FOR EDUCATIONAL PLANNING (IIEP). *New educational media in action: case studies for planners.* Paris, Unesco IIEP. 1967. 3 vols.

Description of 23 studies carried out in: Algeria, Australia, Colombia, Honduras, India, Italy, Ivory Coast, Japan, New Zealand, Niger, Nigeria, Peru, Thailand, Togo, the United States, and Western Samoa. The techniques used are radio, television, films and programmed instruction.

UNIVERSITIES OF GLASGOW AND STRATHCLYDE. *Audio-visual services annual report 1969-1970.*

The two audio-visual centres operate autonomously but under joint control. At Glasgow, television is used as a direct audio-visual aid, especially in biology: there are few requests for replays or recordings of television courses. At Strathclyde, television is mainly used for pre-recording courses, particularly in mathematics. The three types of recording ordinarily asked for are: the traditional lesson, the experimental demonstration, the structured programme. The teams from the two universities have jointly published a booklet dealing with sound and image in higher education.

The Universities Association for Tele-Studies introduces itself. *Die Deutsche Universitätszeitung* (Bonn), no. 6, March 1971, p. 177.

Article on the press conference that followed the Association's first assembly, in February 1971. A general Association of Tele-Studies is to be formed. This act will have important consequences for university activities, since it involves introducing audio-visual methods; henceforth the universities will have to maintain close links with governmental and broadcasting associations. Contacts have already been established with the *Deutsche Institut für Fernstudien* of the University of Tübingen, etc.

VALERIEN, J. *Les circuits fermés de télévision dans l'enseignement supérieur scientifique*. Ecole Normale Supérieure de Saint-Cloud, Centre Audio-Visuel. 1967. 33 p. (Centre Audio-Visuel publication R-25; mimeographed).

VALERIEN, J. Une expérience d'utilisation de la télévision en circuit fermé dans l'enseignement secondaire du premier cycle (Ecole Alsacienne). Compte rendu 1946-1965. *Sang neuf,* no. 14, 2nd quarter 1966.

The use of radio in science education

by John C. H. Ball
Schools Broadcasting, Voice of Kenya,
Nairobi, Kenya

Summary

Possibly the most difficult, and most important, skill required of an education broadcaster is the ability to understand a subject as his listeners understand it. Taking the word 'understand' in the widest sense, this means knowledge of the listener's way of life—his environment, his language, his culture, and the radio reception conditions where he lives. The technical and aesthetic production of a programme must derive directly from this knowledge. This is of first importance in developing countries, where the tendency has been to import or adapt material from developed countries. The uncritical use of this material has obscured the need for basic research into the learning processes and the environment of the local peoples.

Where radio is used for science education in the classroom, careful thought must be given to the structure of the programme as well as its content. The devices of the radio producer—sound effects, fades, music, dramatization—must be precisely thought out and pre-tested. Where a practical element is built into radio lessons it should be exactly described and timed. The use of printed support material for teachers and pupils requires not only the precise integration of sound and sight, but also a knowledge of the 'visual literacy' of the listeners.

There is reason to doubt if the tape-slide technique, known as 'Radiovision', is properly the field of the broadcaster. This type of package can be produced by audio-visual aid units and requires no broadcasting expertise. Is the radio organization diverting skilled broadcasting staff away from their proper role into work which could be better left to others?

The rapid spread of the inexpensive tape-recorder, especially of the cassette type, is having a profound effect on radio broadcasting. Insufficient thought has been given to the respective roles of these two—the taped lesson and the broadcast lesson. To say that 'All schools broadcasts are better used in tape-recorded form' is to confuse the two techniques. If this statement is true, then radio is simply an intermediate agent between the broadcasting studio and the tape-recorder plugged into the receiver in the school. We then have to consider if a unified broadcast is suitable for the stop-start technique of the tape-recorder, which gives the teacher the 'complete control' now thought to be desirable.

While so many problems surround the use of radio for science education in the classroom, its use for out-of-school science education is more straightforward but equally important. This utilization includes programmes for adult listening, with such self-explanatory titles as 'Science in Action', 'Science News', 'Nature Notebook'; background information programmes for secondary schools and university students; supplementary programmes as part of science correspondence courses; and special health education programmes. These last named could make use of the successful techniques of the radio advertiser to identify and encourage the development of basic health habits for the prevention of disease. That radio is a powerful information medium is recognized by all governments; science educators should use it to inform people on the facts of pollution, population and food.

The success of radio as an educational medium is related to the infrastructure at the listening end—the classroom, adult centre, farm forum—and to its ability to inform, interest and involve the listener.

Introduction

Writing on schools broadcasting [1][1] Charles Armour states: 'The future of radio will be founded on the transmission of high quality programmes, which meet current needs of teachers and their pupils'—a truism which could also be applied to almost any industrial product intended for sale.

The starting point for the industrial producer and the educational broadcaster are the same—the consumer. There is no point in producing something which no one will buy—or broadcasting something which no one will listen to. Possibly the most difficult, and most important, thing for a broadcaster is to see the subject as the listener sees it. Somehow a link has to be found between the way in which he understands the subject and the way in which the listener approaches it. And this in turn depends upon the individual listener and his own previous knowledge. There is no doubt that many 'pop' disc-jockeys find this link with great success. They understand their subject and their listener. Their suc-

1. For references, see page 89.

cess can perhaps be measured by the cubic capacity of their fan mail.

While hesitating to suggest that the success of those responsible for science education by radio should be measured in the same way, the lesson is one which we should not ignore. Perhaps radio science does not lend itself to the Top Ten, but it should aim at the response one might expect from a Number Eleven. We cannot communicate anything by radio unless the listener will listen, and the listener will not do this unless we can hold his interest. We should not be unwilling to learn techniques from those whose aim is sales rather than science. Much thought and research has gone into skilful 30-second advertising spots on commercial radio. These *do* sell products—if they did not, money would not be spent on them. Why, for example, should we not develop the same technique in teaching health science? A skilful 'plug', repeated two or three times a day at peak listening time could make a whole population aware of simple but fundamental health needs.

There are two points here—that radio can reach a whole population and that science teaching by radio should cover the whole population, from the primary school child to his grandfather. This does not mean that radio can do everything—it cannot. As a carpenter selects different tools to do different jobs, so we can select radio or television or books or films or a combination of audio and visual aids. When we select radio we should not confine its use in science education to programmes for schools. If we accept that education does not end with the taking of any particular examination or the arrival at any particular age, then we must consider science programmes for general listening by adults to be as important as programmes for children in the classroom.

Adult programmes can be used with group radio listening, which has been very successful, particularly in rural/agricultural education, in countries as diverse as Canada, Poland, Niger and India [24]. Agriculture itself may be classed as the application of science, based as it is today on the results of scientific research into new strains of plants, soil analysis, pesticides and fertilisers. Adult science programmes can also be part of the normal scheduled output of a radio station. Good examples of this type of programme are the BBC World Service series on 'Science in Action' and 'Nature Notebook', as well as 'The Farming World'. All these provide up-to-date information on wild life pollution, the environment, industry, medicine and so on. Thus a recent programme in 'Nature Notebook' discussed the ecological effects of the steady destruction of tropical rain forests by timber cutting, farming and charcoal making (and warfare in South East Asia)—a topic of particular importance to West and East Africa. Radio has its part to play in 'Doomwatching', in that it can carry information and comment from any part of the world to any other part, crossing all international frontiers. Reports on discussion at the United Nations Conference on the Human Environment, held in June 1972 in Sweden, were broadcast by radio stations all over the world. Such reports were science education on a vast scale and yet were as valid a use of radio for this purpose as simple 'Junior Science' series for primary children in Britain or Kenya.

If we accept science education by radio in this wide context, then we must look more closely at radio itself, before going on to consider types of programmes, structures, visual aids, radiovision and the relationship between radio and tape-recorders.

Radio itself

Radio is cheap compared with television, both at the production end and at the receiving end. Radio studios cost much less to build and equip than do television studios, while the radio receiver costs far less to make and to buy than the television receiver. A radio can be powered by a few cheap dry cells, whereas television requires a mains electricity supply or a set of expensive batteries. There must be few villages in Kenya without one radio set, and a very large number of villages with several sets. More than 3,500 primary schools, 550 secondary schools and 32 teacher training colleges are known to have radio sets. There are no television sets in primary schools and few in secondary schools and colleges. Thus, while the present scope for educational television is small, that for educational radio is considerable. This is particularly important in developing countries, where not only is there a limited distribution of printed information, such as newspapers, but also a high rate of illiteracy. In theory, therefore, because radio requires only the ability to hear, it has a wide audience of literate and illiterate. This is recognized by all governments who use their national radio systems for information, entertainment and education—sometimes in that order of importance. Many countries transmit radio programmes internationally with varying aims—education, entertainment, but more frequently to put over a particular point of view—their own. There are millions of multi-band radio sets able to receive short-wave transmissions from all over the world, and millions more receiving medium and long-wave transmissions locally.

To have the 'hardware' available is essential—and there is no point in broadcasting radio programmes if there are no radios to receive them. But even the presence of sets does not necessarily mean that we are 'communicating' with people. We have to persuade them, first, to switch on and then we have to hold their interest and attention so that they do not switch off. This is a universal requirement of radio—we must reach and hold the attention of our particular audience, whether we are broadcasting a religious service, a 'pop' session, a sports commentary or a science lesson. The problem is how to achieve this in the face of many difficulties ranging from poor reception conditions to programme structure and content.

The argument sometimes advanced that radio is specially suitable for Africans because they have an 'oral tradition' is an oversimplification. Europeans, with centuries of 'written tradition' also listen to the radio. Furthermore, oral traditions have an important visual element—facial expressions and hand gestures—which are absent from radio. Far from being a 'natural' way of listening, the 'voice from the box' is a sophisticated technique which has to be learnt by the listener as well as by the producer. One important factor in this is the clarity and character of the radio voices, qualities which assume much greater importance than when speaker and listener are face to face.

Undoubtedly the biggest problems face the education radio broadcaster, and we will have to examine these problems in relation to science education. But it is well to keep in mind the wider problems of education into which those of radio fit. Spaulding correctly summarizes the situation [2]: 'The unsophisticated enthusiast often believes that new educational techniques will quickly and cheaply resolve the current problems of education. The enthusiasts usually *underestimate the complexities of education and the immense problems inherent in changing* the way we go about education so as to permit the new technologies to contribute effectively...' To put it another way, it can be like trying to move an elephant with a matchstick.

Radio reception

As only the sense of hearing is involved with radio, the clarity and quality of what is heard is fundamental to the understanding of the listener. Reception on long and medium wave is usually satisfactory, although it can deteriorate in certain geographical conditions. Reception on short wave, because of variable conditions in the ionosphere, can adversely affect speech clarity. This is of special importance in science programmes, when it is essential that the words can be heard clearly.

R. L. Hilliard writing on radio broadcasting [3] puts it in this way. 'The conscientious radio broadcaster aims to transmit sound in such a way that the listening public will be able to understand speech clearly and enjoy the reproduction of music. This seems a rather simple objective, yet the achievement of it is often difficult and, unfortunately, at many stations is never accomplished. After leaving the transmitter, sound is subject to all sorts of losses and quality degradations before it reaches the listener's ear. Therefore it is necessary to keep the technical quality of the sound as high as possible at all times while it is under the control of the station. The station must not relax its efforts to provide "clean" audio, especially because much of the audience may be listening on inexpensive table-model radio sets and tiny transistor portables. For example, although newspaper photographs are necessarily rather coarse in comparison with original photographs or magazine reproductions, expensive cameras, careful processing and delicate engraving methods are used to ensure that the clearest possible pictures, within the limits of the reproduction process, result.'

This idea of quality is not necessarily the same as the aesthetic quality which may appeal to the producer. A beautiful programme, involving several voices and the subtle use of sound effects, may emerge as a confusion of sounds at the receiving end. By quality here we mean quality directly in relation to audibility. It is of prime importance that the producer of programmes be aware of reception conditions in the area to which he is broadcasting and that he take these into account.

Another factor which must not be overlooked is the external noise which may be occurring at the receiving end—noise from other classes, children in the playground, street sounds and traffic. The producer cannot control these, but if he is aware of them he can take them into account in speech quality, programme structure and techniques of production.

Radio production techniques

In certain circumstances, commonly used radio production techniques can be a hindrance to hearing and understanding. An example can be taken from the use of sound effects [4]. Sound effects are used to add depth and interest to words and to help the creation of pictures in the mind of the listener. But as the listener can only interpret what he hears in relation to what he already knows, clearly the sounds must be ones he already knows, clearly the sound of a ship's siren will not create a picture of a ship in the mind of person who has neither seen one nor heard the sound of the siren *coming from it*. Just to have seen a big ship is not enough—seeing alone does not tell him the sound it makes. Likewise, hearing alone does not create the picture. *Sight and sound* must at some time previously have been brought together—in reality, on film, on television.

The problem of using sounds in a radio lesson does not end here. Listeners will interpret an unknown sound in relation to a similar known sound. This can produce a completely different mental picture from the one intended by the producer of the programme. That the sound may be used with words which appear to make its context clear may not be enough. In tests carried out in Kenya, 22 per cent of the children tested were unable to identify the lowing of a cow. Some thought it was an elephant, a leopard, a man, trees, engine of a car, and 7.3 per cent thought it was a hyena. Of a group of teachers from a rural area who were tested, 30 per cent thought the sound was made by a hyena. One teacher's method of identifying the two sounds, cow and hyena, was 'If the sound goes down at the end it is a cow; if it goes up at the end it is a hyena'. The lowing of a cow is an acceptable sound for a peaceful agricultural setting on radio and the call of a hyena an equally normal one for a natural history programme, but when the identification is reversed, the presence of a hyena in a farm, and a cow in a game park may not be the mental pictures which the programme producer wishes to create. Again, in a natural history lesson it might be assumed that the roaring of a lion would create a mental picture of that animal. But this is much more likely to occur in the minds of children in Europe and America than in East Africa. Continuous exposure to radio, television and films from a very early age familiarizes the European and American child with the lion and his roar, although the majority of them may never have seen a real lion or heard it roar.

In the Kenya tests mentioned above, the roar of a lion was included. Only 40.7 per cent of the children tested recognized the sound. More significantly, an analysis of the individual school results showed that in the urban Asian and multi-racial primary schools the recognition ranged from 100 per cent to 84 per cent, whereas in the rural African schools the range was from 22 per cent to 0 per cent. A major factor in this difference is the exposure of urban children to television, films and visits to the Nairobi National Park—facilities not available to rural children. An African child living in a rural area is less likely to recognize a lion's roar than a child in London, New York, Tokyo—or Nairobi. In addition, wrong identification can be a definite blockage to understanding—children 'identified' the lion's roar as a pig, dog, car, aeroplane, radio, motor-cycle, tractor, machine for coffee, people singing, beating a bull and breaking dry leaves.

It should never be assumed that because we, as adults, with our wide terms of mental reference, understand a word or sound, children will likewise understand it. This does not mean that sounds should never be used in a science broadcast, but that when they are used, due thought must be given to them. They can be very helpful if properly used and identified by words. Thus a recording of a foetal heartbeat can be used if the listeners are told what it is. Such a sound could add interest to a biology programme, because it is a sound which the listeners are never likely to hear in any other way. What sounds are used and how they are used will depend on the country and on the producer's knowledge of his listeners. Thus sounds associated with industrial processes are more likely to be understood by children in Western Europe than by children in rural Africa.

Sound effects are given at length as an example of a common production technique which can have the opposite effect to that which is intended, a hindrance rather than a help to understanding. Other production techniques also require care. The voices chosen have a major effect on understanding. This applies not only to their pitch: too low and too high pitch are often distorted by small loudspeakers—but also to their character and personality. The success of two junior science series in Britain and East Africa was partly due to the personality of the radio voices used.

The use of dramatic inserts is also a technique which requires more thought than is sometimes given to it. Students will accept studio actors playing a dramatic part from a novel, but will they also accept them playing the parts of real scientists (Lister, Pasteur, Rutherford, etc.)? Younger children may do so, older children may consider it rather silly and prefer a straight narrative account. A really good radio narrator or reader may be able to tell the story of Lister far better than any fictitious dramatic inserts in a programme. Thought must also be given to the relevance of drama or play-acting to the culture of the listening children. Is the western type of dramatic presentation 'understood' by all children? This problem may not occur in Asian countries (India, South-east, Asia, Japan, etc.) which have their own traditional drama forms. But in many parts of tropical Africa, education radio has derived from the models of western Europe because it was started by Europeans during the colonial period. What does the dramatic presentation of the lives of long-dead European scientists mean to the rural African child? What politicians call 'colonial relics' remain in education broadcasting in Africa, where much more local, *practical* research is still needed into the methods and techniques of radio presentation. The word 'local' does not refer to national frontiers, for within many African countries there is a great diversity of environments, where the way of life, even the food, of the children is completely different. Thus Kenya has tropical, savannah, desert and semi-temperate zones—and these present problems in a national radio network, particularly in such topics as natural history and health science.

'The ability to bring a wealth of outside experiences into the classroom at the flick of a switch excites the imagination. But the definition of what this should be, its planning, production, broadcast and use in schools must be part of the overall curriculum research and development process' (Spaulding [2]). There is urgent need for local, practical research into the use of radio as an instrument for education, and 'on-the-spot' application and re-testing of the results of that research. This is a *field-worker project,* not for itinerant 'consultants' and 'experts' whose 'reports' and 'recommendations' gather dust in ministerial pending trays.

We should include in the study of radio production techniques, the actual language of the broadcast. 'Often because of their limited vocabulary and inadequate background experience in a particular area of study, children are at a loss to follow a programme and appreciate its contents.' In this caution on the use of language, Gwen Allen [5] is referring to broadcasts in English to native English-speaking children, and her caution undoubtedly applies to all children listening in their native tongues. A more difficult problem arises when children have to listen in a language which is not their mother tongue and which they have to learn as they go along. In Kenya the medium of instruction in primary (and secondary) schools is English. Primary science broadcasts, including health and natural history, are confined to Standards 6 and 7, by which time the children have, in theory, been learning English for five years. However, as the children learn their English from African teachers, who in turn were taught by African teachers, mispronunciation of the language is not an uncommon occurrence.

These mispronounciations are mainly of two types—those arising from the fact that they have never heard the correct one and those arising from the absence of certain English sounds from their mother tongue. One example of each will suffice. From the name alone children were unable to recognize the Pied Crow, a bird common throughout East Africa. This was because they knew the bird as a 'krau', pronounced in the same way as that other common animal, the cow. Many Bantu language speakers have difficulty with the English letters l, r, and nd, resulting in hode (hold), had (hand), ovaly (ovary), brood pleasure (blood pressure), lough locks (rough rocks), vuture (vulture). These mispronounciations produce similar mis-spelling in written work. A few examples, taken from tests in connexion with English Language lessons for upper primary children in Kenya, are furo (floor); mborise (polish); ngirass (glass); geren (clean) and masels (muscles).

Differences in pronounciation can cause a complete blockage in understanding during a radio broadcast, especially a science lesson. Much more practical, on-the-ground research is needed with this problem in countries where it is considered necessary to use a single broadcasting language to communicate with peoples who also speak a large number of different tribal languages. Lot Senda [6] puts the point clearly: 'Research into the application of mass media techniques to education cannot be carried out in the abstract.

Greater resources for research and experimentation with new techniques should be made available and a systematic analysis of the learning process as it applies to developing countries carried out.'

Radio science education for general listening

Having seen that commonly employed radio techniques can sometimes hinder understanding, what of the structure of science programmes? This will depend on the aim of the programme and the audience to which it is directed. These can range from general listening to classroom instruction.

The aim of general listening programmes is to supply information on many aspects of science. Titles such as 'Science in Action', 'Science News', 'Science and Industry', 'Nature Notebook', indicate the type of material used, covering research and application. Their structure includes talks, interviews, discussions and the use of recorded archives. Most of the world's major broadcasting organizations hold fascinating collections of recordings of the voices of men and women who were personally involved in many great scientific achievements. Often these are not available to the small, less fortunate, radio stations except on a commercial basis, which they are unable to afford. There may well be a role here for an international organization to play, by financing such transactions in the furtherance of science education. Radio is already making a significant contribution by educating people on the dangers of pollution of their environment. Creating awareness of these dangers is also creating a public opinion which can be strong enough to force action on governments and industrialists.

All the above types of programme are 'pure radio', that is they do not involve any other sense than that of hearing. They do not require pamphlets, posters, film strips or slides. They are therefore well within the scope of all radio stations, provided that they can obtain the material from which to build their programmes. This again is an area where the larger and richer broadcasting organizations can help the smaller and poorer ones, by making taped material freely or cheaply available. Even favourably placed individuals can help. A practical example of such help is a British science correspondent, who regularly provides taped material at nominal cost to Schools Broadcasting, Kenya, where it is built into science news programmes for secondary schools and general listening.

That this type of general science programme can find a place in university studies is shown in the use made of radio by the Open University in the United Kingdom, which will be discussed later under 'Radio and Science Correspondence Courses'. It can also provide useful background material in secondary school science courses. An interview with an eminent scientist or a research worker involved in some new project or an argument in which two scientists put varying points of view, can provide starting points for class discussions.

Radio science talks have been broadcast in Zambia for evening listening, and include such topics as water, the physics of air, electricity, the scientist and food, the electronics industry. Those on air, prepared by F. P. Nalletamby, included a stapled booklet distributed through the Extra-Mural Department of the University.

There is also a place for radio in mass health education. If a toothpaste manufacturer can make people aware of dental care in order to sell his toothpaste, he is making a significant contribution to national health. Whether we use his toothpaste or some other brand is not important (except to him)—what is important is that we use toothpaste. The skilful employment of advertising techniques on radio could repay careful study for application to all aspects of health and personal hygiene. It is generally recognized that the prevention of disease is the most important single factor in national health, particularly in tropical countries. Vast sums of money are spent on curative treatment, only to have the same patients return a few months later with the same diseases. The conventional type of radio health talk may 'reach' a literate urban population, but is totally ineffective with a rural illiterate population. And yet radio may be the only way to reach these people. Short health 'plugs' of 30 seconds' duration, regularly repeated, might have an impact and assist in the prevention of disease. This technique could also be used in accident prevention and road safety campaigns.

All the general and 'advert'-type science programmes referred to above present few technical or structural problems for the radio producer. The problems really begin when we try to use radio for direct instructional purposes in the classroom.

Radio science education for the classroom

It is generally known that children, especially younger children, can only concentrate for a short period in a radio lesson, without their attention wandering. 'We know, if not from an observation of conventional teaching, then from programmed learning studies, that factors of, for example, boredom can interfere with the reception of the input signal.' (P. J. Hills [7].) The period before boredom sets in may be as short as three minutes. For programmes longer than this, devices have to be built in to recapture attention and hold interest. Such devices usually mean doing something. They have the dual purpose of breaking up the radio lesson into shorter pieces and introducing practical work. If we accept that personal experimenting is an essential part of a child's science education, then a practical element must be built into a radio science course. This is not an easy thing to do.

'The studio presenter cannot use the reactions of his audience to guide him in adjusting the pace and level of his teaching during a programme. This must be thought out beforehand, conscious decisions must be made and then adhered to.' (Geoffrey Hall [8].) When a child is asked to do something during a broadcast, not only must the instruction be clear and unambiguous, but the time he needs to do the experiment must be carefully assessed. As this will not be the same for every child, a mean has to be worked out which is closer to the slower than the quicker children. This can only be done by classroom testing, using a tape recording of the proposed activity. Even then observation is necessary during actual broadcasts and possible adjustment of timing before re-broadcasting. We have to apply to the use of the medium, radio, the same scientific methods we are endeavouring to encourage in the students—observation, deduction and the testing of those deductions by experiment. Because a radio science lesson cannot be stopped, turned back and listened to again, it must be precisely structured in language and content. An example of instructions for an experiment is given in Appendix A.

In the 1960's the BBC introduced a series entitled 'Junior Science' for primary classes. The feature of this series, which included teachers' notes and pupils' pamphlets, was the way in which it involved children in activities during the broadcast. They were asked to do things with simple apparatus—rulers, rubbers, pencils, matchboxes, cotton reels, paper clips. That the series successfully met a need can be seen from the fact that in its hey-day it sold over 300,000 pupils' pamphlets each term. Its achievement was twofold—it stimulated children to think about the scientific explanations for many things in their everyday lives and it gave confidence to teachers, who had no science training, to handle simple science. Arthur Vialls, BBC Science Producer, writes 'It's entirely due to his (Harry Armstrong, the originator of Junior Science) work that any science at all is being taught in some primary schools today (1972)'. Geoffrey Hall [8] writing in 1971 on the success of some television and radio science series in Britain, states '... one wonders if they would have an audience at all, but for those earlier series, which introduced many primary school teachers to simple practical science for the first time and showed them exactly what to do within a very limited field. Those were the series that built up their confidence in handling scientific themes and conducting practical science in a much freer way.'

'Beginning Science' [9], a radio series with notes and pamphlets, for tropical Africa, owed much of its early thinking to 'Junior Science'. The series, developed in the mid-60's has undergone changes, modification, and is still in use in Uganda and Kenya, while a Kiswahili version is broadcast in Tanzania. All programmes contain practical elements—look at a picture or diagram in the pamphlet, listen to a special sound, do something with simple apparatus. Pupils' pamphlets, of which over 100,000 have been issued, are now selling (1972) at around 60,000 copies a year. Before the introduction of this series, science teaching in Kenya primary schools was negligible. Now some 60 per cent (about 4,000 schools) of all primary schools teach science with the aid of this radio series and its support material. In fact, the teachers' notes and pupils' pamphlets have become the basic, and often only, science textbook in a large number of schools.

The success of this radio series (and 'Junior Science') has been due to two factors—its starting points are in the everyday experiences of the children and it has created confidence in teachers to handle the subject.

It is not difficult to find starting points—when a child moves a pencil point across a page a mark appears, why does this happen? When he puts sugar into tea, the sugar disappears, what has happened?

When he cuts his finger the blood soon gets thick and stops flowing, why does this happen? 'The programmes must fit into the previous experiences of the children and at the same time extend those experiences into new areas of investigation, inspiring follow-up activities through which the children can manipulate new ideas and information so as to make them their own.' (Gwen Allen [5].) The topics selected must have their starting points in the children's existing knowledge and must give scope for practical investigation and follow-up. An example of such a topic is water—an integral part of every child's daily experience. It can lead to learning expeditions over a wide range—man and water (biology and health), plants and water (botany), the water cycle (geography), water and power (hydro-electricity), floating and sinking (ships and trade).

The building of confidence in teachers is a slow but vital part of a radio series. This is especially so at primary level where the teachers have had no science training of their own. In countries where all teachers have been through a period of secondary education before taking up teaching, they are likely to have had some science education, however slight. But in countries where the majority of primary teachers have themselves had only a primary education, as in many African countries, they have received no science education whatsoever. To give them confidence in teaching science from a radio 'starter' takes a long time and needs great patience. This can involve broadcasting the same series, at the same times, for several years. They have to be shown 'exactly what to do within a very limited field'. Herein also lies the danger of trying to go too fast. Innovators often make the mistake of trying to change in a few months an education system which has grown up over fifty years. 'Chalk and talk' methods of teaching science are deeply rooted and are easy for the teacher to handle. The new 'discovery' methods involve more time and more work for the teacher. Pushing them too fast is much more likely to cause a complete breakdown than to achieve any advance in primary science education. Such a breakdown, with its inevitable loss of confidence among teachers, could set everything back by several years. It is dangerously optimistic to say, as does a 1964 report [10] on modern technological approaches to education in East Africa—'The skills needed to create and produce interesting and educationally effective radio and TV programmes can be learned in a relatively short time' (in two- to four-month courses). It is equally dangerous to assume that confidence and expertise at the receiving end of radio and TV can also be achieved quickly.

'Junior Science' and 'Beginning Science' were developed at a time when, on the one hand, there was the cry to individualize teaching to allow each child to develop along his own lines, and, on the other hand, the cry to make more use of the mass media of radio and television which would involve hundreds of thousands of children seeing and hearing the same things at the same times. Both series were an attempt at a compromise solution to the problem—how to combine mass instruction with individual learning. 'Beginning Science' is still on the air and is still needed. 'Junior Science' is no longer broadcast, having fulfilled its aim, and has been replaced by 'Discovery' (age group 9-11 years), which represents the next step. This is no longer a complete series, instead separate topics are dealt with in units of three or four programmes. The aim is to stimulate children into observing, asking questions and trying to answer their own questions by experiments and research. A pupils' pamphlet and teachers' notes are supplied. The programmes do not treat each topic completely, but select a few and varied aspects, in the hope that different children will be interested in different things.

Vialls feels that this type of approach is more likely to be followed in the future: 'From our point of view I foresee a change in general structure in that we shall probably drop the idea of year-long series of programmes by making contributions in particular areas as and when need or opportunity arises and limiting that contribution simply to the number of programmes which seem to provide the best coverage—the contribution will be very localized and very carefully chosen to take the greatest possible advantage of both the need and of the medium and in this we hope to get a great deal of co-operation from the Schools Council (Integrated Science Project) who are well disposed towards this operation, thus we hope to overcome at least some of the difficulties of communication with teachers.' Schramm [11] also believes that future production 'is more likely to be short subject-matter units than entire courses, because they will be easier to orchestrate into classroom experience or individual study', although he does not make it clear whether he is referring to actual broadcasts or to taped programmes. (The relationship between radio and tape recorder will be discussed later.)

Many countries, including France and Australia, do not use radio for the type of science education programme discussed above. Broadcast primary science is confined to the television screen. Recent inquiries (1972), through Unesco personnel, showed that little or no radio science was being used in Tchad, Malawi, Niger, Mauritania, Cameroun, Rwanda, the People's Republic of the Congo and Zaire. Zambia and Botswana have used radio for science, as mentioned in the previous section.

Of the Australian states only Western Australia still broadcasts science programmes to primary schools; other states now use television for teaching this subject. With the exception of a series of five programmes on sound, for which illustrated suggestions for follow-up are provided in the pupils' pamphlet, the other twenty-seven programmes for the year deal with separate topics, science club activities and question box. 'The broadcast topics have been chosen to help you to learn how to work as scientists do, that is by observing things about you, by conducting your own experiments, and by keeping and comparing records of what you discover. There will also be programmes showing how the scientist applies his knowledge to help us in our everyday life—building skyscrapers, supplying electricity, preserving our wild flowers and so on. There will be two regular magazine-type programmes, the broadcasts on club activities and the question box proved so popular last year that it has been decided to have one each month. In alternate fortnights there will be a current affairs programme divided into two sections: one about what's going on around us in the world of nature study and the other dealing with noteworthy items of news about physical sciences.' (From Broadcasts to Schools, Western Australia, 1972.) These radio programmes are not only a good example of the topic rather than series approach, but also of a variety of presentation methods instead of all programmes being alike.

The BBC 'Discovery' series adopts a similar technique. Thus a programme on micro-organisms, called 'Unseen World', intended to stimulate children to do their own experimental work on such things as moulds, is followed by an optional direct teaching programme on the discovery of penicillin. Another unit begins with the basic sources of energy ('It starts from the Sun') and follows with direct teaching, historical-type programmes on Watt's improvements of the steam engine and the history of the conversion of energy up to and including Whittle and the jet engine. 'I would see the future of radio as a medium for teaching science in the presentation of material designed to present the historical background of basic scientific work' (Arthur Vialls).

Radio has been used successfully by several countries to teach health science. All Australian states have two health science series on radio, one for juniors and one for older primary children. Programmes are of 15 minutes' duration and are supported by pamphlet material. The junior series covers such topics as Your Body, Health and Growth, Health and Cleanliness, Growing Older and Accidents. The last programme in each unit is 'Question Time', to which children are encouraged to write and have their questions discussed and answered. The senior series includes units on food and nutrition, the prevention of infection and community services for health. Both the junior and senior series pupils' pamphlets are also workbooks, with questions to answer and places in which to write them. Appendix B is a specimen page from the 1972 pupils' booklet for health broadcasts to senior primary schools. Radio is also used for health science in New South Wales and Tasmania, for secondary schools. Topics at this level are of a more advanced nature, such as 'the relationship between mental health and physical health'. 'Each programme deals with a separate topic and each presents aspects of a topic that will raise questions in the minds of teachers and pupils. The natural follow-up of each programme would be discussion or more detailed explanation of questions raised in the programmes.' The presenter of the programmes 'speaks to children, psychologists, doctors and others: comment is broken up by music and source material from books and newspapers', (from NSW, Australia, Secondary Teachers' Notes, Radio and Television, 1972). We have already mentioned the importance of breaking up information-type programmes into shorter pieces by various devices—in this case, music, interviews and source material.

Radio is used in Kenya and Uganda (in English) and Tanzania (in Kiswahili) for health science lessons to primary schools. The Kenya and Uganda series are similar and employ the same techniques as in 'Beginning Science', that is, visuals and activities during the broadcast. The way in which these visuals have been produced will be discussed later (page 83). The programmes deal with the human body and with the diseases which can attack it. Thus a study of the breathing system leads on to respiratory diseases; a study of the digestive system and excretion gives the background to such diseases as bilharzia, hookworm, dysentary and intestinal worms. Wherever possible, activities are used in the broadcasts to show that health education really is about people. A specimen layout of teachers' notes for this series is shown in Appendix C.

A final example of the ways in which radio can be used for instructional science education is in the field of natural history or nature study. For several years the BBC has broadcast a very popular series entitled 'Nature', for 8-10-year-olds. The topics are wide-ranging and are not confined to nature subjects in Britain. A programme on lions in Kenya was included with others on local birds, insects and earthworms. An excellent pupils' pamphlet, including colour photographs, adds to the attraction of this series. Recently an attempt to include physical science programmes among the nature programmes has met with a mixed reception; teachers and children alike apparently prefer to keep the nature study flavour. At a time when so much is being written and said about the environment and the disappearance of wildlife, nature study programmes can make a most valuable contribution to children's knowledge and understanding of the other animals with which they share this planet.

This has special application to children living in East and Central Africa where 'big game' still exists in the wild state. Kenya and Uganda have a natural history radio series, supported by twelve posters (black and white because colour printing was beyond their financial means) and detailed teachers' notes. The aim is to make children aware of the mammals, birds, reptiles and insects to be seen in their own countries and to encourage them to observe and note what they see around them. The programmes act as 'starters' to individual observation and interest. The survival of many animal species, in the face of increasing pressures from human population and agriculture, may well depend in the future on the strength of a public opinion which has developed from natural history education in schools. In this respect, especially in developing countries, radio has a vital role to play.

Mention has been made above of pamphlets, notes and posters as integral parts of radio science, and we must now consider this visual material. It is, however, worth noting how visual material, originally designed as part of a schools radio course, can become useful in its own right. Just as the pamphlets of the 'Beginning Science' series in Kenya have become a textbook, so the health and natural history posters have been widely used by all types of health and wildlife educators who are therefore benefiting from material which would not have been produced had it not been for the stimulus of radio.

Radio and printed support material—teachers

As we have seen, some radio programmes are purely audio experiences, requiring no other sense apart from that of hearing. Other radio programmes make use of printed support material. The need for this support material has led many broadcasting organizations into the publishing business on a big scale. The range and complexity of these publications is to a large extent dependent on the financial resources of the organizations producing them and on the ability of listening schools to buy them. This is also an area where aid money could be well spent. It is the developing, poorer countries, whose need for printed support material is greatest, who are also least able to afford it. Radio science pamphlets and natural history posters are widely used in East Africa today and this is partly due to the British Council which provided the early finance to enable this material to be prepared and printed. Similarly, a small cash grant from the USAID made possible the reprinting of a health booklet, so that every primary school in the country received a copy as part of a radio health course. Thus in both cases a few hundred pounds spent in the right place at the right time produced educational benefits out of all proportion to the amount spent.

The involvement of broadcasting organizations in the publishing business seems certain to continue. 'The present tendency is towards the production of an integrated broadcast pamphlet and notes for the teacher' (Armour [1]). New developments in visual aids 'increasingly depend on more and more sophisticated packaged software which is provided for the teacher and which can only come from current research and development efforts' (Spaulding [2]). For radio in the teaching of science 'to be useful in this sense, the programmes must be part of a package including pupil's pamphlets, simple experiments to be done during or after the broadcast and questions to be answered and discussed' (G. van Praagh [12]).

How are these packages to be put together? There are clearly two parts—material for the teacher and material for the pupil. The prime need of the first is to guide the teacher on how to use the broadcast. Its content will depend on the type of teacher for whom it is intended, and on the teacher's own educational background. A trained secondary science teacher will not need to be taught science, and may even resent it, but he may need suggestions on how to handle the broadcast in class. The non-science primary teacher will require to be taught more science than is put over in the broadcasts to his pupils, to enable him to attempt a meaningful follow-up. This in its turn can influence the choice of topic—it is inadvisable to try to teach a topic to the children which will require a disproportionate amount of study by the teacher. This also raises the question as to how far the broadcasting organization should go in its educational endeavours, in relation to other educational bodies and facilities. The answer will once again depend on local conditions—the role of radio science will be different in a developed and in a developing country. But there are common factors—information about broadcasts should be available to teachers well before the broadcast dates and the notes provided should relate closely to the broadcast and pupils' materials.

Advance information is sent out in the form of schedules, showing the titles and times of the broadcasts. From these schedules, teachers can decide which series they wish to take, and so order the notes and pamphlets from the broadcasting organization. They can also fit the series into their class timetable. This in itself is often a difficulty and has been partly responsible for the recent, and growing, development of supplying tapes of broadcasts. Possession of the tape enables the teacher to fit the programme into his timetable at a point where he feels it is most useful. There are several important points to consider later, on the relative roles of broadcasts and tape recorders.

Once again, the production of teacher and pupil support material varies from country to country. In the U.K., each radio series has its own support material. In East Africa, the notes for teachers on all subjects (except primary science) are included in one booklet, of magazine size, and are issued free to schools which are registered as having radio sets. There is a separate book for primary and secondary schools. All subject notes are put in one book for reasons of finance and distribution. In France and Australia different subjects are also included in one book, although there are several books.

The layout of notes, particularly to primary teachers, can be very important. Appendix C, shows a type of layout which has proved successful for health science. The lesson includes blackboard work, posters, activities and suggestions for follow-up. The programme itself ends with a question which leads on to the next lesson. The use of a formal blackboard summary may appear 'old-fashioned', but is a compromise with reality. Follow-up time to radio science lessons is always less than we hope for. The children may not be given time to do all the observing, experimenting and recording necessary to explore the topic fully. It may therefore be better for them to have something in their notebooks, to refer to later, than nothing at all. We must also consider that radio programmes which require a good deal of follow-up will lose much of their effectiveness if time for this is not available. This presents a dilemma—an educationally valuable programme requiring follow-up which it does not get or a less valuable programme requiring less follow-up. A solution may be the inclusion of 'optional' programmes as described in the 'Discovery' series, which allow more than one week for follow-up on a 'starter' programme. 'Certainly, for full value to be obtained from any programme, considerably longer than a week or two will probably be necessary, otherwise the children, having begun to develop ideas and make suggestions, will have to be torn away to look at or listen to more ideas new and different to them, even though, to us adults, they are in the same range of thought. Consequently concentration will be destroyed and an opportunity for sound learning lost' (Gwen Allen [5]).

Unfortunately, the provision of advance information and detailed notes does not ensure that teachers use radio lessons. The trouble frequently lies with the training colleges, where little instruction is given in the use of the radio—'... a very great problem (in Britain) is the sheer lack of communication with teachers—it is incredible but true that teachers are still leaving the training colleges (1972) with no information whatsoever either on our output in radio or television or on ways of using this output to greater effect' (A. Vialls).

The situation in East Africa is no better. Trained teachers are posted to schools all over the country. In the face of isolation and the inertia of older staff, it is all too easy for them to fall back on the old ways of teaching, which in any case may get better results in the factual examinations the children have to take. Much expensive training is therefore wasted. Consideration must be given to providing far more money for special itinerant college staff, who will constantly visit schools to guide and encourage newly trained teachers to put into practice what they have learnt. Such follow-up work is of equal importance with the training itself—its absence is a serious weakness in present efforts to raise standards in schools.

The development of radio stations with local rather than national coverage, to which teachers are seconded by local education authorities (as in the United Kingdom) to help the broadcasters, is a step in the right direction of involving teachers more in the production and use of radio education.

Radio and printed support material—pupils

The preparation of the most effective type of visual for use with particular children is properly the field of the visual-aids expert. But as we have seen, broadcasting organizations produce such a vast amount of printed material, that broadcasters must be aware of some of the problems involved in this work. This is especially so in developing countries where radio staff have to be 'jacks-of-all-trades', preparing teachers' notes and pupils' pamphlets as well as writing scripts and producing programmes.

It cannot be assumed that everyone will 'see' a picture in the same way—what Cassirer calls 'visual illiteracy' [13]. Gwen Allen writes—'Only as children accumulate a variety of experiences can they begin, among other things, to interpret pictures (still or moving) and finally use their knowledge in abstract. These things cannot be taught, they are part of the process of development for which provision must be made, and the children must start at that point in the process appropriate to their needs.' Several writers have investigated visual illiteracy, including S. Spaulding in Costa Rica and Mexico in 1953 and S. Fonseca and B. Kearl in Brazil in 1960 [14]. Spaulding concludes that an illustration as such has no assured educative value and may even be a distracting influence, if its content has not been presented in terms of the past experience of the viewer. Fonseca and Kearl conclude: 'The study confirms our assumption that the ability to interpret many kinds of pictorial symbols is a learned skill and has, in this sense, much in common with the ability to interpret verbal symbols. Young people, illiterate or with limited years of schooling, did a significantly poorer job in interpreting pictorial symbols than older respondents with more education. These two variables—age and education—are the key to one's opportunities for increased learning and are, at the same time, main forces influencing one's ability to comprehend pictorial symbols. . . . Age also relates to the ability to comprehend pictorial symbols, but to a lesser degree than formal education. . . . Cross tabulation of age and education showed a high correlation between the two variables, but indicated that in general education has a larger role in increasing ability to interpret pictorial symbols.'

Similar tests carried out in Kenya by A. C. Holmes in 1962 [15] confirmed the above conclusions. An example is given in Appendix D. The hypothesis, pictures and results of the test are shown. When we used the same test with small groups of children in 1967, the correct answer for 'perspective hills' was 9.1 per cent and for 'perspective road' was 16.7 per cent. These results would indicate that caution is needed in the use of perspective drawings in science pamphlets. Nor is it only perspective which causes problems. For health science, enlarged drawings of insects are often used. Yet tests carried out by B. Shaw in Kenya in 1969 [16] showed that only 55 per cent overall of those tested recognized an enlarged drawing of a house-fly and only 38 per cent overall recognized the enlarged drawing of a mosquito. These tests also showed that recognition of drawings did improve with education, as seen by Fonseca and Kearl.

Further thought must be given in science visuals to the style of cut-away diagrams. These are often used to show the insides of animals and machines. What style of presentation should be used and how much detail should be included? 'Comprehension is reduced either by excessive unnecessary detail or excessive deletion of detail' [14]. To try to find out a little more about the most effective health science visuals for Kenyan children, tests were carried out in 1969 in nine primary schools [17]. The visuals supplied had a definite influence on the drawings which the children were asked to do after the test. An additional test showed that this influence was further increased when direct references to the visual were made in the broadcast. Also as part of the test, the children were asked to *write down* what they remembered from the broadcast. An analysis of their written work showed no significant difference between the three groups, two with visuals and one without. Even the additional test, with direct references to the visuals influenced children's drawings, had no effect on written work. Visuals influenced drawings but not written work. This seems to be in line with the conclusions of educational psychologist Ake Edfeldt, that stimulation over bisensorial paths must be parallel. He goes further and asserts that simultaneous audio and visual stimulation which is not parallel can be actually harmful rather than helpful.

When visuals are used in conjunction with science broadcasts, not only must those visuals relate exactly to what is being said but also, whenever possible, direct references should be made to them. To be looking at one thing and hearing about another can only cause confusion.

'Many abstract concepts in art and science could be made clearer and conveyed through analogies, models and comparisons, not only in a visually exciting way but also in a way that was visually much clearer than if only words had been used, and in which eye and mind absorbed these new concepts simultaneously' (F. Henrion quoted by Fifield [13]). But this is by no means universally true, quite apart from the need for parallel stimulation mentioned above. Pictorial analogies have been shown to be of doubtful value. An example of pictorial analogies is shown in the health science poster in Appendix E. This poster was produced as part of a WHO project and was widely distributed in East Africa. It was used as support material for health science broadcasts to primary children on food and nutrition. The effectiveness of these picture analogies was tested in a radio series on nutrition to upper primary classes [18]. Building foods were likened to building a house; protective foods to locking the house against thieves. energy foods to the burning of a fire. Pictures of a house, padlock and fire were displayed alongside pictures of the appropriate foods on the posters. Some figures from these tests are given in Appendix E, and show that the use of these analogies (words and pictures, and words alone) do not appear to have helped those children to remember the work of the foods in the body, any better than children in the control group which did not have any analogy references. Other observers, using these posters with adults in villages, found that many were unable to see the association between body building and house building; some thought the padlock was to be eaten and others thought that the energy foods were to be burnt to make the fire, and some wondered if 'eating some building food every day' meant killing a cow or a goat every day.

Verbal analogies are commonly used by teachers and are an attempt to illuminate the unknown by the known. The class teacher can see from the reaction of his pupils whether or not the analogy has worked, and can try others if it has failed. These interreactions are not possible in a broadcast. Therefore the analogy must not 'miss', and it *will* miss if the object to which comparison is made is unknown to the child or is known in a different form. 'The inside of your lung is like a big tree which has been cut off and turned upside down' presents several problems in understanding, especially if 'tree' to the child means a palm or a stunted thorn bush. Of children who were asked to write down what they remembered from a science broadcast which contained three analogies, 23 per cent recorded a tree/inside lung analogy, 9 per cent recorded a 'lung like bags' analogy and 7 per cent recorded a 'ribs like cage' analogy. These figures show only that they had remembered something about a tree, bag, cage—they do not show how far they had been able to use the abstraction or how far the analogy had helped in understanding about lungs and ribs. The inability of children to understand the shape and function of the mammalian backbone when it was compared to a 'pipe' was explained by the fact that 90 per cent had never seen a pipe, except a smoking pipe, which was not what the scriptwriter had in mind. To try to illuminate the unknown in terms of the unknown can hardly succeed. The few examples given above of visual and verbal analogies indicate that this method should be used with caution in radio lessons. A meaningless analogy can increase confusion in a child's mind rather than help understanding.

Consideration should be given to the use of colour in publications for pupils. Colour can be valuable in natural history, particularly for birds, insects and plants, and for programmes on light. That it need not always be 'better in colour' was shown by children using the health poster shown in Appendix E. Tests revealed that children using a colour version did only marginally better than children using a black/white version [18]. Where a food was easily identifiable by shape (bananas) colour made no difference, but where shape was similar (oranges, tomatoes) colour was a help. Colour photographs of foods are now being used, but no information is yet available as to their value in relation to colour drawings of food. In the final analysis, whether or not colour is used depends largely on finance. Colour reproduction is extremely expensive in relation to black/white. Once again this is an area where aid to broadcasters in developing countries could be usefully spent. The cost, in aid terms, is very small and could help to provide, for instance, colour natural history posters which are well beyond the budget of many small broadcasting organizations.

The size of visuals is important. If intended for use during the broadcast they must either be on a poster large enough to be seen from the back of the class or must be supplied to each child. Individual pamphlets are expensive but can be self-financing if sold to schools. The combination of instructional pamphlet and work book, in which the children write their answers as they work through the series, is probably the most effective material support to broadcasts. A financial disadvantage of the work book is that it can only be used once, whereas ordinary pamphlets can be used many times and with different classes. In developing countries, where visuals have to be issued free because children cannot afford to buy, the individual work book is beyond the resources of many at the moment. Thus re-usable pamphlets and wall posters have to be used.

The provision of health science visuals in Kenya went through three stages. First, teachers were asked to copy onto the blackboard enlarged versions of small diagrams included in the text of the teacher's notes. This proved unsatisfactory, as teachers either did not make the drawing at all or made it so badly that it was valueless as a teaching aid. The second stage was to include a double-page, pull-out sheet in the middle of the book of notes. The problem here was that the diagrams, although useful, were much too small. The third stage was to increase the size of diagrams to fit onto posters 70 cm by 45 cm. These posters were made more durable by using thicker paper and binding all twelve together at the top so that they could be hung on the wall and folded over as necessary. Each step was also governed by financial resources—as more money was available, more was possible. As a last step, the set of posters was included in school equipment lists for schools to buy. But it took four years to reach this point —a point which would have been reached in a shorter time in a developed country.

Radiovision for science education

By definition radiovision means the simultaneous use of a radio broadcast and a film strip or slides. Possibly it originated as the radio man's reaction to educational television.

The film strip (35 mm) is bought from the broadcasting organization, together with a booklet of teachers' notes. Audio signals are incorporated in the broadcast to tell the teacher when to turn to the next picture. This method—broadcast direct into the classroom—is used in Australia, where radiovision is still fairly new. Few programmes have yet been produced (1972) and none on science. Schools are allowed to tape these broadcasts from their radio sets, but taped copies of the programmes are not supplied by the Australian Broadcasting Commission. Radiovision is used extensively by the BBC in Britain and by Radio-Télévision Scolaire in France, including science programmes for both primary and secondary schools. The BBC output includes two units on electricity (mainly historical, e.g. 'Volta and the battery') and two on 'inside the body' and human reproduction. There is little doubt that radiovision has proved itself to be a very effective addition to educational broadcasting. It fits into the generally accepted belief that radio teaching achieves more when combined with something else —visual aids and activities. But the spoken word, visuals and activities must all be closely integrated. We have said that looking at one thing and listening about something else can be confusing. This applies very much to radiovision. When the children are looking at the picture on the screen, what they are hearing from the radio must be about the picture. A disadvantage of radiovision can be that it inhibits child participation, something which we have considered very important in other types of science broadcasts. To try to get round this the BBC produced a series on electricity and magnetism in 1970 in which they 'aimed to make certain of the frames in the filmstrip working documents from which the children could make up their own experimental set-ups and carry on their experimental work from there'. No information is available (1972) as to the success or otherwise of this technique.

The respective roles of radio and tape-recorder will be discussed later. However, as regards radiovision, 'I am quite sure that the future of science by radio for schools in this country will be profoundly influenced by radiovision. It has many notable advantages—a colour picture of very high definition, the fact that a programme and its pictures *can be repeated* in whole or in part at will, *complete control by children or by the teacher over the use of the programme,* even occasionally the use of a picture without the sound that goes with it' (Vialls). Add to this Armour's view 'a radiovision programme is a radio broadcast for schools *to tape* and use with its accompanying 35 mm filmstrip or slides'; also add the fact that tapes of the sound programmes can be *bought* direct from the broadcasting organization, and we have quite a new situation. The one certainty about *radio* is that it cannot be stopped, turned back or repeated. So what we now have is not *radio*vision but *tape*vision.

If the sound tape of the programme can be bought, why should it be broadcast as well? As the programmes cannot be used without the filmstrip, what therefore is the school audience in relation to those who cannot—or do not want to—use the programmes because they do not have the visuals? The number of filmstrips sold should indicate the audience, and the number of those who buy the tape should, by subtraction, give the number who presumably tape from the radio. This figure may then be related to the air time used on a national network for this minority audience. Furthermore, the material provided in a radiovision package, both audio and visual, hardly requires *broadcasting* expertise. Most large training colleges, university departments of education and audio-visual aid centres could produce this type of material perfectly well. The only radio element in radiovision is the ability to broadcast the audio content on a nationwide scale and so enable a large number of schools to tape it free of charge. Whether or not this can be justified depends on the figure mentioned above—how many schools tape from the radio in relation to, firstly, the number who buy the tape, and secondly the number of schools who do not use this series.

This is in no way to decry the value of this teaching method of sound and vision. But if it is intended solely for taped use, as indicated by the BBC (and by Radio-Télévision Scolaire—'La radiovision unit intimement des documents sonores qui arrivent par les ondes et que de nombreux maîtres enregistrent sur bande magnétique, et des diapositives, une par séquence sonore, qui donnent des images de grande taille, en couleurs, que les enfants ont le temps d'observer et qui peuvent être reprojetées, réétudiées après l'émission'), then we may question whether this is really a field for radio, or whether it could be done better by an audio-visual aids centre selling the package.

It might be possible to argue a case for radiovision in developing countries, where radios are available but tape recorders are not. However, when we consider that a filmstrip projector, with power supply, is essential, we at once limit the audience to the type of school which is also most likely to have a tape-recorder. Taping direct from a radio also has its problems where reception conditions are poor and variable. Furthermore, is it justifiable to occupy 30 minutes of air time on a national network for the benefit of a few hundred students? So once again we must consider if the radio (tape) vision teaching package is the field of the audio-visual centre and not of the broadcasting organization. As we have seen, there is plenty of scope for science education by radio, both in general and instructive programmes. It might be better for radio to concentrate on things it is uniquely able to do well, and leave to others the preparation of educational material in which its special techniques, cheapness and universal coverage are not needed.

Probably the most thoroughly researched and tested use of radiovision (or tapeslide as it is also called) has been carried out in Sweden by G. Markesjö and P. Graham, of the Royal Institute of Technology. They have used a combination of radio programmes, slides, workbooks, teaching guides and tapes to teach electronics to upper secondary school students. Their system includes techniques for integral feedback—'Results from diagnostic tests, exercises, attitude tests, examinations etc. are used as checking elements (or censors). Suitably data processed, these part results can be transformed into recommendations for teachers, students and producers of courses.... A quick feedback permits a "process control" of the instruction and an individual form of assistance is necessary for having the intended effect' [19].

Radio and the tape-recorder

Radio is a purely listening experience, and it must contain enough incentive, of one kind or another, to persuade the listener to keep listening. 'It must be remembered that a schools broadcast is still a broadcast and if it is to hold its audience it must be vital and interesting. The child in the classroom may not be able to switch off the radio but he can quickly lose interest and 'switch off his ears' (R. Aspinall [20]). This statement is equally valid for education programmes intended for general listening. Pure audio learning has its own problems of programme content, presentation and studio production, but all have to be solved within the context of hearing and the interpretation of these sounds by the brain.

An entirely new set of problems is involved when other senses—sight and touch—are combined with radio listening, and we have already discussed this above. It may be considered that the need for these other sensorial paths reveals the limitations of radio as a teaching aid. However, it is a valid use of the medium in relation to its unique ability to reach a wide audience at very low cost. But the introduction of the tape-recorder has created a new situation in relation to radio, a situation which most writers on education broadcasting confuse. Thus 'with a decentralized educational system such as ours, it is impossible for the BBC to fit in with school timetables. The best we can do is to avoid obviously bad placings and then let schools know in advance what our plans are to be, so that if they value broadcasts, they can take steps to arrange their timetable accordingly. With sound broadcasts the tape-recorder *can solve* most of these timing problems....' (Hall [21]). And again, 'one technical advance which I am sure is going to produce a radical change and possibly a great improvement *in the way in which school broadcasts are used* is the tape cassette. The machinery is cheap, the cassette is cheap and it is exceedingly easy to use; never before have we had this combination of advantages' (Vialls). But it all depends on what you mean by 'the way in which school broadcasts are used'. The basic difference between radio broadcasts and tape-recorded programmes is that one can be stopped, started, replayed at will and the other cannot. Over one the teacher has control, and over the other he has not. In printed notes to teachers (radio series 'Life Cycle', Summer 1972) the BBC states, '*All* school broadcasts *are better used* in tape-recorded form. This is particularly true of programmes (with a science bias) which are likely to provoke a great deal of discussion; it is thus more than desirable that you and the children should have complete control over the listening experience.' The notes go on to extol the many advantages of the tape-recorder—the teacher can familiarize himself with the programme *before* he uses it, the programme can be more easily fitted into the timetable and so on. Two words are worth noting from the above, and they are 'all' and 'before'.

It is not possible for a teacher to familiarize himself with a radio programme before it is broadcast, unless it is repeated a second or third time. So 'before' means that he is expected to have a tape of the radio programme to which he listens in out-of-school time and decides where and when he wishes to use it. 'All' indicates that this is the normal procedure for the use of a broadcast. So if this material is best used from a tape in the classroom, why broadcast it? Why not supply the tapes direct to schools? It has been argued that one advantage of radio over tape is that radio is flexible. Radio programmes can be revised and changed before being re-broadcast, whereas the contents of a supplied tape are 'frozen'. However, this advantage disappears where a radio programme is integrated with visuals. Changing the programme means changing the visuals, which have already been sent to the schools. There seems to be only one reason for broadcasting instead of supplying tapes—and that is, by broadcasting the material it becomes available for all schools to tape free of charge from their own radio sets. The role of the broadcasting organization is therefore the role of a transmitting agent and need not be that of an originator of the material.

Broadcasting primarily for taping raises the question as to whether or not the unstoppable broadcast is suitably structured for a stoppable role in a classroom. A radio programme is designed for listening to as a unit and may not lend itself to the 'complete control over the experience' which now seems desirable. This end could better be achieved by a programme specially designed for such control. There is confusion between these two types of programmes. Thus 'Individual programmes are on the whole conceived as unified entities and should satisfy the needs of a class *as complete broadcasts*. Nevertheless, the wide range in the ability of listening groups makes it *essential* to control the pace of the experience *by stopping the tape* to answer questions and assess assimilation and understanding. Carefully timed interruptions can encourage a high degree of involvement on the part of students and provide variety for those who find concentration on a purely aural source difficult. In the case of science programmes it allows experimentation and practical work to follow immediately on suggestions *within broad-*

casts. Indeed only a portion of the broadcast may be necessary to provoke the work that the teacher wishes to develop. Programmes on nature study or biology can be played back to coincide with the development of classroom specimens' (B. Chaplin [22]). There is apparently confusion here between 'complete broadcasts', 'stopping the tape', 'within broadcasts'—what are we talking about, radio broadcasts or taped programmes? It is difficult to see how a 'unified entity' can also include 'carefully timed interruptions'. Better results might well be achieved by broadcasting the 'unified entity' type of programme and supplying quite separately the 'carefully timed interruptions' programme on tape.

Furthermore, if the main reason for broadcasting a radio lesson is to enable it to be taped free of charge, then should these programmes not be transmitted at convenient off-peak times? Is there any point in transmitting programmes in school hours, when the teachers will be busy in class and unable to listen to them or tape them? Why not, instead, broadcast extracts from the programmes before or after school time, so that the teacher can listen (and in conjunction with notes already sent to him) decide whether or not he wants to order the tape and pamphlets? This type of broadcast could serve as sample or demonstration material, thus giving teachers the freedom of preview and choice, now considered so important in the use of audio and visual material for class teaching.

Education broadcasters must consider very carefully the situation created by the omnipresent tape-recorder. In developed countries nearly every school has a tape-recorder and a very large number of teachers have their own machines. Audio material, particularly in science where a practical element is so important, might be better issued on tape, with radio previews at suitable times to allow teachers to select and order. The preparation of the material could be done by audio-visual centres, leaving expert broadcasters free to concentrate on programmes better suited to radio—the type of general science education programmes described above (pages 75-79).

The situation in developing countries is rather different. Here the tape-recorder is confined to secondary schools and training colleges. Thus in Kenya, of 3,500 primary schools known to have radios, less than 100 have tape-recorders. In such a situation, radio still has an important part to play. Carefully prepared programmes, as described above, can make a valuable contribution to science education. Even if tape-recorders become widely available, their use to tape direct from the radio will depend on improvements in reception conditions. The supply of taped programmes to schools is financially out of the question at present. It is also mechanically doubtful in view of the almost total absence of maintenance facilities for tape-recorders in rural areas.

Radio and mathematics

One science subject which hardly appears to lend itself to radio teaching is mathematics. Various series have been tried on television, for example, the BBC series 'Maths Today' (1971) and the Radio-Télévision Scolaire 'Les Mathématiques' (1972). Both series are for secondary schools. The BBC series includes teachers' notes (96-page booklet), pupils' package of study sheets and worksheets, and 8 mm film loops. Despite all this material, intended to provide the basis of a two-year course in mathematics, the notes contain a caution—'... it cannot be too strongly emphasized that it will nevertheless involve teachers in a great deal of work and re-thinking, especially if their knowledge of "modern maths" is slight. Do not undertake it lightly; it is certainly not a "soft option". And please remember, the intention is that pupils should spend all, or nearly all, of their mathematics time on the course. If they do not the results are likely to be disappointing.'

In 1965, the BBC used a radio series for secondary students called 'Background to Mathematics'. This involved the use of a 27-page pamphlet for the students, to which references were made during the broadcast. The treatment was partly historical (Roman, Egyptian, Babylonian numerals) and partly the application of mathematics. Thus the binary system was related to its use in modern computors. This series was used in Kenya in 1966, with the same pamphlet but adapted scripts. There is no information available as to its success.

In 1971 a series entitled 'Looking at Mathematics' was broadcast in Kenya. This involved a 55-page students' pamphlet, and was again partly historical and partly the application of mathematics. Numerous references to the pamphlet were made during the broadcast and exercises were included for follow-up work. Three of the titles in this series of eight programmes were 'Numbers and Measurements', 'Some Great Mathematics (Pythagoras, Euclid, Archimedes, Napier)',

'Time and the Calendar'. Again, feedback on this series was negligible. This may be due to the fact that it dealt with the application of mathematics, and in an examination-orientated school system most teachers concentrate on teaching the essentials of the subject in relation to the examination, and have no time to spare for 'background' or applications.

For the direct teaching of mathematics, radio would seem to be an unsuitable medium, although it has been used as an element in correspondence courses. The Centre National de Télé-enseignement in France offers three background mathematics programmes as part of a correspondence course.

There does, however, seem to be a place for radio in the application of mathematics and as background to the subject. Such programmes would need visual material related precisely to the broadcast, would preferably include activity, and would include follow-up suggestions for the children to explore for themselves afterwards. This could take the form of a workbook. Such background programmes could help to develop the interest of pupils in a subject which for many, despite 'modern maths', is still difficult and dull. This type of programme might perhaps provide that incentive which makes all the difference to the learning process.

Appendix F shows an extract from a pamphlet issued with a series of programmes on the application of mathematics, involving radio, visuals and activity.

Radio and science correspondence courses

Whereas television has been widely used to teach science as part of a correspondence course, little use has been made of radio in this context.

The Open University in Britain, as described by M. J. Pentz [23], used 36 television programmes and 37 radio programmes, all of half-hour duration, in its Foundation Course in Science. Of the use of television, Pentz states that it is 'an essential component of the system, rather than a desirable appendage, because of the vital importance of experiment in science. We do not believe we can give our students a proper understanding of science without involving them, through the medium of TV, in experimental work which they cannot do at home. . . . For these reasons, we set out to *integrate* the TV programmes with all the other components of the system.' On the use of radio he states 'the radio programmes will, in the main, be used to enrich and unify the course. For example, we may bring a group of experts into the studio to discuss some wider aspects of the scientific subject matter—technological, social or political. Or the radio programmes may be used to draw attention to the relationship between the ideas and methods of one part of science to another. Or they may be used to supplement or correct parts of the course which have not been sufficiently clear, or to bring the course up to date on recent scientific developments.' This is certainly a clear statement of the respective roles of television and radio in such a course. Furthermore, the types of programme envisaged for radio are well suited to the medium. It is worth noting, however, that both television and radio together are designed to occupy only 5 per cent of the total learning time in a week. (See also Kaye and Pentz, 'Integrated Multi-media systems for Science Education which achieve a Wide Territorial Coverage,' in the present volume, page 143 *et seq*.)

Teaching by radio and correspondence is practised in Australia, where correspondence schools and the 'School of the Air' work through two-way radio to teach children who live too far away from schools to attend. However, science is not taught in this way, at least as far as New South Wales is concerned, although experiments are being carried out there in the use of taped science lessons at primary level. The Principal of the Correspondence School states 'the facilities available for education in this State are such that the use of radio for the teaching of science is unnecessary. Television is used for the teaching of this subject.' In France, the Centre National de Télé-enseignement offers four radio programmes on the biological sciences as part of a correspondence course for students and adults—'... ces émissions apportent un complément oral à l'enseignement dispensé par correspondance aux élèves. ...'

The Institute of Adult Studies of the University of Nairobi organizes a Correspondence School which contains a radio element. The main science course, at junior level, is in biology and currently has some 600 students enrolled, but a newer course in physical science has only 34 students so far. These figures can be compared with an enrolment of over 2,000 in an English language course. The radio programmes in science consist of two fifteen-minute broadcasts each week, and these are repeated once during the week. Although the radio element of these courses is considered important (it is estimated by the course organizers that 65 per cent of those taking the course listen to some of the broadcasts) it is possible for a student to pass the examination without the use of the radio lessons. In other words, the correspondence material is complete in itself, while the radio lessons are a bonus which will help the student, but are not essential. There appears to be no way round this situation in countries where radio reception is variable and where radio by correspondence course broadcasts cannot be allocated good listening times for a minority audience on a radio service with only one or two channels.

Clearly, correspondence schools in most countries must rely on tapes for the audio part of their courses, and not on radio, although in Zambia radio has been used for science lessons in a correspondence course. The radio becomes more useful where stations have several broadcasting channels and can use one mainly for education, or where there are local stations covering a limited area.

The radio element of a science correspondence course is subject to all the limitations on the use of this medium for teaching an essentially practical subject. Its universality has to be balanced against the need to supply written material and visuals. Consideration must be given as to whether or not the issue of tapes might not do the job better. In developed countries the preference may be for tapes, in undeveloped countries where tape recorders are few, radio may still be useful. Possibly its role as described above in the British Open University remains the most effective, to enrich and supplement, and not to be an indispensable element of the course.

Some conclusions

'Certainly, the Pied Piper of television and radio often hypnotizes the unwary into believing that these mass media can somehow suddenly educate all our children painlessly and cheaply.... The ability to bring a wealth of outside experiences into the classroom at the flick of a switch excites the imagination. But the definition of what this should be, its planning, production, broadcast and use in schools must be part of the overall curriculum research and development process ... [24].

The starting point must be a clear definition of objectives—what is it we want to do and then what medium shall we choose to achieve this aim. If it is to be a multi-media approach, then where does radio fit into the pattern? And as we have seen, we must consider carefully the role of radio in science education to ensure that we are making the best use of it. The original use of radio as a purely audio experience has been changed by the need to build in a practical element for science lessons and to include a range of visual aids, pamphlets, posters, filmstrips and slides. The advent of the tape-recorder has introduced a new factor, the implications of which have not yet been clearly assessed and resolved. 'The technology of broadcasting has much to offer within the complex structure of formal and informal education. Educators must learn how to use the medium, however, and learn how to integrate broadcasting into an overall educational strategy. This will often mean changing the way we do things in a traditional classroom, if a broadcasting-orientated system is decided upon. It will also mean that broadcasters will have to collaborate with learning psychologists and educators in developing new styles of instructional broadcasts' [24].

Such closer co-operation between educators and broadcasters must include much more practical 'in-the-field' research into radio and the learning process. Although 'one of the main difficulties in the investigation of any teaching/learning interaction is that only the input material to the student, and the output material from the student can be observed ...' [7], such input/output evaluation of radio science lessons, in relation to the use of radio as a whole for educational purposes, is negligible. The amount of expertise, air time, pupils' time, teachers' time, visual aids and money wasted on ineffective, unsuitable and useless radio lessons over the years must be considerable. The need for practical research into the real effectiveness of educational radio, particularly in developing countries, is urgent if the medium is to play a useful part in science education. This research must be directed at all levels from the definition of objectives to the organization of the listener, whether in school or out of school. The effectiveness of radio 'as an educational medium depends largely on the infrastructure at the receiving end' to encourage continuous, sequential listening and interaction on the part of the audience. This infrastructure may be the classroom, the village forum [25] or the correspondence course. Yet common to all there must be an incentive to listen—'you can lead a horse to water but you cannot make it drink'.

'In the glamour of TV, we have underused radio. We have little doubt that it can do many educational and informative jobs as well or almost as well as TV, and at about one-fifth the cost. Should we not review the tasks and the support we have assigned to radio? I am not suggesting that we should do less with TV, but rather that we can do more with radio' (Schramm [11]). It is not only a question of 'doing more with radio' but also of deciding what we want to do and of knowing how to do it.

It is time now that radio educators rethought their role in the educational system—asking, first, what can radio do that other media cannot, with special reference to the single sensorial path, hearing; and second, to review most critically the relationship between radio and other teaching aids, that is, its role in the multi-sensorial path of hearing with sight and touch.

REFERENCES

1. ARMOUR, C. School radio: its future in Britain. *Educational Broadcasting International*, vol. 5, no. 1, March 1971.
2. SPAULDING, S. *Prospects in Education*, vol. 1, no. 3, 1970. Paris, Unesco.
3. HILLIARD, R. L. *Radio broadcasting*. New York, Communications Arts Books.
4. BALL, John C. H. Using sound effects in schools broadcasting. *Educational Broadcasting International*, vol. 5, no. 3, September 1971.
5. ALLEN, Gwen E. Broadcasting—an integral part of learning. *Educational Broadcasting International*, vol. 5, no. 3, September 1971.
6. SENDA, L. Viewpoint. *Educational Broadcasting International*, vol. 5, no. 3, September 1971.
7. HILLS, P. J. Science teaching and educational technology—Part II. *Teaching Science in Secondary Schools*. Compiled by the Association of Assistant Masters and the Science Masters Association.
8. HALL, G. Some thoughts on science and schools broadcasting. *Educational Broadcasting International*, vol. 5, no. 3, September 1971.
9. BALL, John C. H. Beginning science: A radio series for primary schools in Africa. *Educational Broadcasting International*, vol. 5, no. 2, June 1971.
10. SCHRAMM, W.; HUSON, R.; RIVERS, W.; KEMPFER, H. Modern technological approaches to education in East Africa. Report of a Mission to Kenya, Uganda, Tanzania. June/July 1964.
11. SCHRAMM, W. The future of educational radio and television. *Educational Broadcasting International*, vol. 4, no. 4, December 1970.
12. VAN PRAAGH, G. Viewpoint. *Educational Broadcasting International*, vol. 5, no. 2, June 1971.
13. FIFIELD, R. Audio visual evolution or revolution. *New Scientist*, vol. 51, no. 762, 29 July 1971.
14. FONSECA, S.; KEARL, B. Comprehension of pictorial symbols: an experiment in rural Brazil. University of Wisconsin.
15. HOLMES, A. C. *A study of understanding of visual symbols in Kenya*. (OVAC Pub. No. 10—London, Cedo, Tavistock Square.)
16. SHAW, B. *Report on the recognition of drawings in Kenya*. London, CEDO.
17. BALL, John C. H.; MAY, J. P. Health education radio lessons for primary schools: some problems examined. *OVAC Bulletin* no. 20, October 1969.
18. BALL, John C. H.; MAY, J. P. Health education radio lessons for primary schools: some further problems. *Educational Broadcasting International*, vol. 5, no. 4, Dec. 1971.
19. MARKESJÖ, G.; GRAHAM, P. *An electronics course with integrated feedback*—Reports PE-6 and PE-9. (Available from Department of Education, Royal Institute of Technology S-100 44 Stockholm 70, Sweden.)
20. ASPINALL, R. *Radio programme production*. Paris, Unesco, 1971.
21. HALL, G. Biology teaching and broadcasting. *Aspects of Education—Ten; A New Look at Biology Teaching*, p. 98.
22. CHAPLIN, B. Educational radio and the use of the tape-recorder. *Educational Broadcasting International*, vol. 5, no. 2, June 1971.
23. PENTZ, M. J. Science teaching at the Open University. *Physics Bulletin*, vol. 21, October 1970.
24. *Science and education in developing states*. Proceedings of fifth Rohovet Conference; Praeger Special Studies in International Economy and Development, 1969.
25. MATHUR, J. C.) NEURATH, P. *An Indian experiment in farm radio forums*. Paris, Unesco, 1959.
    PICKSTOCK, M. The value of radio in rural broadcasting. *Educational Broadcasting International*, vol. 5, no. 1, March 1971.
    RIITHO, V. Radio in family planning education in Africa. *Educational Broadcasting International*, vol. 5, no. 4, December 1971.

Other publications on educational radio:

CASSIRER, H. Two-way radio in rural Senegal. *Educational Television International*, vol. 4, no. 2, June 1970.
ERDOS, R. F. *Teaching by correspondence*. Paris, Unesco, 1967.
HALESWORTH, B. Radio—the Cinderella medium. *Educational Broadcasting International*, vol. 5, no. 3, September 1971.
HARRIS, B. R. Broadcasting and the teaching of science in secondary schools. *The School Science Review*, vol. 53, no. 182, September 1971.
Science for primary schools—5: Using Broadcasts. Association for Science Education; pub. Murray, 1970.
TAYLOR, E. J.; BARNES, N. Some reflections on educational research. *Educational Television International*, vol. 4, no. 2, June 1970.
WILSON, M. Starting educational broadcasting in Afghanistan. *Educational Television International*, vol. 4, no. 4, December 1970.

Appendixes

A. BEGINNING SCIENCE   Sound (Kenya)

Published by Schools Broadcasting, Ministry of Education, Nairobi

Extract—Teachers' notes

Lesson 2. How sounds travel and how we hear them

The Aim

The aim of the lesson is to show how sound travels from where it is made to our ears and how our ears receive it, i.e. how we hear.

The Apparatus

Apparatus needed for this lesson:
1. Each child must have a book.

The Blackboard

The key words are—VIBRATE, EAR, ECHO, SOUND WAVES.

The summary points are:

1. Sound can travel *through* the air, *through* solids and *through* liquids.
2. Sound *takes time* to travel. It takes time for a sound to travel out and then bounce back as an *echo*.
3. Sound travels *through* the air in a whole series of little pushes—it is called a 'sound wave'.
4. Sound waves reach our eardrums, which vibrate. These vibrations pass along nerves to the brain and so we hear.

Other words for the blackboard are—solids, liquids, bounce, wave movement.

Experiments

There is one experiment during the broadcast. The children are asked to pick up their book, hold it against one ear and with the other hand scratch on the cover of the book. [How to do this is shown by picture 1, page 4 of the Pamphlet.][1] They should be told to look at this picture *before* the lesson.

Sounds. As in Lesson 1 these are very important. At the beginning of the lesson the children are reminded of sound and vibration by hearing again one of the musical instruments shown on page 3. Later they hear an echo. And at the end of the lesson they hear two sounds one a *high* sound and the other a *low* sound. They are asked to think about these before the next lesson.

The Follow-up

In the last lesson the children learnt that where sounds are made there must be vibrations.

1. Sound Travels Through Air

In Unit AIR, the children learnt that air presses on everything from all directions. So if anything vibrates it must make the air round it also vibrate. How these vibrations pass *through* the air and how we hear the sound is explained below.

2. Sound Travels Through Solids and Liquids

We should expect that as sound is caused by vibration, anything which can vibrate can carry sound—and this is so. Most substances are 'elastic' (some more than others) and so they can be made to vibrate and will carry sound through them. For example, if you shake up some nails in a *closed* box, you can still hear the sound of the nails hitting each other inside the box. The sound has travelled through the sides of the box.

In the experiment [on page 4, picture 1], the children can hear the scratching of their finger *through* the book. Page 8, picture 7, is an experiment to show how sound travels through the wood of a table top. There are two other experiments to show that sound travels through substances other than air. They are—

(a) Page 5, picture 5. When the fork is hit gently it vibrates. The vibrations are carried by the stretched string to the ears, where they sound like a bell.

(b) Page 8, picture 6. Sound travelling through a stretched string is also the explanation of this telephone. The air near the mouth of the boy who is speaking vibrates. The vibrating air makes the bottom of the tin vibrate and that makes the string vibrate. The vibration travels through the string and makes the bottom of the *other* tin vibrate. This finally makes the air near it vibrate and this vibration reaches the ear of the boy who is listening. Remember that the string must be stretched tight in this experiment.

Here is another experiment you can do. Place a watch on the floor. Rest one end of a four-foot stick on top of the watch. Put your ear to the other end of the stick. You will be able to hear the ticking of the watch through the stick.

[Page 4, picture 2 shows a swimmer under water.] He can hear sounds made under the water some distance away. Water, and other liquids, are 'elastic' and so sound can pass through them.

Extract—Broadcast script

2. How sounds travel and how we hear them

Lesson 3. The Human Body

In the last lesson you learnt that sounds are made when things vibrate.
Do you remember this sound?
Listen—
(*Instrument No. 3 - Lesson 1*)
That sound came from the grass instrument on page three of your pamphlet. By moving the thumb across it I made the

---

1. Square brackets indicate omission.

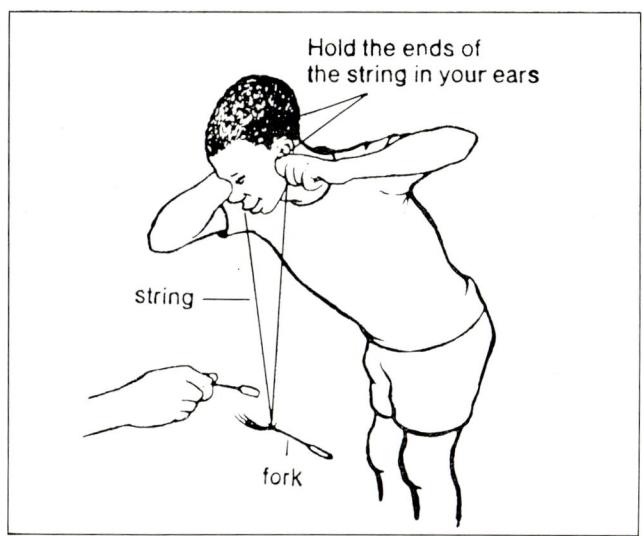

5. Hit the fork and hear the vibrations *through* the string. [Page 5.]

7. Put your ear on a table. Ask a friend to scratch the other end of the table. Try the same experiment with a desk, a door and a chair. This is an experiment to show that sound travels *through* solids. [Page 8].

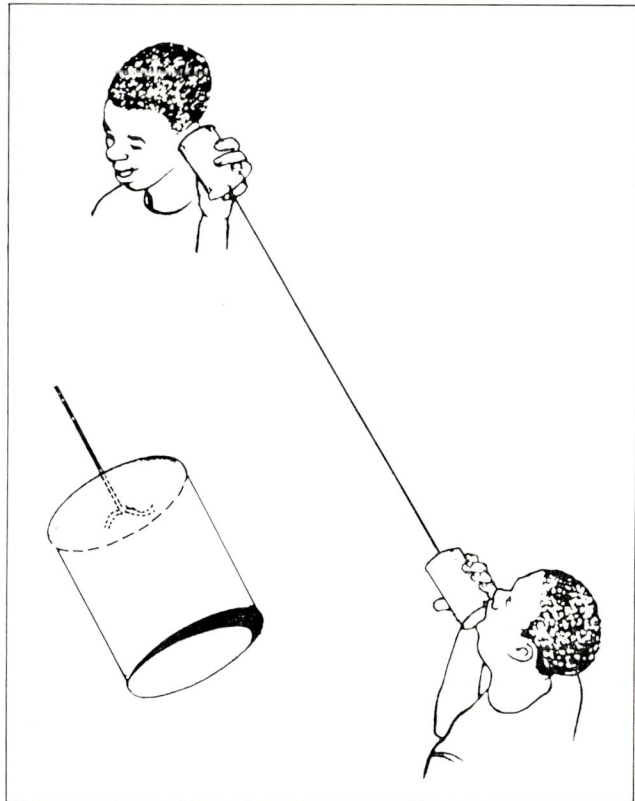

6. Make your own telephone from two tins and a 6 metres length of string. Pull the string tight while you talk. [Page 8.]

long pieces of grass vibrate. The sound made by them travelled to your ear and you heard it.

The two questions we must try to answer today are—how did it travel to your ear and how did you hear it?

First then—how did it travel to your ear?

To answer this I want you to think back to your lessons on Air. You learnt that Air is a real thing. You can feel it—it takes up space. It has weight, it presses on everything from all sides.

When something moves, the air round it must also move.

So when the grass instrument vibrates like this—

(*Instrument No. 3*)

—the air round it also vibrates.

And these vibrations travel through the air, and we hear the sound. But can they travel through other things—can sound travel through solids and liquids?

Well, you can find one of the answers quite easily, by a simple experiment. I want you to do it now. Look at page 4 of your pamphlet.

Pick up your book with one hand and hold it over one of your ears. Press the book firmly over your ear—just as you see in the picture.

Now with your other hand—cratch on the outside of the book. Scratch with a fingernail on the book.

Can you hear the scratching *through* the book? (Pause for answer.)

Please put the book down now.

You heard the sound which your fingernail made on the book. The vibrations passed *through* the book to your ear.

Sound travels very well *through* solids. And it also travels *through* liquids. Those of you who can swim know that you hear sounds when you are *under* the water.

What I want you to remember is this—sound can travel *through* solids and liquids—as well as *through* air. *But it must have something to travel through*—it cannot travel through *nothing*.

B. HEALTH — For primary schools (Australia)

Published by the Australian Broadcasting Commission, South Australian Branch, 1972

Pupils' pamphlet
—Specimen page

Grades VI and VII. Fridays: 1.40-1.55 p.m.

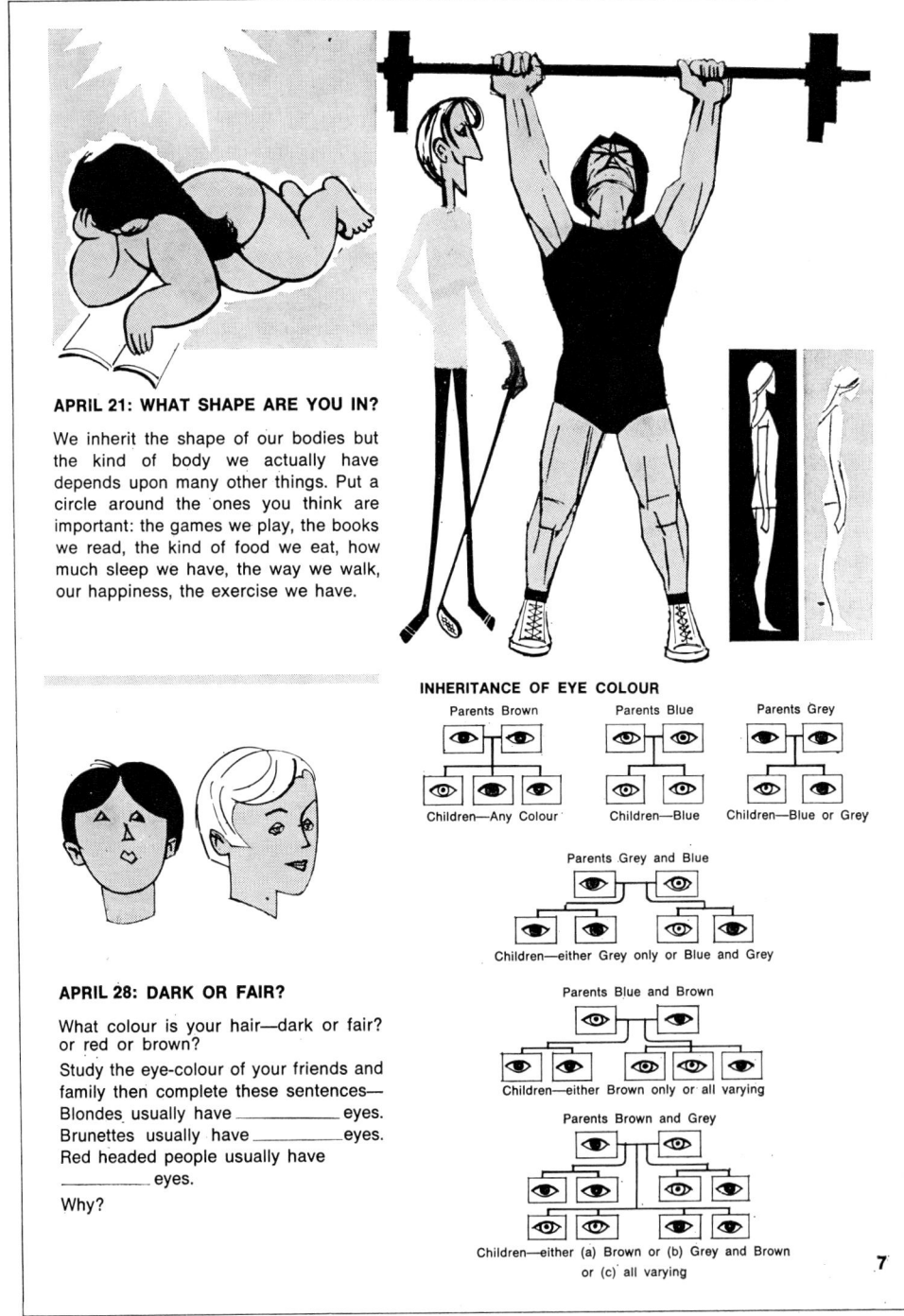

APRIL 21: WHAT SHAPE ARE YOU IN?

We inherit the shape of our bodies but the kind of body we actually have depends upon many other things. Put a circle around the ones you think are important: the games we play, the books we read, the kind of food we eat, how much sleep we have, the way we walk, our happiness, the exercise we have.

APRIL 28: DARK OR FAIR?

What colour is your hair—dark or fair? or red or brown?
Study the eye-colour of your friends and family then complete these sentences—
Blondes usually have _____ eyes.
Brunettes usually have _____ eyes.
Red headed people usually have
_____ eyes.
Why?

INHERITANCE OF EYE COLOUR

Parents Brown — Children—Any Colour
Parents Blue — Children—Blue
Parents Grey — Children—Blue or Grey
Parents Grey and Blue — Children—either Grey only or Blue and Grey
Parents Blue and Brown — Children—either Brown only or all varying
Parents Brown and Grey — Children—either (a) Brown or (b) Grey and Brown or (c) all varying

C. HEALTH EDUCATION BROADCASTS   For primary standards 6 and 7 (Kenya)

Published by Schools Broadcasting, Ministry of Education, Nairobi

Teachers' notes—Specimen layout

Lesson 3. The Human Body

Thursday 10th February

Repeated Friday 11th February

The Aim

To describe the Human Skeleton and show how it is moved by Muscles, Tendons and Joints.

The Blackboard

Key Words—

Other Words—

Summary points

SKELETON; BONES; MUSCLES TENDONS; JOINTS.
Skull, backbone or spine, ribs, pelvis.
1. The Skeleton supports, protects and makes movement possible.
2. Muscles move the bone and are attached to them by tendons.
3. A Joint is where two bones meet.

Visual Aids—

Before the lesson, put up the Skeleton Poster (No. 2) and be ready to point to it, and to the Muscles and Movement Poster (No. 3), during the broadcast.

The Radio Question

At the end of the broadcast the children are asked to think about these questions—what makes your body move? How do you tell your muscles to move? These questions should be used as preparation for next week's lesson on the Nervous System.

The Activities

During the broadcast the children will be told **to do** three things. They are:

1. Put a hand on each side of their head and feel the bones of the skull.
2. Put the right arm on the desk. Put the left hand on the large muscle between the elbows and shoulder of the right arm. Move the right arm up and down (bending it at the elbow) and feel the contraction of the muscle.
3. Holding out an arm and bending and stretching it, as an example of a joint movement (elbow joint).

The Broadcast

The Skeleton consists of the skull, the bones of the trunk and the limb bones. The three functions of the skeleton are support, protection and movement, and each is illustrated by the activities in the broadcast.

The names and positions of the bones are shown on the Poster. The spine or backbone is specially mentioned because the 33 bones make a pipe or tube, down the middle of which passes the spinal cord. This illustrates one of the functions of the skeleton—protection. The backbone protects the spinal cord from damage. Likewise the 24 ribs protect the lungs and heart, and the skull protects the brain.

Another function of the skeleton is to give firm support to the body. All the other parts of the body are held up by the framework of bones, as can be seen on the Skeleton Poster.

A third function of the skeleton is to enable the body to move. Muscles are attached to the bones by tendons—when one muscle relaxes the muscle on the opposite side of the bone contracts and pulls one end of the bone and so moves it; when this muscle in turn relaxes the other contracts again and the bone returns to its original position. The diagram below shows how you can make a working model of the arm muscles and elbow joint with wood, rubber (e.g. pieces from a bicycle tube or elastic bands), string and nails.

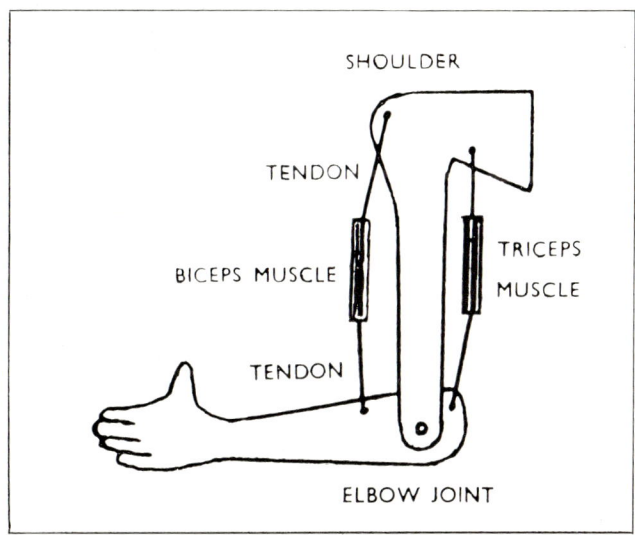

The movement of the arm muscles is shown on the Poster —and you should also point out other muscles of the body from the big ones in the legs to the tiny ones which move the eyes. Movement is made possible by joints, where two bones meet. The joint is surrounded by a capsule (or bag) containing an oily fluid. The ends of the bones are covered with a firm, elastic material (called cartilage) which protects the ends of the bones—i.e. stops them rubbing together and getting worn out. Detail of the elbow joint is shown on the Poster—Muscles and Movement.

Teeth are made of a kind of bone, with a hard outside cover called enamel. When a baby is born it has no teeth, but they appear between the age of 6 months and 2 years. These are called milk teeth and number 20. During childhood they all come out and are replaced by 32 permanent teeth which begin to appear from about 6 years.

We point out at the end of the lesson that all people, of whatever race or colour are **basically the same**. They all have a head, two legs, two arms, two arms, skeleton, muscles, heart, lungs, stomach, and so on.

The final question leads on to the next lesson—The Nervous System.

After the Broadcast

Oral Work—
1. Why are bones called the framework of the body?
2. What other important work do the skull, backbone and ribs do?
3. How are your bones moved?
4. How are muscles attached to the bones?
5. What is a joint? What work does it do?

Written Work—

Write 10 lines on—'Why do we have a Skeleton in our body?' The answer should cover the three functions of the skeleton.

Practical Work—

Ask the children to point out on their bodies where some of the bones and muscles are, and to give the names of them. They should also do again the experiments in the broadcast.

References to Other Lessons

In the Teachers Notes for 'Beginning Science', Term 2, Energy, there is a drawing showing many of the muscles in the body, front and back.

D. VISUAL AIDS (Kenya)

Extract from *A study of understanding of visual aids in Kenya,* by A. C. Holmes [15]

4. Perspective

HYPOTHESIS That where understanding of the convention of perspective is important to comprehension of the picture, correct interpretation will be more difficult.

This hypothesis was tested with the use of two symbols depicting a hunter, an elephant, a tree and a buck. These symbols are part of a set used by Dr. Hudson of the Institute of Industrial Research in Johannesburg in making similar tests.

In each case the arrangement is that the hunter on the left of the picture is aiming his spear at the buck on the right hand side of the picture, with the elephant and the tree drawn much smaller to indicate its distance in the centre background. On Sheet 'A' lines of hills and foreground were added to indicate the different planes of the picture and on Sheet 'B' the illusion of distance was further aided by the fact that the man and the elephant were both placed on the same road.

TABLE 4—PERSPECTIVE—MAN SPEARING BUCK		
	Perspective Hills	Perspective Road
Total number questioned … … … …	806	738
Correct answers … … … … …	53	61
Incorrect answers … … … … …	753	677
Percentage correct … … … …	6.6 %	8.3 %
Percentage incorrect … … … …	93.4 %	91.7 %

Correct Answers
Man spearing buck—
deer—antelope, etc.

Incorrect Answers
Man spearing elephant
Don't know
Man
Hunter
Man Hunting

Correct Answers
Man spearing buck —
deer—antelope, etc.

Incorrect Answers
Don't know
Man spearing elephant

COMMENT
These figures support the assumption that the use of pictures of this type have little value as teaching aids.

E. USE OF VISUAL ANALOGIES WITH BROADCASTS (Kenya)

ANALOGIES

Method

Nine primary schools, three in each of three different areas were selected as being typical of their area. The three areas were Nairobi City (urban), Machakos (a small township 45 miles east of Nairobi) and Kiambu (a rural area 15 miles west of Nairobi). For analysis each group consisted of one school from each of the three areas. The children were all in the sixth year of primary education and were mixed boys and girls. Only the answer papers of children who sat both parts of the test were used and these were reduced by random selection from about 115 to 100 papers in each of the three groups.

The visual aids

Five food names in letters 1.5 cms. high were written on each poster. Small names, where these were the same as the large names, were painted out. The posters were used as follows:

Schools—
  Group I   = Full poster, including analogy pictures (house, lock and fire).

Schools—
  Group II   = Poster without analogy pictures (house, lock and fire).

Schools—
  Group III   = Poster without analogy pictures (house, lock and fire).

The broadcast

Two five-minute broadcasts were recorded on tape. These were in the simple, direct teaching method used by Kenya Schools Broadcasting for health education in primary schools. The two 'broadcasts' were the same except that Tape 'A' included descriptions of the analogy pictures, e.g. 'You can think of your body as a house which is being built … etc.'; 'We put bars on the windows and strong locks on the doors to protect it from the thieves. In the same way, we protect our body from the attacks of disease … etc.'; 'Think again of the house we have built and which we have protected with bars and locks. The house still needs a fire in it for cooking and to warm it … etc.'. These references concerning analogies were left out of Tape 'B'. In both tapes the children were told to 'look at the pictures' when each food group was talked about. The foods named in the broadcast were the fifteen (i.e. five on each)

The poster (actual size 45 cms. × 70 cms.)

written in large letters on the amended posters, and were the same verbally for all nine schools.

Tape 'A' was used for Schools I and Schools II. Thus Schools I had both pictures and broadcast descriptions of the analogies. Schools II had only the broadcast descriptions (no picture of the analogies) and Schools III heard Tape 'B' which did not include the analogies, and they were not given any pictures concerning analogies.

The tests

The appropriate poster for the school group was put onto the blackboard and left for fifteen minutes, during which time the lesson was introduced and the tape played. The poster was then removed and the questionnaire handed out. There was no teacher follow-up to the broadcast.

A simple questionnaire was devised. This stated, 'In the radio lesson you learnt about three kinds of food.' There followed three identical sections with the following three sentences in each. 'One kind of food is called———food. The names of the different foods of this kinds are———. The work of this kind of food in the body is———.'

After collecting the children's answers, the poster was again put up on the classroom wall. A second visit was made to each school two weeks later, the poster removed and identical question papers given out. The children were not told about this second test in advance.

Analysis

Table 'A' shows that in remembering the work of the foods, Schools III without the analogies did better than the two groups which were given them. Likening the work of foods in the body to house, lock and fire does not appear to have helped Schools I and II in comparison to control group, Schools III. Nor is there any significant difference between the two tests. Schools I who had the analogy pictures in front of them for two weeks did score better in the second test but still did not reach the level of Schools III who had declined slightly by the second test. Our doubts from earlier tests about the value of analogies were thus confirmed. Analogies are used by the peoples of Kenya but are usually confined to a given area. Therefore they would appear to be a hindrance rather than a help in a lesson broadcast to the whole country.

Table 'B' indicates that the verbal use of analogies, supported by pictures, with Schools I eliminated confused answers in Test I, immediately after the broadcast.

By Test II some confusion had crept in. Confusion was highest in both tests in Schools II where the analogy references were verbal only. Verbal analogies used alone caused more confusion than when the work of the food was given without them. It would appear that *if* analogies are used, supporting them with pictures is the most effective way of doing it.

Table "A"
Correct answers to the questions on the work of the food in the body

SCHOOLS I (Analogy Pictures *and* References in Broadcast)
Building Foods
  *Without Analogy   32 ⎫ =52 %   41 ⎫ =64 %
  †Including Analogy 20 ⎭   23 ⎭
Protective Foods
  Without Analogy   26 ⎫ =43 %   34 ⎫ =51 %
  Including Analogy 17 ⎭   17 ⎭
Energy Foods
  Without Analogy   33 ⎫ =52 %   32 ⎫ =49 %
  Including Analogy 19 ⎭   17 ⎭
Average   **49 %**   **54.7 %**

SCHOOLS II (Analogy References in Broadcast only—*no* Pictures)
Building Foods
  Without Analogy   40 ⎫ =58 %   35 ⎫ =55 %
  Including Analogy 18 ⎭   20 ⎭
Protective Foods
  Without Analogy   40 ⎫ =52 %   44 ⎫ =61 %
  Including Analogy 12 ⎭   17 ⎭
Energy Foods
  Without Analogy   49 ⎫ =61 %   43 ⎫ =60 %
  Including Analogy 12 ⎭   17 ⎭
Average   **57 %**   **58.7 %**

SCHOOLS III (*No* Analogy References or Pictures)
Building Foods   64 %   59 %
Protective Foods   57 %   60 %
Energy Foods   60 %   57 %
Average   **60.3 %**   **58.7 %**

* Work of food correctly stated but house, lock, fire analogies *not* mentioned.
† Work of food correctly stated and linked to house, lock, fire analogies.

Educational Broadcasting International
© 1971 Centre for Educational Development Overseas.

Table "B"
Confused answers on the work of the three food groups

SCHOOLS I (Analogy Pictures *and* References in Broadcast)
Building Foods
  *Without Analogy   0 ⎫ =0 %   8 ⎫ =9 %
  †Including Analogy 0 ⎭   1 ⎭
Protective Foods
  Without Analogy   0 ⎫ =0 %   8 ⎫ =8 %
  Including Analogy 0 ⎭   0 ⎭
Energy Foods
  Without Analogy   0 ⎫ =0 %   6 ⎫ =6 %
  Including Analogy 0 ⎭   0 ⎭
Average   **0 %**   **7.7 %**

SCHOOLS II (Analogy References in Broadcast only—*no* Pictures)
Building Foods
  Without Analogy   20 ⎫ =22 %   14 ⎫ =19 %
  Including Analogy 2 ⎭   5 ⎭
Protective Foods
  Without Analogy   14 ⎫ =15 %   8 ⎫ =11 %
  Including Analogy 1 ⎭   3 ⎭
Energy Foods
  Without Analogy   9 ⎫ =10 %   13 ⎫ =14 %
  Including Analogy 1 ⎭   1 ⎭
Average   **15.7 %**   **14.7 %**

SCHOOLS III (*No* Analogy, References, or Pictures)
Building Foods   5 %   7 %
Protective Foods   1 %   6 %
Energy Foods   7 %   8 %
Average   **4.3 %**   **7 %**

* Foods named, *without* reference to house, lock, fire but placed in wrong work group (e.g. meat=energy food)
‡ Foods named, and linked to house, lock, fire but placed in wrong work group.

F. MATHEMATICS (Kenya)

From *Looking at Mathematics,* Pupil's pamphlet (Junior Secondary School level, Kenya 1971). Published by Schools Broadcasting, Ministry of Education, Nairobi.

End of Lesson 1: 'Numbers and measurement'

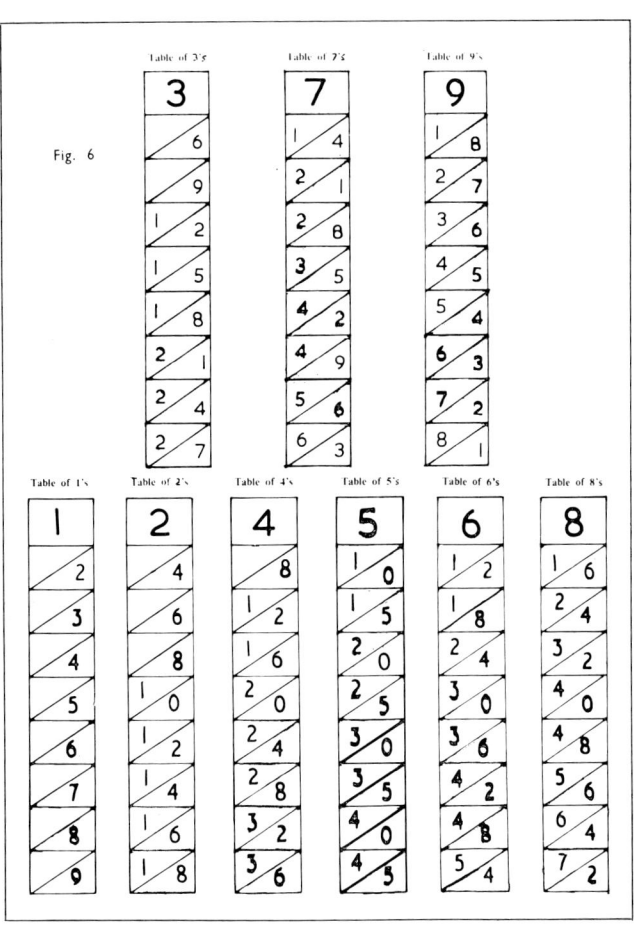

Fig. 6

In Figure 6 you will see strips which I'd like you to prepare for next week's programme. You will need 10 strips of paper, or card, and if you can make them 9 inches long by 1 inch wide you will find them easier to use. Each strip represents one of the multiplication tables from 1-9 and we will use them to show a method of long multiplication devised by a man called Lord John Napier in 1617.

THE SELECTION WHICH FOLLOWS FROM THE SCRIPT OF LESSON 2 SHOULD BE READ WITH REFERENCE TO THE FOLLOWING ILLUSTRATIONS FROM LESSON 2 'SOME GREAT MATHEMATICS'.

Extract from the script of Lesson 2
('Some Great Mathematics')

Teacher   He was John Napier, born in Scotland in 1550 and his main claim to fame was his work on calculation. I'd like to show you his method for multiplication based on what are called 'Napier's Rods'—or

Achungo	'Napier's Bones', probably so called because they were made from ivory or from animal bones. Have you got the strips I asked you to prepare at the end of Lesson 1 in your pamphlet?
Achungo	Yes, I have. I couldn't get any stiff card but I've got paper strips—will they do?
Teacher	Yes, of course, but you must be careful not to tear them. Now, shall we try to use them?
Achungo	Yes, please. How would you multiply say (PAUSE) 2754 by 297?
Teacher	Well, keep your 'bones' and your pamphlet—page 7—beside you and we'll try to solve that problem (SLOWLY). You will see that we have picked out the rods which have the top numbers in the order 2754. Now, what are we multiplying by? 297. Lay your rods—the 2, 7, 5, 4, rods beside each other as shown—(PAUSE)—notice how the diagonals seem to be continuous—and go first of all to the 2-row. The numbers on that row are in pairs between each diagonal and we combine each pair so that we get 4 + 1, 4 + 1, 0 + 0, and 8 or 5-5-0-8. (PAUSE.)
Achungo	I see that. What's next?
Teacher	(SLOWLY.) Next we go to the 9-row and add the pairs between diagonals, which gives us 1, 14, 7, 8, 6. (PAUSE.) What do you notice? (PAUSE.) We've got a 14. When we come to write this down as a proper number we give the tens part of 14 to the number before 14 which in this case, is 1. The number we get here, therefore, will be 2-4-7-8-6 (SLOWLY). Finally, we go to the 7-row and by adding between the diagonals we get 1-8-12-7-8 and by transferring the tens part of 12 we have 1-9-2-7-8. Now look at your pamphlet and see how these figures are arranged for adding up.
Achungo	So the answer is 817,938.
Teacher	Yes. After this programme, perhaps you'd like to check up using your modern method of calculation. I'd like you, too, to try some more of these when you've got time. There are two in your pamphlet to try, but you can work some more of your own.

Extract from Pupil's pamphlet for Lesson 2

```
Napier's Rods                    2754 x 297

      | 2 | 7 | 5 | 4 |
  x1  | 2 | 7 | 5 | 4 |
  x2  |4/1|1/4|1/0|0/8|   5508
  x3  |   |   |   |   |
  x4  |   |   |   |   |
  x5  |   |   |   |   |
  x6  |   |   |   |   |
  x7  |1/4|4/9|3/5|2/8|  19278
  x8  |   |   |   |   |
  x9  |1/8|6/3|4/5|3/6|  24786
```

2 Line:
    Adding numbers between diagonals gives 5 (4 + 1), 5 (4 + 1), 0 (0 + 0) and 8 or 5508.
9 Line:
    Adding again gives 1 (0 + 1), 14 (8 + 6), 7 (3 + 4), 8 (5 + 3), 6 (6 + 0). The tens part of 14 is added to the first 1 making it 2 and the whole number becomes 24786.
7 Line:
    Adding gives 1 (0 + 1), 8 (4 + 4), 12 (9 + 3), 7 (5 + 2), 8 (8 + 0). Transferring the tens part of 12 to add it on to 8 we get the number 19278.

To complete the sum write down the original numbers with the multiplier under the given number like this:

$$2754$$
$$297$$

We now write the answer (5508) which we got from the 2-row with its last digit (8) under the 2 of the multiplier.
The answer (24786) from the 9-row under the 9 of the multiplier and the answer (19278) from the 7-row under the 7 of the multiplier; and in the normal way.

```
   2754
    297
   ----
   5508
  24786
  19278
  ------
 187938
```

It is a bit complicated but that's only the first time we've done it. With practice you'll get much quicker.

Question (follow-up work)

1. Using Napier's Rods calculate the answer to:

    (*a*) 275 x 34;
    (*b*) 3217 x 422.

(Designed by R. H. Thompson.)

Learning media

Theory, selection, and utilization in science education

by Arthur I. Berman
Institute for Studies in Higher Education,
University of Copenhagen,
Fiolstraede 24, DK-1171 Copenhagen, Denmark

Summary

If we were to devise a classification scheme for learning media, we might consider, along with the students, the importance of format. Our immediate questions would be: is it a diagram; is it moving; is it three-dimensional? The carrier of the display is of interest mainly to the teacher: is it produced by a 35-mm slide or by a plasma panel? One may agree with McLuhan—the medium *is* the message—if by this we mean the involvement of the student with the personally made line drawing, or with the lecturer on the screen who speaks to him alone. For media-oriented education, this is an essential point.

But science education demands more, namely, precision of display. The textbook drawing is no substitute for the holographic 'wraparound' image of the moon rock. The teacher's description of the life cycle of a flowering plant is no substitute for the time-compressed colour motion picture of the event. How to effect this display is the teacher's domain. One can, for example, delineate six distinct advantages of overhead projection over slide projection. One can discover several ways in which televised instruction is superior to live lecturing. The role that holography plays in science education is of unique interest: image reproduction through a microscope that is sharp at virtually all depth levels, shock-wave interference, sharpening of blurred photographic images, and, above all, an 'isomorphic' rendering of an original scene that hardly can be described merely by printed words.

Programmed learning is not a medium but a *strategy*. Nevertheless, its principles are so fundamental to the employment of learning media that they should be included in any relevant discussion. The essence may be conveyed by five interlinked words: goalformulation - stimulus - response - confirmation - validation. Some teachers concentrate solely on the stimulus function of media in the belief that the rest of education will take care of itself. How many film makers are willing to validate pilot runs with real students before wrapping up their final product? How will a teacher edit a filmstrip that has failed to activate his students? These are the real media questions.

To use media effectively one must be guided by an understanding of the requirements of the modern student. What stimulates him? What maintains his interest? How does he learn? But we must also ask: what do we want him to learn? Here we draw on the philosophy of modern science and search for the 'pointer readings' of educational operationalism. We ask the present-day student not to recite Newton's laws but to calculate the trajectory of a rocket, and give him all the means to do this efficiently.

But what are the means—the seminar with its emphasis on free expression and immediate validation of one's own ideas, yet slow-moving and indecisive; the lecture with its emphasis on concepts and principles, but cast in a somewhat authoritarian, and hence a bit outdated, mode; the computer with its emphasis on dialogue shaped to the input knowledge of the student, but lacking, after all, that trace of human compassion that may make all the difference? Or can we call on educational technology to produce the ideal strategy, combining the best elements of each, and supplying various forms of learning media that do more than stimulate, that are sensitive to and adjustable to the learning needs of each individual?

I. Media classification schemes

A. *General categories*

In physics a medium is defined as an interim substance through which a force acts or an effect is produced. In information theory the term is extended to mean an agency through which information is transmitted.

A very basic definition of media might be made by a separation according to the stimulation of sense organs so that we may speak of tactile media, olfactory media, etc. As nearly all media in education are sensed by the ears or eyes, audiovisual media has come to be used as a generic term, although learning media would seem to be more appropriate especially as the emphasis is placed on the end intended.

Finn categorizes media by the following scheme [1] [1]:

1. For notes, see page 138.

1. Tool level. Devices that aid the teacher to continue his normal teaching operations.
2. Data level. Information storage devices (microfiche, video tape, EVR cassette, etc.).
3. Behavioural control level. Programmed learning devices.
4. Meaning level. Devices that relate to meaning (motion picture film, slides, etc.).
5. Research level. Devices for research activities (dialect recording, high-speed photography, etc.).
6. Systems level. A media package, such as computer-based learning, to achieve precisely defined objectives.

Educational media were divided into low and high technology domains by Berman [2]. The information input (software) of a *low-technology* medium is under the direct influence of the teacher. The medium is relatively inexpensive, highly responsive to student feedback and may be edited by the teacher without the aid of specialists. A *high-technology* medium is relatively expensive, requires expert assistance for information updating, and generally subordinates the teacher's role. Because of the overall cost and inflexibility, high technology media tend to develop a greater authoritarian bias.

Examples of low-technology media are printed matter including expository texts and programmed manuals, slides, tape recordings, and overhead projections, especially if teacher-generated. Examples of high-technology media are computer-based learning systems and packaged video-courses. Into either category may fall teaching machines and motion picture projections, depending on the extent to which the local teacher's functions are displaced by the medium. A teacher is generally able to select at will appropriate textbook sections, slides, or single-concept films. He is not, however, at liberty to edit filmed courses, or tinker with the 'innards' of an AutoTutor.

B. *An educational media taxonomy*

A media classification scheme emphasizing informational hierarchies is given below [3]:

I. Real objects, people, and events, kinetic and static.
II. Tactile representations, three dimensional (e.g. statues, toy electric trains, raised maps, solar system model, etc.).
III. Visual representations.
   A. Kinetic
   1. Three-dimensional veridical: (a) Isomorphic sequential replication (holographic). (b) Stereoscopic sequential projections (polaroid and bichromatic motion pictures).
   2. Two-dimensional veridical: (a) Continuous projections (overhead projections with models or polarizer-analyser pair). (b) Sequential front or rear projections (motion pictures and television).
   3. Three-dimensional diagrammatic (computer-generated and hand-animated stereoscopic displays).
   4. Two-dimensional diagrammatic: (a) Continuous projections (overhead projections with polarizer-analyser pair). (b) Sequential front- or rear-projected animations (motion pictures, television, computer displays).
   B. Static
   1. Three-dimensional veridical: (a) Isomorphic replication (holographic). (b) Integral photographic (lenslet: spherical and cylindrical). (c) Stereoscopic (polaroid and bichromatic projections and bioptic viewers).
   2. Two-dimensional veridical: (Paper photographs, single, duplicated and projected, rear-illuminated transparencies, single-frame front and rear projections).
   3. Three-dimensional diagrammatic: (a) Integral photographic (lenslet: spherical and cylindrical). (b) Stereoscopic (polaroid and bichromatic projections and dioptic viewers).
   4. Two-dimensional diagrammatic: (paper drawings, single, duplicated and projected, telescribing, rear-illuminated transparencies, single-frame front and rear projections and overhead-projected transparencies and programmed-learning devices).
IV. Auditory representations
   A. *Non-verbal: analogic*. Music, tones, sound-effects.
   1. Multi-channel sound (optical or magnetic film track through theatre projection systems).
   2. Quadriphonic sound (magnetic tape or phono disc through four parallel amplifier-loudspeaker units).

3. Dual-channel sound: (a) Dichotic (tape, disc, or FM stereo radio through dual amplifiers into stereo headphones): (b) Stereophonic (same as (a) but into stereo loudspeakers).
4. Monophonic sound (tape, disc, radio, and TV audio into loudspeaker or headphones).
B. *Verbal: digital*. Narration, dialogue.
1. Compressed or 'discerned' speech devices.
2-5. Same as 1-4 in A (add language laboratory use in 5).
6. Telephonic.
7. Telegraphic.
V. Simulations: Analogic and digital computer simulators (e.g., nuclear reactor simulator), voice and music synthesizers, artificial taste and smells, and games.

Note that programmed-learning devices are not included under a separate heading and may be considered to lie within virtually all areas of II and IV although their emphasis is in IIIB4. In the visual domain polychromatic/monochromatic divisions are avoided as polychromatic agents are capable of monochromatic displays; furthermore, polychromatic capability in all visual media is widespread; in common use, projection television is generally monochromatic and holographic displays are nearly always monochromatic and veridical. The invention of the pulsed ruby laser has opened up the possibility of moving-object holography [3]. Overhead projectors have two-dimensional veridical capability for small slides especially when an optical device is placed above the stage, or for large positive transparencies such as may be made with a view camera; otherwise, its primary use is for diagrammatic display. Note that further subdivision of the diagrammatic display may be made (true outline of veridical, schematic, abstract or free impression, graphical as continuous or histographic, or simply verbal lettering) but this is avoided by considerations of convenience and space.

The hierarchical arrangement in the taxonomy is based on bit-rate requirements for information transfer. From the student's point of view the medium is inherent within the stimulus. A video projection is basically indistinguishable from a cinema projection, a 35 mm slide projection from an overhead transparency projection, a tape-recorded sound from a disc-recorded one. Unless differentiating factors were in evidence (the brightness of an overhead projection, or steadiness of a 16 mm film projection, for example) the student would not discriminate media sources. On the other hand, the agency through which the information is transmitted is certainly unique. There are differences of educational importance that may not directly concern the student such as cost, convenience, flexibility, or availability. These educational criteria would enable us to further subdivide the taxonomic representation to establish in detail the devices (agents) appropriate to each.

The arrangement also ignores the ratio of signal-to-noise that may be unique for each agent. How should we classify, for example, a blurry video colour image in relation to a sharp black and white motion picture? Or a scratchy stereo record to a monophonic magnetic tape played through a 'Dolbyized' amplifier? Other questions about media may arise: How does a scene viewed through binoculars, providing an enhanced stereo effect, represent a unique medium? Is the view of an electron microscope stage a video display?

Finally, we ought to consider in detail the psychological effects that established McLuhan as a centre of controversy when he introduced the concepts of low- and high-definition media ('cool and hot') and his hypothesis describing the relative 'involvement' induced by each [4]. The question of media combinations, producing, in the Broadbent [5] sense, an effective single medium, is one to be considered in a later section.

C. *Media functions*

Learning media have several definable functions [3]:
1. *Recording,* to permit events taking place in the present to be reproduced in facsimile at a later time.
2. *Display,* to permit groups of individuals to receive information at convenient times and locations.
3. *Manipulative,* to permit the observation of events which are normally unobservable to the senses.
4. *Stimulative,* to arrange the environment esthetically or otherwise uniquely in order to increase motivation.
5. *Evaluative,* to interpret responses of students in relation to a variety of variables including personality and potential for learning, and to permit teachers to evaluate their own performance as well [6].

Media can be used to transmit students' responses directly in class, for example, by holding up colour-coded cards in reply to multiple-choice questions, or by more elaborate devices that send an electric signal from a student's position to the teacher's desk [7].

The use of specialized media in educational institutions is widespread. Figures for the United States in 1970, for example, indicate that almost 100,000 public schools own approximately one million projection screens, 700,000 record players, 600,000 headphones, and a half-million overhead projectors [8].

The reluctance of educators to use innovations has been well documented [9]. Even the education industry has proved to be as conservative as the institutions [10]. As for students, many are convinced that at least high-technology media are 'depersonalizing and run contrary to the emerging social structure' [11]. Nevertheless, although progress is slow, technical innovations are finding their way into the educational milieu at all levels.

The following section emphasizes the *carrier* of the medium display, of principal importance to the teacher and administrator.

II. Media subsystems

The devices that are listed below are familiar to the reader for the most part. For general details at an introductory level on the use of many of them, a modern reference such as Gerlach and Ely's *Teaching and media* [12] is suggested. Practical suggestions on the use of media will be found in Garrison's *1001 media ideas for teachers* [13]. Specific commentary of special interest on media and their use follows in the subsections below:

A. *Filmstrip and slides*

The ordered sequential arrangement of filmstrip material is a major advantage when a well-planned and tightly structured presentation is required but this could also be its major disadvantage. Many teachers prefer the conventional (35 mm) slide projector because of the freedom and flexibility that it offers, particularly the ability to create one's own input with an inexpensive camera.

The limitations of room darkening have been alleviated to some extent by: (1) the use of the Kodak Sunscreen, a highly reflective surface that permits viewing under normal room lighting conditions provided the viewing angle is close to normal, and (2) the use of heat-resisting materials in slide manufacture (Harcourt Brace and Javonovich have been developing this type of slide) to permit the transmission of a high-intensity light flux. The widespread use of filmstrip materials may eventually yield to miniaturized storage and electronic display systems of the EVR type as discussed in E below.

B. *Overhead projection*

By far the most widely employed and versatile device for static visual display has been the recently developed overhead projector. Its advantages relative to slide or single-frame projection are (1-4):

1. Screening materials are simple and inexpensive. Contrary to popular belief, a grey (or even a pale blue or green) screen will produce the most pleasing image. A plastic or matte sheet are excellent; the more expensive beaded type or metallic screen are unsuited. Surprisingly good results are obtained when the operator rotates the projector ninety degrees and illuminates a large area on the side wall of the auditorium. It will greatly enhance the clarity of the image, a positive factor in learning; however, of course, students must not only sit away from the wall but from the front as well, where the projector will interfere with their viewing. In normal front-screen operation, the projector is lowered and the screen is tilted forward, normal to the optical axis. A projector too near the screen tends to produce a brown fringe in the image and one too far, a blue fringe. Several manufacturers have now been able to correct this (e.g., Besseler and Leitz) by providing a variable adjustment between the light source and stage. (See Section VA for further details.)

2. The teacher may operate the projector while facing the class, either in a sitting or standing position. This has immediate psychological advantages in establishing a more direct contact with the students, and certainly when sitting, a less authoritarian posture without loss of credibility. Some teachers feel, however, that the competition between them and the visual medium for

the attention of the students is less satisfying to them than when using the conventional chalkboard.

3. Between one and two orders of magnitude of light flux may be projected on the screen, compared with the slide projector, without overheating the film materials. Thus, room lights may be left on for note taking and especially during the discussion mode of teaching. Brightness and high contrast is such a problem, even with a grey screen, that the teacher is advised to experiment with the interposition of a colour filter or with wall projections. Unfortunately, a common complaint among teachers is the high concentration of light at the stage by the Fresnel lens, a problem that is dealt with most easily by interposing a blue-green plastic filter between the operator's eyes and the stage, or by redesigning the optical system to reduce glare:

The high light flux has enabled one manufacturer (3M) to design an overhead projector that is cooled by convection rather than by forced air blowing. The light is focused down on the stage and reflected up by a Fresnel mirror below the transparency. If smooth flexible transparency materials are used they will make good contact with the surface by electrostatic attraction, thereby avoiding a double image. Contrast is lower, scratches are more visible, and overlaying of transparencies is severely restricted. Nevertheless the general result is quite satisfactory and the short focal length and absence of blower noise are decided advantages. More conventional single-transmission projectors also come in short focus models (permitting operation close to the screen, an aid in seminars and discussion sessions) and with blower-noise control.

4. The capability of creating a very effective transparency in a few seconds, preferably with a thermal copier, and alter it immediately on the basis of student feedback, is perhaps its most important advantage. As was noted in B above, its projections are limited to diagrammatic representations unless: (a) an optical unit is placed on the stage for full-sized projection of small slides, or (b) positive film materials, view-camera size, are projected. However, (c) an array of, say, 6 × 6 cm slides is effectively projected to present simultaneously alternate views or aspects of a single subject. Many teachers have found that the concurrent use of two or three projectors gives a multimedia theatre effect in an auditorium allowing the equivalent of several hours of blackboard writing to be projected at once, but with all the attendant distraction disadvantages, however.

Except for biological specimens, geological samples, etc. requiring veridical photographs or projections for much of their study, the alternative of providing neat, simple, hand-made teacher-produced materials offers many interesting learning possibilities. Experiments comparing hand blocklettered freehand diagrams to commercially made transparencies of comparable information showed a student preference for the more informal (and personal?) product so long as clarity and neatness were maintained throughout [13a].

5. The ability to alter at will by addition or substraction of portions or all of a given projected transparency is a particularly useful advantage. Overlay projection is especially common in biology where complex parts of an organism can be built upon a simple framework, or in general sciences where the labels on a diagram can appear last, as well as details unwanted in the initial inspection. The usual procedure is to tape the overlays on the edges of the cardboard frame on which the transparency is mounted and then swing them into position. A simpler procedure for overlays as well as for paper or cardboard 'revelation masks' would be to omit the cardboard frame entirely and tape guides on the stage that are made with rulers to permit positioning and registration (see Fig. 1, below). Omission of the frame saves time, expense, and storage space. In this way a set of transparencies can be placed in a single file folder and protected and stored accordingly.

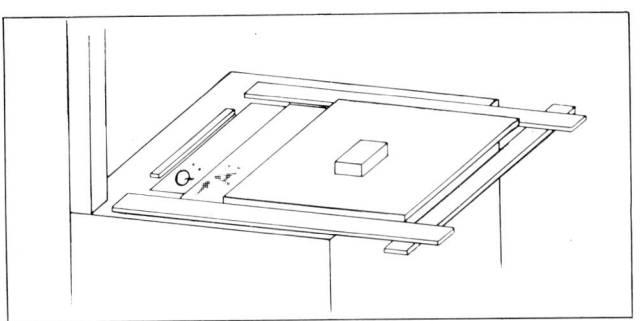

FIG. 1. A device made of rulers for positioning unmounted transparencies and supporting a revelation mask.

Blocking out sections of a transparency with the mask not only aids in concentrating attention, but is particularly useful in programmed learning group instruction where a single transparency could include material of a half-dozen frames, equivalent to a dozen slides if responses are in sequence with the stimulus material.

6. The large format of overhead transparencies permits direct reading by the teacher as well as full-sized reproduction in many copies for student use.

Probably the main drawback of the overhead projector relative to the blackboard relates to its frequent misuse, a condition that can easily be remedied by a competent teacher. Students cannot readily cope with a mass of non-linear information that is suddenly flashed on the screen from a prepared transparency, particularly when verbal material simultaneously flows from the teacher. They generally prefer the restricted pace of the blackboard, suggesting to the teacher that he had better dispense with prepared transparencies entirely and work instead with felt pen on acetate, less neat but more spontaneous. The alternative solution is to retain the prepared transparencies and:

(a) uncover them section-by-section with the revelation mask;

(b) limit the development on transparencies to those course objectives that require knowledge of basic structure, and make more use of printed materials for independent study of the course details;

(c) pass out copies of the transparencies in advance so that students may become familiar with them prior to class attendance, and

(d) ensure that verbal auditory exposition reinforces the visual channel by delaying commentary until students can assimilate the visual information, furthermore arrange that verbal elements within each channel approximately parallel each other. Some teachers have found it effective to eliminate lettering and some details in the prepared transparency and insert them by hand while speaking. This also encourages students in the active role of overt note-taking during the exposition.

C. *Motion pictures*

1. *Single-concept films.* The most exciting recent innovation in the employment of motion-picture media in science education has been the widespread use of the single-concept film in the continuous-loop cartridge.

The use of cartridge projectors as well as self-threading motion-picture projectors has removed the major obstacle to the employment of film media by teachers: equipment handling.

The single-concept film usually omits a sound track permitting a teacher to extend his creativity by offering his own audio cassette-recorded sound commentary to the showing. The importance of this type of activity in enhancing the teacher's self-esteem ought not to be ignored. He enjoys a similar role as producer of his own transparencies. In both of these media, modern technology can provide the means for a teacher of limited artistic talent to create an attractive, efficient, and personal end product for his student.

There is a very large catalogue of single-concept films available in science (for example, by Encyclopaedia Britannica and Ealing), particularly physics [13b] and biology. Motion pictures are especially suited to biology in visualizing growth studies by time-lapse photography, moving organism studies by slow-motion photography, and microstudies by photomicrography.

The super-8 cartridge projector with its 150 watts of power can actually produce an acceptably bright, fairly stable, and relatively sharp image from the rear of an auditorium, offering good competition to a projected television image.

The 16 mm projection system is, of course, the professional one to use for large-group instruction. New self-threading machines make life easier for both teacher and students. Projectors, such as the Kodak Ektagraphic MFS-8, may be pre-programmed to combine movies with still-frame projection without loss of image brightness.

The cinematic medium itself has advanced esthetically since the early 1960s when new experimental techniques were being developed such as split-screening and zooming. Peter Robinson's tri-screen film *To be alive* was for many the main attraction at the 1964 New York World's Fair. Robinson has since contributed much of his skill to the making of science education films. Most media specialists would agree that the central point of interest at Montreal's Expo-67 was the emergence of the new techniques and new uses of cinema.

2. *Computer-generated films.* With the production of kinetic-diagrammatic films by computers an important area opened up in all fields of science. The computer can calculate and display such complex problems as the tumbling of a satellite in the earth's field. It can calcu-

late and display in halftone objects that exist only in parametric form, as well as intricate mathematical constructions—even a computerized stereoscopic view of a tesseract has been made. Harmon and Knowlton [13c] have noted these and other applications in science such as the graphical representation of dynamic three-dimensional electric or hydrodynamic fields, cochlear motion in the inner ear, the structure of world-wide weather (pressure, temperature and precipitation patterns), and explosive wave fronts. Cotterill and his group in Denmark's Technical University have produced a computer-generated film on the statistical changes in potential and kinetic energy and position of individual molecules in a crystal matrix during the melting process [13d].

D. *Audio-tape recording*

1. *The cassette recorder*. The most important recent innovation in sound recording for science education use has been the cassette recorder and playback unit. Its prime features are convenience and low cost. There is some tendency to dismiss convenience as a very secondary factor in education. Yet the problem of *in*convenience inhibited the widespread use of classroom education films through the 1930s and 1940s, and is still a factor in the acceptance of many teachers of overhead projectors.

Professional and semi-professional reel-to-reel tape recorders are easier to thread than most film projectors; nevertheless, it is inconvenient to store the equipment and the tapes and, most important of all, virtually impossible to pass among students as learning packages for home study. The low cost of cassette units comes about not because they are especially simple in their mechanism, but because they are manufactured by many dozens of competing firms for a world-wide mass market apart from the education field.

It is for this reason that several companies have been able to engineer cassette recorders and high-density chrome-based recording tapes of professional quality that compare in performance with the better reel-to-reel recorders in spite of the low tape speed and narrow track width. The costly materials of high-quality, such as ferrite recording and playback heads, may be manufactured in smaller sizes. An electronic circuit has been developed by Dolby, referred to earlier, that is widely used in the best recorders; it virtually eliminates the tape background noise disturbance that has plagued the industry.

2. *Stimulation potential of high fidelity*. The result of these engineering advances, as far as the educator is concerned, has been a manifold increase in faithful reproduction relative to background, i.e., an increase in the ratio of signal-to-noise. This raises the question: To what extent can the information cues in media of high S/N (signal/noise), in impeccable reproduction and presentation of sound, for example, actually stimulate and motivate a student to a greater extent than the real thing? This is not trivial. The cinematic medium has more to state, by sharp close-up, arrangement of stimuli, and general style, than a live stage play. Some composers write music for dichotic reproduction of sound, separate information channelled through each ear, as with stereo headphones, as distinct from stereo loudspeakers which mix the sound channels to reproduce concert-hall realism.

The general subject of background noise, which would include distractions of all types irrelevant to the signal information, as well as reinforcement and interference of simultaneously competing media, is, of course, an important research area in information theory. Among educators, the study of the esthetic employment of media of high S/N for unusual and varied effects as a means of stimulating and motivating students who have grown up in an age of television, is a research area worthy of greater attention.

3. *Headphones*. There is little doubt that the extensive use of headphones in education has succeeded in concentrating attention and isolating a student from background distractions. The reader can make this experiment for himself. The learning advantage of stereophonic over monophonic listening has yet to be proved, yet the realism and esthetic enhancement of multichannel sound does seem to have a stimulating effect for most listeners, somewhat analogous to the video use of colour, even when irrelevant to essential information. Vannevar Bush notes how reproduced music can give a more accurate impression, 'free as it is from audience noise or distortion of instrument balance due to bad acoustics in a crowded concert hall' [14]. There is a very effective device that interpreters use called a *bidulle* ('gadget') where a large conference table is fitted with headphones for each of the participants who then receive perhaps a more intense link with the

speaker, who might himself be speaking into the microphone, than under ordinary conference conditions [15].

4. *Compressed sound.* Experiments have been made with speech compressors that succeed in reducing the time of presentation of a lecture or recorded discussion by higher-speed playback of a tape recording and simultaneous frequency compensation for the increase in the sound pitch that would take place [16]. There is a great deal of interest and some success in the use of this device, also called a sound discerner [17].

E. *Television*

1. *Direct video displays.* The importance of television as an effective educational medium, as in the case of motion pictures, relates very much to the entry mode of students into the education system, a subject that will be discussed in IIIA below. Prior conditioning has made the medium acceptable and in many cases desirable.

Educational research fails to support the view that courses taught exclusively by television are better than live instruction as measured by the performance of the learner [18]. In a remarkable review of 91 research studies covering four decades of teaching by virtually every means—live lecture, seminar, television, etc.,— Dubin and Taveggia reported, 'We are able to state decisively that no particular method of college instruction is measureably to be preferred over another, when evaluated by student examination performances' [19]. Their conclusions have provided a stunning blow to researchers who have attempted to use standard cognitive evaluation techniques as a means of verifying the effectiveness of their preferred teaching method. Affective criteria would appear to offer a more positive basis for drawing conclusions, were it not for conflicting effects that distort data in opposing directions. First, the *placebo* (or Hawthorne) effect, after the medical research term for a drug having no pharmacological value but administered as a control in testing experimentally the efficacy of a biologically active preparation—the supposition here is that students will think the innovation, like the calcium tablets, ought to help in any event. Opposed to this is the 'disorientation' effect, wherein a student's preference would be for the 'time-tested and true'. It would be a risk to attempt to predict an evaluation of a teaching system without a prior evaluation of the input data, the average of which varies markedly from decade to decade if not from year to year [20]. The most serious disadvantage of direct (as distinguished from stored) video presentations, as with live lecturing to large groups, is the lack of satisfactory continuous student feedback or even its possibility [21]. In deference to the auditorium lecturer, however, a measure of silent feedback does occur for him in terms of the pulse of the class, their facial expressions, for example, restlessness or enthusiasm, all of which can have a contagious effect in a large group. He can respond to this and alter his pace or his mode of presentation; at the very least, he can pause until they have finished their note-taking.

A second disadvantage of the video medium is the limited range of visual verbal presentation. The live lecturer has an extended range of media available at his fingertips from a wall of blackboards to a battery of projectors. Transference of all this to the confines of the television screen is usually unsatisfactory.

On the other hand, the video medium contains the potential for course enrichment through science demonstration close-ups, film-clip insertions, stop motion, increased-frame cinematography and amplification of sounds as well as magnification of visual stimuli, etc., which has, in fact, encouraged the placement of monitors within auditoriums during live lecturing. This not only extends the range of the science lecture down the microscope barrel or into the glass beaker but permits the lecturer to focus on Scroedinger's equation on a page of the text if he wishes.

In one other respect the video medium retains a marked advantage. The TV lecturer presents an almost hypnotic aspect by maintaining eye contact with an imagined viewer, who in turn receives the illusion that the lecturer is speaking to him alone on a very personal basis. The attention-maintaining quality of the medium is enhanced by the brightness of the image relative to the surroundings and a general absence of distracting influences that often accompany live lecturing [22].

On balance, the advantages and disadvantages of the opposing media are consistent with a null hypothesis (nsd, or no significant differences) outcome on evaluation, taking into consideration both cognitive and affective values [23]. Despite the apparent null result, McKeachie, in his 1963 review, considered that a more careful evaluation of the statistical evidence showed television to be definitely inferior to live lecturing 'in communicating information, developing critical thinking, changing attitudes, and arousing interest' [24]. In the intervening decade we ought to reflect on the change both in students' attitudes and television style before permitting past data to influence present decisions.

2. *Stored video displays.* (a) *Advantages and disadvantages*
The recorded televised course presentation is on a completely different level from live instruction: the care and preparation might be justified to be perhaps an order of magnitude greater merely because the total viewing audience may be several times this. This would mean greater use of film clips, more elaborate preparations, etc. The availability to students is unlimited in time. The pause-and-repeat mode now adds to the flexibility to permit not only note-taking and self-paced instruction, but questions and discussions if operated in small groups with and without a teaching assistant. This makes it suited to group-paced programmed learning [25]. Gropper and Lumsdaine reported success in terms of learning in group-paced programmed learning by television requiring overt responses to students in a control group exposed to a conventional television lesson [26].

Crucial to the employment of group-paced programmed learning in any form is the accumulation of research data that can justify its use, especially if one considers that self-pacing is one of the basic tenets on which the programmed learning movement rests. In a series of experiments, Carpenter and Greenhill reported that there were no substantial differences for university students in performance among three pacing methods: self-paced teaching machines, self-paced programmed textbooks, and externally paced filmstrips [27]. The experimenters noted a wide tolerance for variations in pacing rate in the population studied, and concluded that self-pacing was not necessarily the optimal mode. Briggs, *et al.* suggested that variable automatic pacing, with shorter time intervals as learning trials progressed, could be superior to self-pacing [28]. Gryde suggested multiple-track TV programming at different levels of presentation as one way to adjust to student differences, where a student could continually alter the track [29]. At the moment there seems little doubt that group-paced programmed learning can be at least as effective as self-paced [30].

Inflexibility is the most apparent disadvantage in video-taping courses or course segments. A media rule-of-thumb principle is that the more difficult it is to compose a presentation, the more unlikely it is that alterations will be made should the need arise either on the basis of student feedback or simply because the information is outdated. This inflexibility could be accompanied by an authorian bias, not only affecting educational values for the student, but creating a role conflict for the teacher [31], as noted earlier in the comments on high-technology media. The inflexibility of video taping is alleviated somewhat by electronic editing methods that can insert new material, erase unwanted material and shift ahead old useful material (the teacher is expected to wear the same tie).

(b) *Video cassettes.* For several years educators have been promised a media revolution in the form of mass-produced video-cassette equipment. There are compelling reasons for this, basically the same as for the audio cassette: convenience and low cost. To these would be added all the advantages already noted for the television medium. The techniques of television broadcasting would be modified by the variability of viewing in space and time on a broad scale, including viewers' homes.

The most widely known system is the one invented by Peter Goldmark of CBS Laboratories called Electronic Video Recording (EVR) [32]. The potential of this system is very great and an enormous effort has been spent in developing it; the ultimate plans for world distribution by its creators and their associates, Imperial Chemical Industries and CIBA, would depend, it would seem, on the interest of the public, including educators, in it and competing systems.

Information is stored optically on miniaturized film, 2.5 mm $\times$ 3.35 mm per frame, rather than magnetically on tape. Coding of the optical information is accomplished by writing with a fine electron beam *in vacuo* thus alleviating problems of crystal granularity and light scattering that affect ordinary optical photographic methods. This results in an order of magnitude increase in resolution over optical methods to 400 lines per millimetre. The storage capacity per frame thus exceeds one million bits. This is four times the information capacity of an ordinary video display and even exceeds that of 16 mm motion picture film under ordinary optical conditions.

Retrieval of the stored information is achieved by means of a flying-spot electron scanner in the home or school unit, which then feeds the signal directly to the antenna leads of the home receiver for conventional display. This type of retrieval system has been used for years by TV broadcasters for decoding and encoding information transferred at 24 or 25 frames per second from motion-picture film to the video screen, the frame rate of which is synchronized to line-current alternations.

The quality is high and the projections may be made in colour by coding a parallel track in black and white, a relatively inexpensive technique. The programme may be paused at any time for single-frame study without the risk of damaging the film. A random-access retrieval system coded on the magnetic audio tracks permits the system to act as an immense storage bank, competing with microcard, microfiche and ultra microfiche systems. It could indeed become a homeviewing device of entire libraries, considering the 180,000-frame capacity of a single seven-inch telecartridge that normally runs a half-hour in its colour mode.

In analogy to the manufacture of phonograph records, precision engineering is employed to create a master to be replicated for a mass market; however, relatively unsophisticated technology is required at the home-viewing end. It is claimed that small children can operate the playback unit without difficulty, opening its potential to every level of education in and out of the classroom. Texts and manuals in science education have already been co-ordinated with EVR [33].

The permanence exceeds that of motion-picture film for sprockets are unnecessary and the raised magnetic audio tracks on the edges inhibit contact of the optical surfaces. Also, heat damage is impossible. On the other hand, artefacts as small as five micrometres will be visible requiring high standards of cleanliness in manufacture and handling.

The artefact problem is eliminated in the RCA Selectavision system, which codes information in embossed holographic form on vinyl tape for retrieval by a laser actually built into the home or school converter. Despite what appears at first to be an overelaborate system, two advantages are clearly obvious: (1) in a hologram, a fraction of a square millimetre of any portion has all the information needed to reproduce the entire visible image—one can cut out a small piece and view the entire scene through it—and so artefacts are indeed invisible; (2) like plastic phonograph discs, they can be mass produced at low cost.

Other manufacturers have developed competing systems that offer promise; however, in the education field, there is bound to be confusion because of the lack of standardization. Sony, for example, has extended its reputation in the magnetic video tape area to produce cassettes in this form in black and white or colour, with the added facility of home recording, impossible in the CBS and RCA systems except by remote processing. Telefunken and Decca manufactured a plastic phonograph-record-type of video disc for quarter-hour video programmes, a challenging engineering problem when one considers that the bit information is a hundred times greater in the video medium than in the audio.

Several manufacturers have developed relatively easy-to-use home and school electronic cameras for television playback. The CBS device is a coloured still-frame or movie camera that can take 12,000 still pictures on a single loading for video display in colour that avoids costly conventional colour photography processing [34]. RCA and Bell have also developed compact electronic units [35].

3. *Microteaching.* Microteaching and microcounselling are two of the more important innovational applications of television to education [36]. A portable video-recording chain is generally used to record small-group supervised teaching sessions. There is attention paid to discipline control, the capacity to evoke student responses, organization of work, and general cueing technique. As an aid to the teacher in self- and guided-evaluation, it has been described as corrective, therapeutic, and dynamic. In science education its value extends to the study of demonstration and laboratory technique.

F. *Holographic and other 3D displays*

1. *Holography.* The educational possibilities of three-dimensional imaging extend into greater areas than one would suppose by casual reflection. Obviously the stimulation and motivation potential of media as fascinating and unique as those outlined below is high; however, the effect would gradually wear off unless there were significant learning areas associated with three-dimensional optics, as indeed there are.

(a) Holography is lensless photography. Reproductions of objects can be seen at very high resolution with virtually all depth levels clearly in focus. Any student who has observed objects under a high-powered microscope will recall the impossibility of keeping a range of depths in simultaneous clear view. A micro-hologram will maintain highly resolved pictorial information throughout the medium under investigation including an extended lateral field [36a]. Consider the possibilities opened up in biological education when observing a hologram of micro-fauna in twenty or thirty centimetres of pond water. Or searching for bubble-chamber tracks in a nuclear physics laboratory.

(b) Students may make holographic interferograms of shock-wave interactions such as those which surround a flying bullet, or aerosol studies of the supersonic jet from a jet engine, the contour profile of a turbine blade, or the vibrations of a loudspeaker. The possibilities are extensive.

(c) Blurred photographs and other displayed media from electron micrographs to radio astronomical readouts may be holographically sharpened by a method considered to be superior to computer methods [37].

(d) The information-storage capacity of a Bragg-effect or deep-volume hologram is quite impressive. In a 20 cm × 20 cm hologram a bit capacity of 90 *billion* is available. (Compare the 250,000 bit capacity of a video image, noted previously.)

(e) The holographic image is substantially isomorphic with the original subject and quite distinctive compared with stereoptic renderings. The observer can move his head and see all visual aspects of a scene provided he still is able to observe through the hologram. Flashes of reflected light from surfaces will appear with head movement. If a magnifying lens formed part of the original scene, the magnified virtual image will appear in depth and take on different aspects as the observer moves. Apart from the immensely stimulating effect of a phenomenon such as this, one may use the medium to examine in detail rare crystals, geological and biological specimens, and intricate molecular arrangements, particularly if the hologram is in the form of a cylinder to permit complete views around an object. The implications in science education of this new medium have barely been scratched [38].

The name hologram is derived from the Greek words *holo* (complete) and *gram* (message). Dennis Gabor conceived of holography in 1947 [39], and began the original work on the idea for which he received the Nobel prize in 1971 [3a]. Holography was relatively dormant until the invention in 1962 of a powerful coherent light source, the laser. When it was introduced to holography by groups led by Leith, Upatnieks, and Stroke [40] a quantum jump occurred in its developments. At present it is a major area of optical study [41].

In holography, coherent light as from a laser (light in which the electromagnetic waves maintain a fixed phase relationship) impinges on an object, and the light that then issues from the totality of object points is made to interfere with a second beam of similar coherent light at a photographic emulsion. The pattern of light and dark fringes that appears after development has no apparent relationship to the original scene. However, if the developed film is then illuminated by similar light the original scene will appear as a virtual image in space when observed through the film, and under specified conditions as a real image.

Holograms are generally made only of still objects, for object movements of a significant fraction of a wavelength during exposure will destroy the record. The technical development of the high-intensity pulsed ruby laser as well as high-speed photographic emulsion has opened up the field of holography of living objects made with exposure times as short at 10^{-8} seconds, and even the promise of kinetic holography [3a].

2. *Integral or lenslet photography.* This method of photography, familiar to the reader in the form of ('three dimensional') greeting cards, or as illustrations on the cover and inside the popular magazine *Venture,* also has important educational applications. As incoherent light is employed, objects under ordinary illumination, as well as self-luminous objects, may be studied. The colour of objects, still or moving, is more readily imaged than the holographic techniques, and image reconstruction in very large formats may be produced for pictorial display.

As long ago as 1908, Gabriel Lippman, who received the Nobel prize in physics for creating the first method of colour photography, proposed the first conceptually-simple lenslet method of recording and displaying 3D images. In integral photography, light reflected by a subject is imaged by each one of an array of tiny spherical lenslets on to a photographic plate, so that each lenslet sees the object point from a unique aspect, forming therefore a unique image point upon the plate. When the developed plate is placed in original register with the lenslet array and illuminated from the rear by ordinary white light, a real point image is formed at the original object point location [42].

3. *Stereoscopic imaging.* Other 3D visualizations such as those appearing in textbooks as diagrams or photographs observed with two-colour glasses, or stereo viewers, or polarized projections on metalized screens viewed with polaroid glasses, have proved valuable in education where a course objective requires that a student form a conceptual understanding of a geometric or physical shape.

G. *Programmed learning*

1. *Extrinsic programming.* At about the time that world education was searching for new guidelines to meet the pressing economic and social demands for scientific education in the 1950s, the concepts of programmed learning were reaching fruition and being put into practice. Although the *modus operandi* is drawn from the behavioural principles of B. F. Skinner [43], programmed techniques of both the Skinnerian extrinsic and the Crowderian intrinsic types can trace their origin to Grecian times [44]. The principles may be summarized as follows [45]. In theory a student's motivation and learning is enhanced by (a) having him participate actively in the learning process by responding to incomplete phrases (or pertinent questions) in the form of sequentially displayed interdependent frames, each forming a unit of the development, (b) advising him immediately of the status of his response, (c) adjusting the difficulty of the sequence so that an average student has about a ninety per cent chance of responding correctly each time, (d) enabling a deviant from the group average to be brought closer in performance to the others by making the completion time of the sequence open-ended for each student and (e) utilizing the cybernetic aspect of the procedure to establish a close fit with the target population on the basis of point (c), by rewriting frames following an analysis of the responses.

The principle of behaviour-shaping or operant conditioning behind this form of programmed learning is somewhat repugnant to many science educators. A typical reaction, recently noted by Piaget [46] took place at a 1959 Wood's Hole conference of mathematicians, physicists, and biologists who were seeking means of recasting the teaching of the sciences. Skinner's propositions were treated rather coolly since 'the particular problem facing the conference was less one of finding the means to achieve accurate comprehension than that of encouraging the development of inventive and inquisitive minds'.

2. *Intrinsic programming.* In an effort to diminish the effect of insufficient background for a student of poorer than average ability, the programmer inserts remedial sub-programmes into the sequence, to which a student would be routed on the basis of his last response. As an infinite number of constructed responses are possible, the procedure is made tenable by merely asking the student to choose one of several plausible responses presented to him, thus determining the route to be followed. A guide to methods of intrinsic programming has been written by Rowntree [47].

It is clear that the overriding consideration in both forms of programming is that the responsibility for learning rests entirely with the programmer-teacher. Whereas the traditional form of instruction, lecturing and required readings can have an existence independent of the receiver, programmed learning depends upon student validation for its existence, a rather new idea in education.

Klaus distinguishes the type of learning in intrinsic and adaptive programming from that in extrinsic programming [48]. The former is stimulus-centred where learning takes place in the acquisition of stimulus material, prior to the student's response. In the latter form the model is trial-and-error; behaviour is learned only when reinforced and the programme is response-centred. Both forms are in use in science programmes and both have particular advantages and disadvantages. There are obvious dangers in their overuse as such values crucial to science education as inquisitiveness, intuitive leaps, inductive generalizations, etc. cannot be readily programmed [49]. Nevertheless, much of science education consists of well-ordered, logical presentations of structured information to be mastered, a task in which programmed-learning techniques have demonstrated high performance levels.

3. *Mathetics.* A recent research report by Balson [50] confirmed the feeling of many educators that the conventional form of Skinnerian programming—very small steps and low error rate—took longer to complete for a group of grade-school pupils than an alternate programme composed of longer frames and higher error rates. Although a null hypothesis applied to the assessment of the comparative cognitive data, the longer time

of completion was of significance. It is a common complaint of bright students that conventional small-step linear programming is generally boring and unsatisfying, contrary of course to theory [51].

For some years Gilbert has maintained that programmes should consist of long frames upon which a student performs several operations to gain mastery, a programming scheme he calls 'Mathetics' [52]. Certainly the current trend in programmed learning has been in the direction of larger 'operant spans' if only to maintain student motivation [53].

4. *Programming formats.* The four formats in which programmed learning appears at present and possibly in the future are:

(a) A teaching machine by which a student is prevented from observing the correct response prior to his own responding. Furthermore, the programme may not continue unless a student enters his response.

(b) A printed manual or book in which the presentation is easily produced inexpensively, but relies on student self-control for its effectiveness.

(c) An electronic computer with an enormous memory and display potential. It may have branching capability by analysing constructed responses, rather than only multiple-choice responses. Furthermore, it can evaluate a student's integrated performance continuously and adjust the programme to this parameter as well as factors such as personality characteristics.

(d) A 'HAL-type' computer, one that does not merely ask questions of the student but also answers random questions posed by the student. It is modelled after the principal character in the film *2001: A Space Odyssey* and presumably will be generally available by that time.

5. *Adaptive programming.* Programming in machine or manual form can be considered a subsystem and generally within the learning environment controlled by the teacher. The case of computer-based learning is different. From its inception, aside from drill and practice modes, it was considered an ultimate learning system, and moves rapidly in that direction. Its important role in science education will not be developed here; however, some trend identification is in order.

As noted above, a well-designed computer programme has evaluation capability of a set of expected forms of constructed responses, and adaption capability to the student's entry behaviour, adjusting both format and content in a design called idiomorphic programming [53].

6. *Strategy.* Of interest to science educators are the possibilities of programming by various methods such as an inductive (egrul) or discovery method, a deductive (ruleg) or expository approach, or a guided discovery approach [54]. An experiment with three groups of mathematics students employed three teaching strategies corresponding to these methods [55]: (a) A number of hints were given to enable a student to derive a sum formula by himself. (b) The formula was presented followed by numerical examples. (c) A relationship was noted between rows of term values, rows of sum values, and the series itself. The groups performed from best to worst in the order of guided discovery, discovery, and expository, suggesting a positive correlation between performance and active student involvement.

Programmed learning in general and computer-based learning in particular have been criticized because of their impersonal nature in leaving the teaching to a device operating in tutorial mode (despite the counter argument that the person behind the programme is as human as the author of a textbook). Filep points out, however, that 'to learn without involvement with another human being, to have no commitment to share his psychological space, may be appealing to many' [56]. He notes also that the 'hands-on' quality of typing or the use of a light pen gives a student feedback through his tactile senses. The participation in the personal process of learning, Filep likens to the gratification experienced by the craftsman of earlier times [57]. O. K. Moore, inventor of the 'talking typewriter' used extensively in early childhood education, maintains that the one crucial criterion of an 'interactional machine', one that interacts with humans, should be its 'congeniality' [58]. Trow considers that the 'assembly-line nature' of traditional schooling is far more depersonalizing than the new media [59].

In a very recent review, Hammond [60] attempts to analyse the reasons for the slow pace of computer adoption in education [61]: cost, institutional resistance, reliability, and computer language are factors. Nevertheless, Dartmouth College has installed a hundred terminals in dormitories and various locations including other schools and colleges. The computer language is simple, the failure rate low, and the interest high. Hammond reports also that S. Papert and his group at MIT avoid the conventional use of the computer in favour of such devices as a mechanical 'turtle' that draws complex geometric figures under the control of grade school pupils who write the programmes themselves. This reflects the trend to expand the computer in science beyond its tutorial mode of digital conversation; computer-generated films, already noted, are a case in point [62].

H. *Simulations and games*

1. *Simulations.* A simulation is a representation, often in miniature, of a large-scale project. A wind tunnel is a good example. An analogue computer—e.g., one that can duplicate the operations and events in a nuclear reactor in terms of pointer readings without high cost or risk of damage—is another. Coker's speech simulator is a computer that could have an important impact in computer-based learning, especially for young children [63].

In *Future shock* Alvin Toffler considers the trend toward simulated environments [64] where computer experts, designers, etc., will join to create experimental enclaves in which customers will don costumes and go through a planned activity sequence that envelopes them in the simulated environment. In Ray Bradbury's *Fahrenheit 451* couples are actor-participants in televised psychodramas surrounded by four-wall video sets. In *The Search* C. P. Snow recommends that students be placed in the role of the original discoverer to excite them about science and stimulate their creativity [65], an idea that resembles the history-of-physics laboratory developed by Devons and Hartmann [66].

Certainly this type of participatory activity, as media involvement, can more easily enable the student to delve deeper into the ideas of science and interrelate them into the cultural fabric of his rapidly changing environment, a subject of deepening concern among science educators within the past few years [67].

2. *Games.* As an instrumental medium, games have had a wide application [68] particularly in the social sciences [69]. *Psychology Today* has published a number of participatory games in areas such as race relations. In the Montessori prepared environment, games are used extensively, including mathematical ones. This medium is finding its way into the natural sciences closely related to the social aspects of science; for example, Toffler refers to a game called 'Future' distributed by Kaiser Industries, that introduces players to various technological and social alternatives of the future and forces them to choose among them. He notes also J. Villegas's game, developed at Cornell University, to elucidate the ways in which technology and values will interact [70]. Sutton-Smith considers the importance of children's activity, including games, in developing an early conceptual understanding of basic principles [71], a subject to which Piaget has devoted much of his life study [72].

Electronic games that can actually be played on a school or home television receiver have recently been developed and manufactured. These may find useful applications in education [73].

I. *Printed materials*

1. *Books.* In spite of the impact of new learning media during the past two decades, the book still reigns supreme as the leading transmitter of information in formal education, although many would take issue with this unverified opinion, feeling that the vicarious visual medium of television plays a far greater overall role in the life of the individual, and furthermore, that there is no real distinction to be made between formal and nonformal or informal education. The distinction here is not meant to be between structured and unstructured, but rather what one normally associates with professional activity and what one does not.

Five factors are responsible for the importance of books: portability, low cost, high scanning and random access capability, tactility, and perhaps most important of all—self-paced individual study. Tactility might be overrated as a factor but a microreader, to some students, is a book only in theory.

Even forty years ago Max Born's *Restless Universe* could excite many a young science student who would flip the pages and view animated sequential drawings along the margins: colliding molecules, precessing electrons, or radiating dipoles.

In the early 1950s the most significant innovation of printed matter for education was the emergence of the low-cost paperback book as a semi-expendable supplement to the hardbound text. Doubleday's Science Study Series, published in a dozen or more language editions, has enjoyed wide use especially at secondary levels.

Prentice-Hall's Momentum Series supplements physics education at the university level.

In the 1970s textbook publishing is subject to two constraints, aside from ever-present cost factors aggravated by the wide availability of photocopiers. These are:

(a) *The need for flexibility and revisability.* Many books are outdated almost as they are printed, because of the rapid expansion of information and its short half-life [74]. Books will be published at a far faster rate and at lower cost with the aid of computers with optical character recognition (OCR) input that can scan a typewritten or even a handlettered manuscript in a few seconds and translate it into computer language for subsequent printout. Other books are printed by photo-offset methods to save time and cost.

(b) *The importance of format.* The success of magazines such as *Scientific American* and *Psychology Today* appears to be due as much to their unusual and attractive format as to their subject-matter and writing style. Large clutter-free diagrams that focus on the essential principles, multiple colours that discriminate concepts, graphs that reveal their content at a moment's inspection are part of the modern publisher's art, for communication without oversimplification is indeed an art.

2. *Other printed materials.* (a) *Fragmentation.* As information becomes more and more fragmentized and interrelated, textbooks may indeed be replaced by multimedia packages, consisting essentially of printed materials but supplemented by holograms, audio and video cassettes, thin plastic recording discs, etc. Materials that are supplied by the Open University have unusually clear and attractive multiple-coloured formats that add to the stimulation and motivation of the students, packaged in kit form for flexibility and convenience.

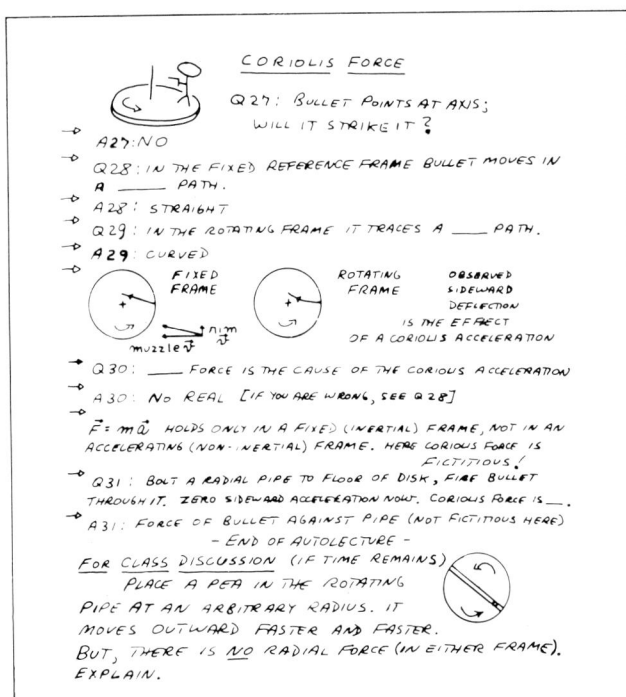

FIG. 2. Micronote in 9x areal reduction.

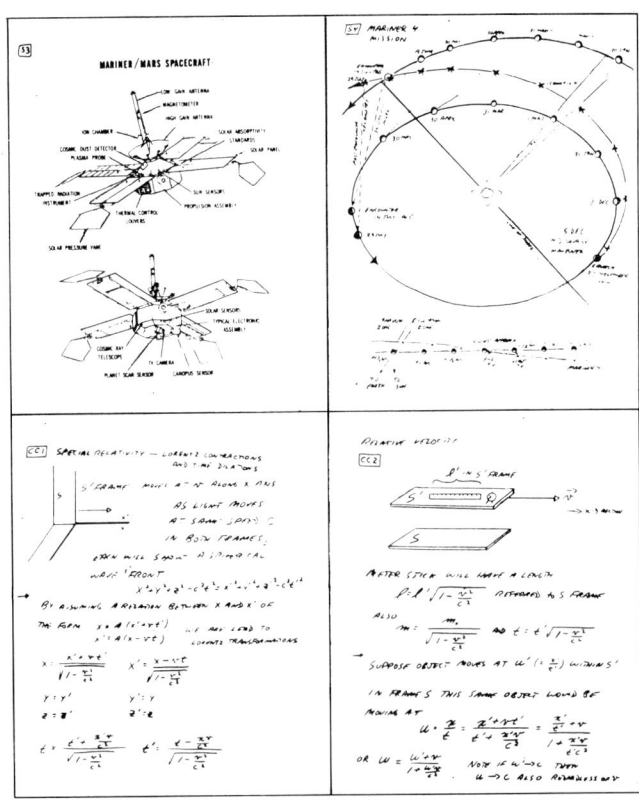

FIG. 3. Micronote in 25x areal reduction.

(b) *Micronotes.* The reducing Xerox copier and new low-cost techniques in precision offset printing have made it feasible to print reduced-size copies of learning materials. These may consist of handout duplicates of overhead transparencies, reduced to 9 or 16 per page. The smaller reduction is uncomfortable to read with the unaided eye if the information is typed and barely within the range of hand-lettered material. Experiments with the use of micronotes [75] have indicated that students find the requirement of a reading glass to be unacceptable; nevertheless, if some notes are available merely to be 'a last-resort' aid to students, a teacher may feel format to be a consideration that is secondary to expediency. In very large classes requiring the collating and handling of tens of thousands of sheets of paper, micronotes printed on both sides of a sheet, 32 or 50 frames per sheet, could prove a blessing, even with a magnifier handout. Recently, the American Institute of Astronautics and Aeronautics appealed to its membership to consider the acceptance of 'mini-printed' technical articles, because of the great cost saving [76]. Oxford University Press has just published a similarly printed two-volume edition, complete with magnifier, of its thirteen-volume dictionary.

(c) *Hand lettering versus machine printing.* There is little question that a neat, clear format is more satisfying than a crude fuzzy one, as anyone who has studied with his professor's mimeographed notes prior to the publishing of his textbook will agree. Nevertheless, books like the semi-handwritten *Nuclear Physics* notes of Enrico Fermi (edited by his students, J. Orear *et al*.) managed to convey a certain personal bond from author to reader, not unlike the inspiration conveyed by a lecturer in an auditorium, or by the intimate close-up image of a television speaker, although the human receiver in all cases may be anonymous.

The criterion for transmission of printed informational cues with a high reliability factor is not necessarily that the format be machine-like but that the message be impeccably clear and neat in the absence of distractions ('noise') of any type. This would apply as well to the production of overhead transparencies [77].

If this rule could be stated as a postulate, as yet unproved, one could then enjoy the convenience of flexibility and revisability in learning media where sharpness and clarity of structure and format can be attained at low cost and with modest effort, yet with no loss of effectiveness.

III. Media selection criteria

A. *Learning theory*

1. *Student input modes.* It is impossible to establish appropriate criteria for media selection and design without being guided by student attitudes that currently exist. Berman has attempted to give a profile of today's young students in terms of a set of aspirations that one would expect to correlate with motivation [78].

(a) *Identity.* The search for self-confidence and recognition in relating one's goals to those of society.

(b) *Understanding.* The search for wisdom and compassion through human contact and for a common bond between the individual and nature.

(c) *Stimulation.* The search for new sense impressions, new relationships, and new experimental approaches.

(d) *Freedom.* The search for the right to choose among various possibilities, and for the awareness and acceptance of the consequences of the choice.

(e) *Commitment.* The search for a role to play, for intensity of purpose, and for critical insight.

Margaret Mead's well-known thesis [79] is that no generation has ever experienced such rapid changes, in communication, the definition of humanity, the limits of the explorable universe, and that the elders, holding the seats of power, are like immigrants in time to a world the young have always known.

William Glasser feels that mankind is entering the fourth of the anthropological phases which he lists as [80]: (a) *Primitive survival society.* Thorough lack of co-operation between individuals struggling against a hostile environment, lasting three-and-a-half million years. (b) *Primitive identity society.* Co-operation within the group, identification with rituals, symbols, and religion, duration: half-million years. (c) *Civilized survival society.* Development of agriculture, conflicts among groups for land acquisition: beginning five to ten thousand years ago to present century. (d) *Civilized identity society.* Role-dominated, self-expressive, individual identity within the group; beginning stages: mid-twentieth century.

The effect of mass media on young people in displaying departures from the ideals and norms of society long before they have been able to make value judgements based upon reflection and understanding has led to the anti-intellectual anarchistic radicalism of the sixties [81]. But on the other hand, a curiously independent, open-minded, pragmatic, optimistic, and potentially creative point of view has developed [82] which, if nurtured, could lead to progress of a better kind.

The effect of these cultural and attitudinal changes upon media style can be seen at once in the new cinema, in a film like *2001: A Space Odyssey* for example. What it is lacking in plot, that may be as incomprehensible to Arthur Clarke and Stanley Kubrick as it is to the audience, is more than compensated for by stimulation, holding especially young people suspended for hours in sheer fascination with the medium, if not the message [83]. It can be seen in commercial television. E. L. Palmer made an 18-month study of children's TV viewing habits: their attention span, areas of interest, eye movements, etc. He and his co-workers, as consultants to the successful children's programme *Sesame Street,* discovered that continuous variety, rapid cuts back and forth between live action and animation, is the style and programming demanded by young people [84]. NBC Vice-President Paul Klein maintains that for young people image is more important than plot: 'There is no need to tell a story with a beginning, middle, and end. They care about people doing things, and all at once.' One particularly successful programme, called *I Spy* depends for its wide appeal not on plot, 'admittedly silly or non-existent', but on the interesting and warm relationship projected by the two principal actors [85].

The current attitudes concerning media and the way in which they stimulate students cannot be ignored by educators without some risk in the loss of learning effectiveness.

2. *Modern learning theories.* The development in this subsection considers learning theory with respect to the role that media can play. Learning is viewed in terms of a cycle consisting of three major phases: (a) the *stimulus* or display phase, (b) the *perceptual,* cognitive, and reactive phase; (c) the *cybernetic:* evaluative, feedback, and regulatory phase [86]. The domain of cognitive psychology appropriate to the study would include such behavioural processes as (a) the *sensory basis* of perception, (b) the *neurological and semantic basis* of understanding, and (c) the *recognition basis* of conceptualizing, all in relation to the motivational process akin to learning.

Maslow set forth a comprehensive theory of motivation in terms of six sets of needs, in somewhat hierarchical order, as important in energizing and directing behaviour [87]: (a) *physiological needs:* oxygen, liquid, food, rest, etc.; (b) *safety:* withdrawal from unfamiliar situations, preference for routine and rhythm rather than disorder; (c) *love and belonging:* desire for affectionate relations with people in general and for group identity; (d) *esteem:* seeking recognition as a worthwhile person; (e) *self-actualization:* need to become in actuality what one is potentially; (f) *cognitive need:* desire to know and understand, to analyse and synthesize. It is of some interest to relate these basic drives with the extended aspirations noted in IIIA1 three decades later.

Melton has categorized seven prototype learning situations [88]: (a) *Conditioning,* such as blinking the eye in response to a signal; (b) *rote learning;* (c) *probability learning:* choosing a correct alternative from a set; (d) *short-term memory:* initial reception and storage of bit or word information; (e) *concept learning,* or learning object properties; (f) *perceptual-motor skill:* making tracking movements continuously; and (g) *problem solving:* discovering a principle that achieves a stated goal.

Gagné has delineated four important modern learning theories [89]:

(a) N. E. Miller's learning sequence may be stated as (i) *motivation,* aroused by, for example, effective use of learning media, built upon the prior experiences of the learner, (ii) a *stimulus:* a cue, discriminated, for example, by colour contrast, (iii) *response:* involvement with the acquired knowledge that may comprehend application or transformation, and (iv) *reward:* the receiving of some satisfaction, preferably immediate, related to the aims desired [90].

(b) B. F. Skinner's learning theory, substantiated by his operant behaviour experiments with animals, resembles Miller's. It emphasizes (i) *shaping,* relative reinforcement of motor responses desired in the learner, (ii) *successive approximation* of stimulus control in which a response, initially shaped by prompting, continues to be given after a prompt has been faded, and (iii) *chaining,* whereby a learning sequence is composed of elements, each of which is dependent upon the one preceding it [91].

(c) Gagné emphasizes (i) seven *distinctive conditions for learning* [92]: signal learning or classical conditioning, stimulus-response learning, motor and verbal chain learning, multiple discriminations, concept learning, principle learning, and problem solving. These specific conditions extend beyond the general more familiar conditions of learning such as contiguity, repetition, and reinforcement; (ii) *cumulative learning*: the principle that new capabilities are built upon certain prior capabilites which must initially be recalled by the learner (e.g., one cannot add fractions without first mastering basic mathematical operations on integers).

(d) Ausabel's theory [93] is based on the principle of: (i) *Subsumption,* or the evolving of a new set of meanings following the subsumption of a new idea into a related structure of already existing knowledge [94]. (ii) Provision of an *advance organizer,* bearing a logical superordinate position to what will be learned in order to place the learner in the proper framework for learning. (For example, before proceeding to develop satellite orbital equations, a student would be primed by being told that the basic subject is the motion of objects under the influence of gravity. An overhead projection of the basic structure: the Newtonian formulation and the Keplerian laws might introduce him to, or remind him of, the plan of attack.) (iii) *Progressive differentiation* of content, where general ideas are presented prior to specific and detailed ones (e.g., the slide projection of photomicrographs of cells and cell groupings before concentrating on the nucleus protoplasm, etc.). (iii) *Consolidation,* the mastery of material upon which understanding of new material is contingent. It is identical to Gagné's cumulative learning (e.g., mastering the concept of torque before analysing the gyroscope). (iv) *Integrative reconciliation,* where newly introduced ideas are interrelated with old ideas and analysed for similarities and differences (e.g., electromagnetic waves using springs, ripple tanks and related to water and sound waves single-concept films).

Ausabel identifies four learning categories: (i) *Meaningful reception* learning: new material, logically organized, is presented in final form and the learner then relates it to his existing knowledge. (ii) *Rote reception* learning: material is presented in final form and is memorized. (iii) *Meaningful discovery* learning: the learner arrives independently at the solution to a problem and relates it to his existing knowledge. (iv) *Rote discovery* learning: the learner arrives independently at the solution to a problem and commits it to memory.

In addition to the theorists examined by Gagné, a media specialist would be well-advised to study the work of the two leading cognitive psychologists, Jean Piaget and Jerome Bruner.

(e) Piaget sequences learning stages chronologically according to the average age of a growing person [95]: (i) *The sensimotor period,* from 0-2 years, is the beginning of organization of visible images and controlled motor responses. (ii) *The preconceptual period,* from 2-7 years, is the interval of categorization of objects, when discriminations are formed. (iii) *The concrete operational period,* from 7-11 years, is the time when internal actions and perceptions are organized into logical operational systems. (iv) *The formal, or hypothetic-deductive operational period,* from 11 years to adulthood, involves abstract reasoning, logical inferences, and the understanding of causal relationships.

(f) Bruner emphasizes three ways of knowing [96]: (i) *Enactive,* or acting on the environment, corresponding to Piaget 'i'. (ii) *Ikonic,* or sensing the environment, corresponding to Piaget 'ii'. (iii) *Symbolic,* or acting through language on the environment, corresponding to Piaget 'iii and iv'.

Throughout his works Bruner emphasizes the importance of structure, the relationship of things: 'The teaching and learning of structure, rather than simply the mastery of facts and techniques, is at the center of the classic problem of transfer' [97]. '... the more one has a sense of the structure of a subject, the more densely packed and longer a learning episode one can get through without fatigue' [98]. 'Understanding of structure enables the student, among other things, to increase his effectiveness in dealing intuitively with problems' [99]. There is a basic consistency here with Ausabel's emphasis on an organizer and progressive differentiation of content.

In his thesis Gagné believes that current learning media—apparently media that are externally produced such as textbooks, films, or filmstrips—do not reflect all these current learning principles. If this be true, the teacher, conscious of them, could be in a position to develop his own media content along the lines suggested in part IV below, or arrange the learning environment so that the student may develop his own learning strategies.

This subsection would be incomplete without considering an area of learning theory that is especially important to science educators: scientific discovery. The ability to make intuitive leaps, to suddenly grasp hold of an integrated rational whole out of scattered parts and produce that 'Eureka feeling' of which Koestler writes [100] is very likely learnable. The extent to which it is inborn is, at any rate, open to question.

For a foundation in learning theory we would have to go back to the Gestalt psychologists, starting with Max Wertheimer [101], who maintained that understanding develops through the study of organized wholes rather than collections of parts, and the field psychologist Lewin [102], who emphasized the importance of the learner's whole experience including his total environment, his geographical/behavioural 'field'. Or the emphasis on 'latent learning' by Tolman [103], who believed that learning that takes place incidentally during reading, conversation, watching television, etc., stored deep in the inner experiences, will suddenly spring forth at the right moment when it is needed.

If these ideas are true then the great emphasis placed on the prepared environment in early childhood learning by the Montessori psychologists and educators will have a crucial bearing on one's attitudes and development in later life. And, regardless of one's inborn 'genius' propensity for creativity and scientific discovery, the potential that lies within everyone could be brought out. It is not without reason that many Montessorians would consider the years from 2 to 6 to be more critical to a person's development than the years spent at the university. Thus, the media of early childhood, the cuisinaire rods, the sensorial objects, the 'golden math beads', are not only not to be ignored, they represent the true building blocks [104].

The 'father' of modern scientific pragmatism, Charles S. Pierce, maintained that every plank of science's advance was laid down by what he calls 'retroduction'— the 'spontaneous conjecture of instinctive reason'. Kepler's discovery of the laws of planetary motion offers one of the best examples of this [105]. The view is similar to Popper's [106]: '... every discovery contains an "irrational element" or a "creative intuition" in Bergson's sense'. But according to Peter Caws: 'In the creative process, as in the process of demonstration, science has no special logic but shares the structure of human thought in general, and thought proceeds, in creation as in demonstration, according to perfectly intelligible principles' [107]. There is indeed hope that modern teaching methods, by providing the best learning conditions, including the effective use of media, will move scientific discovery out of the area of mystery and into the more tangible area of operationalism.

B. *Operational objectives*

1. *Educational operationalism.* The specification of learning objectives in establishing course curricula is a relatively recent innovational approach, designed to provide a deeper scientific base in educational theory and practice. Although every course has at the minimum some expressed aims, albeit vague, inquiries into examinations and courses show marked discrepancies with stated aims. For example, science teachers maintain that a prime goal of the study of their subject is to have their students think as scientists; yet an investigation by the Institute of Physics in England into A-level papers in 1966 indicated that 85 per cent of the questions could be answered merely by rote memorization [108].

The interest in the specification of educational objectives reached an apogee with the publication in 1965 of the dual taxonomic categorization by Bloom and his co-workers [109], and has since shown no tendency to decline. The six hierarchical levels set down in their cognitive domain are: knowledge, comprehension, application, analysis, synthesis, and evaluation. Crucial to the application of their taxonomy are the testing procedures by which attainment of the specified objectives are evaluated. Far more difficult is the problem of testing attitudinal objectives, a subject of their second work, not only because attitudes have an intellectual component that is difficult to differentiate, but because, by their very nature, affective values do not lend themselves well to objective assessment.

This less-than-two-decades concern with what we might describe as 'educational operationalism' is directly derivable from the pragmatic orientation of twentieth-century physics which had its beginnings in the 1870s in the early writings of Charles S. Pierce. Pierce's work on scientific pragmatism was not only the precursor of the religious pragmatism of William James and the moral pragmatism of John Dewey, both of whom played significant roles in modern educational theory, but led the way to the log'cal empiricism of the Vienna Circle that forms the basis of all of modern physics, and indeed all of modern science.

2. *Formulating educational objectives.* Learning objectives are written for four main purposes [110]: to (a) inform and influence citizen groups; (b) provide institutional guidelines for making decisions about broad educational programmes; (c) provide guidelines for

teachers in making decisions relating to general learning programmes; and (d) provide explicit information about content in instructional materials, particularly programmed materials, and examinations.

The semantic distinction to be made between an aim and an objective is that an aim is the general statement of intent that provides the direction to a teaching programme, whereas an objective concentrates on a particular point in that direction. An aim expresses why a particular topic is being taught while an objective defines what specific learning values will have been achieved by the teaching process.

The strongest argument for specifying learning objectives came from educators concerned with assessment [111]. For other educators, prime importance attaches to the relationship between teaching and learning methods and defined objectives [112]. Finally, others, especially students, who place emphasis on communication are particularly concerned with clarification of objectives.

Bloom's hierarchical model recognizes three classes of objectives: cognitive, affective, and psychomotor. The cognitive levels already noted proceed in pyramidal form. Educators concerned with the ends of education attach greatest importance to the higher levels, the levels of the researcher, in contrast to the lower levels, the domain of the skilled professional. Bruner [113] emphasizes the importance of a cognitive-structure link in teaching that extends through all levels. This presumably requires the continuing education of teachers and the development of complementary appropriate teaching media. Although this approach has received a positive reception in secondary education, it has only very recently been accepted at higher educational levels. (For example, the Institute for Studies in Higher Education of the University of Copenhagen was established in 1970 for this end.)

The five elements of Bloom's affective domain are: receiving or attending, responding, valuing, organization, and characterization by a value or value complex (into an internally consistent system as an essential element of an individual's character). We have noted the presence of cognitive linkages to the affective elements. The reverse also is true, for example, an understanding of quantum mechanics involves not only mastery of general principles and problem-solving methods but the development of positive attitudes toward inquiry, independent thinking, etc. Not only knowledge of a subject, but attitudes that a scientist has about the workings of nature and the ability to discover laws and develop mathematical insight is expected to be transmitted by a teacher to a student.

Operational objectives are also called behavioural objectives because their assessment is measured directly in terms of a student's behaviour, the 'pointer readings' of operational physics as applied to education. In Mager's well-known thesis [114] a behavioural objective is characterized by (a) delineation of a specific type of behaviour that will be accepted as evidence of attainment, (b) description of the set of conditions under which the behaviour is expected to occur, and (c) a quantification of learner performance that will meet conditions of acceptability.

From the point of view of science education, it is important to be clear on the distinction between Bloom's form of taxonomy—a classification scheme that emphasizes the development of intellectual skills upon appropriate knowledge of a given content area, that is, a content model—with a process model that emphasizes objectives [115]. Such a process model was given by Fenton [116]. Although it was developed for a social studies curriculum, it is directly transferable to the physical sciences: (a) Recognizing a problem from data, (b) formulating a hypothesis, (c) recognizing the logical implications of a hypothesis, (d) gathering data, (e) analysing, evaluating, and interpreting data, and (f) evaluating the hypothesis in the light of the data.

Göte Klingberg of the Gothenberg School of Education presents a number of classification schemes as alternates to Bloom's. He then cites examples of verbal forms that may be employed to assign descriptions of educational objectives. He lists six aspects: (a) the reproductional: e.g., enumerate, define, describe; (b) higher cognitive: e.g., discriminate, discuss; (c) emotional: e.g., pleased with, indignant at; (d) conative: e.g., interested in, strive towards, rejects; (e) creative: e.g., proposes, improvises; (f) functional: e.g., accepts, is active in. As one example he chooses mathematics-oriented goals: under (a), the reproductional aspect, one could, for example, define the concept 'per cent'; (b) choose a method of calculation suitable for a specific purpose; (c) be pleased to have solved a mathematical problem; (d) be interested in mathematical solutions; (e) find several ways of solving mathematical problems; and (f) make use of percentages in price comparisons during purchases.

The 'contract approach' illustrates the seriousness

with which educational objectives are now taken. Here a student and faculty adviser prepare a formal written document to guide a student toward clearly specified academic goals [118]. Schure emphasized the importance of indicating the precise stipulation of educational objectives within a systems framework to develop 'approaches that combine learning, faculties, materials, and technological innovations, under controlled conditions, so as to achieve optimum learning conditions' [119].

C. *Teaching modes and media selection*

In this section the various strategies in which media are used are described, and the relationship between the spoken and written word is considered together with the problem of reinforcement. The section concludes with Kemp's decision-point diagrams reprinted from his article, 'Which medium?'.

1. *The teaching function.* The principal function of a teacher is to arrange the proper environment for learning to take place. According to Turner there are three attributes of teaching [120]: (a) a dyadic is present—a 'reciprocal, interdependent responding between at least two persons', one holding greater authority and control of reinforcements; (b) a gap is present between the performance standard of the teacher and the performance of the student; (c) the teacher selects the responses which he implicitly or explicitly predicts will close the gap. If the prediction is true, the teaching eventuates in learning. As for empirical studies which have attempted to prove the efficacy of one teaching method over another, we have already referred to the work of Dubin and Taveggia, McKeachie, and Costin, and we could have added McKeachie's more recent works [121] together with that of Beard [122] and McLeish [123] and still fail to come up with confirmed conclusions. In his *Research in teacher education*, B. Othanel Smith points out that '... there are few, if any, skills of teaching whose superiority can be counted as empirically established. All approaches, except correlational studies, have yielded knowledge of little worth to teacher education. And even correlational studies give only rough approximations ...' [124]. In *Teaching and learning,* MacKenzie, *et al.,* outlined the functions required of teaching: (a) presenting information, (b) communicating objectives, (c) structuring the field, (d) explaining difficult concepts and problems, (e) motivating the student, (f) developing critical thinking, (g) developing alternative reference frames, (h) altering attitudes, (i) encouraging originality, (j) developing self-evaluation, and (k) developing problem-solving abilities [125].

2. *Lecturing.* Contrary to popular opinion from both educators and students the lecture as a mode of teaching is here to stay. Its advantages over reading, its chief competition in the teaching/learning arena, are: (a) the inspiration it can offer through a living person [126], who, (b) by facial cues, nuance in voice, and the spontaneity and potential for answering questions and otherwise adjusting instantly to student feedback, can establish a learning environment that books cannot duplicate; (c) as reading is a private affair, students prefer the variety of a group encounter which attending a lecture may afford; (d) the modern lecturer has a battery of multimedia devices at his disposal from colour-video monitors to wall-to-wall overhead projections that can stimulate the most unmotivated student to make each lecture an unforgettable experience (and sometimes does).

The most serious disadvantage of lecturing is its dependence on the less effective mode of auditory linear verbal stimuli [127] with often inadequate visual reinforcement. J. R. Hartley points out that the amount of material usually contained in the hour lecture is simply 'too much for the student to handle even if he is a competent note-taker' [128]. Reading is all visual, randomly scanned material, only linear in part, with innumerable varieties of repeat modes and learning sequences. This is a degrading factor, unfortunately, for the highly structured sequence of the lecture is designed to prevent this very random selectivity of the unknowledgable student. On the other hand, the lecturer too could be a degraded source if the hand and facial gestures, originally information cues, were to degenerate into distracting mannerisms.

3. *Seminar/tutorial modes.* Probably the most widely quoted maxim in all educational literature was uttered by the most unlikely of persons, General Garfield, in the most unlikely of places, a dinner of Williams College alumni on 28 December 1871 [129]: 'Give me a log hut, with only a simple bench, Mark Hopkins on one end and I on the other, and you may have all the

buildings, apparatus, and libraries without him'. There is no doubt that the small group—either the 'student-centred' tutorial of up to four students [130] or the 'subject-centred' seminar of five or more—offers the unique educational values of group encounter that no formal lecture can duplicate.

The small group offers the potential for free discussion with equal participation rights, an open view of the methods of reasoning or judgement rendered by the student, the optional relinquishing of authority by the leader (the tutor or seminar moderator), students in learning dialogue with each other, and a forum for a critique of the structured material in the textbook and lecture. For the somewhat passive student it offers a clear outlet. If the group is too large the same values would pertain by forming a subgroup or 'buzz session' where a more passive member is less apt to be constrained by inhibitions.

The small-group interaction with its emphasis on active engagement rather than passive listening, a key educational value, has led to the view that seminars are indeed 'more important' than the lecture [131]. According to Costin's recent review of the findings of numerous research studies, discussion may be superior for problem solving but not necessarily an improvement over the lecture for the acquisition of facts and principles [132], a conclusion in general agreement with McKeachie [133]. The seminar/tutorial can indeed be time-consuming, and knowledge acquisition in a completely unstructured mode can also be expensive, a consideration that cannot be ignored.

The obvious medium to use in the small group is the overhead projector, completely flexible and completely responsive to feedback. It displays ideas, presents outlines, summarizes points, all from a sitting position with the room lights on and has retrieval and duplication capability. For the recording and playback of discussions for later analysis and synthesis, as well as for teacher-training diagnostics as noted previously, the audio tape recorder and video recording chain can be invaluable.

4. *The laboratory.* Far distant from the lecture where solutions to problems are presented as factual information, or the seminar/tutorial where problems are presented and solutions are hypothesized, the laboratory is the place where problems are presented and solutions are conceptualized and validated. Therefore, the laboratory is the most time-consuming, most expensive, and very possibly the most gratifying and important area of learning activity. The degree of importance is related to the specified learning objectives that are desired by the learner and teacher, which hopefully will be identical. The group-dynamical advantages, with students learning from each other, is at once apparent. Yet the opportunities for independent creativity are limitless.

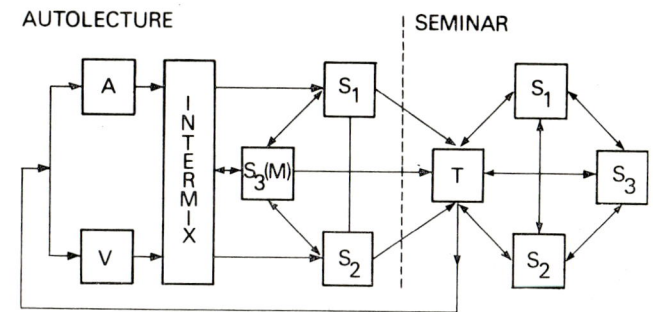

Fig. 4. Seminar/autolecture communication model
S: Student
M: Moderator (here S_3, but may be a teaching assistant)
T: Teacher
A: High fidelity audio cassette system
V: Wide-screen overhead projection system

Laboratory media is, of course, generally tactile, supplemented by computer displays, printouts, meter readings, etc. Albert Baez [134] has successfully used an audio-cassette/overhead-projection medium—sometimes called an autolecture, but better called an 'autolecture medium'—for cueing students engaged in laboratory activity. An autolecture, to be discussed in IVC, is a 'seminar/tutorial interaction activated by a structured expository/programmed-learning mix through a cybernetic high-fidelity medium, the audio-cassette/overhead-projector, operated in unlocked synchro mode' [135]. The use of the single word is preferred so long as its meaning is understood.

5. *Intermedia reinforcement.* Information theorists have been concerned with decoding, encoding, and processing the linear and non-linear content of visual-auditory-tactile media prior to the development of a similar interest in the subject by educators. Many teachers are surprisingly unaware of the effect of bitrate overloading, intermedia and intramedia interference, signal-to-

noise ratio, etc. on learning. Many students, not surprisingly, are not.

The question of *attention* was a principal interest of James: 'Focalization, concentration, of conciousness are of its essence. It implies withdrawal of some things in order to deal effectively with others' [136]. Broadbent's filter model of information transfer [137] referred to earlier in terms of single-channel effectiveness of multi-channel media is based on his experiments in which different sets of digits were presented to subjects dichotically for later recall. He found that the subjects tended to organize the information flow by ears, recalling correctly from one ear while blocking information from the other. He proposed that the brain contains a selective filter tuned to accept the desired message one channel at a time and reject all the others. Other researchers, particularly Treisman [138] believed the model to be an oversimplification, and that under certain conditions multiple channels can act as reinforcing agents.

For the teacher it is a genuinely challenging task to perform the sequence properly: (a) project a detailed visual stimulus, (b) remain silent, (c) begin talking, (d) extinguish the visual stimulus. The sequence is adjusted intuitively to the class average. Those students with a blockage problem, or those who are less motivated, or simply slow learners, or those who possess a limited attention span, or who have other things to think about at the moment, etc., will produce instant or time-delayed feedback information to the teacher that may or may not act as a correction signal to the teaching process (more usually it will be impressed as a stigmatic device upon the learner). This simple cybernetic-loop model from receiver to communicator represents the essential distinction between communication and information transfer, or the distinction between teaching and telling.

The question of interchannel interference between competing media modes has received considerable attention in recent years [139]. Otto [140] and Bourisseau [141] found that pictorial and verbal stimuli elicit different responses in learning, raising the possibility of interchannel and even intrachannel interference rather than reinforcement. This view was supported by Van Mondfrans and Travers [142]. Travers also maintained [143] that simplified representations (e.g., diagrammatic) may be superior in terms of learning effectiveness to realistic (veridical) ones. Diagrams, of course, may concentrate on the essential points and omit distracting details. This would lend added support to the widespread diagrammatic use of the overhead projector in preference to veridical slide projections.

The determining factor in the research on intermedia information transfer as to whether interference or reinforcement will occur is relevancy. According to the cue summation principle, learning will increase as the number of available cues increases [144], but they must be relevant cues. There is little question, on the basis of available evidence, that a reasonably sophisticated learner (e.g., a university student) will accept auditory verbal cues that are related to visual pictorial material, together with some degree of reinforcement provided by the addition of relevant visual verbal cues [145]. The variability that is possible in detail, time allotment, and signal-to-noise ratio is so wide that in practice only careful monitoring of feedback by the teacher will insure a maximum degree of reinforcement.

In spite of the conflicting data in the early 1960s concerning the question of reinforcement versus interference in intermedia displays, we can cite the extensive studies of Hsia in providing some further element of credibility to the combined use of audio and visual channels in an intermedia mode [146]: perfect between-channel redundancy (identical information transmitted in perfect synchronization by auditory and visual channels simultaneously) eliminates possible interference effects and reinforces one channel by the other. If the amount of information is below the maximum capacity of the central nervous system, no between-channel interference can possibly occur, even if the information via the auditory and visual channels is entirely different. Hsia also notes the research of Mudd and McCormick [147]: provided that the information is related, auditory cues of various dimensions appreciably decrease the time involved in a visual search task.

6. *Media selection criteria*. Throughout this review we have noted the various uses of media under several learning and teaching conditions. Briggs has outlined a set of eight specific procedures in selecting media on the basis of prescribed course objectives [148]:

(a) Selecting and defining instructional objectives and expressing them in terms of expected behavioural outcomes.

(b) Sequencing the objectives in a manner such that component or prerequisite learning takes place prior to more complex learning.

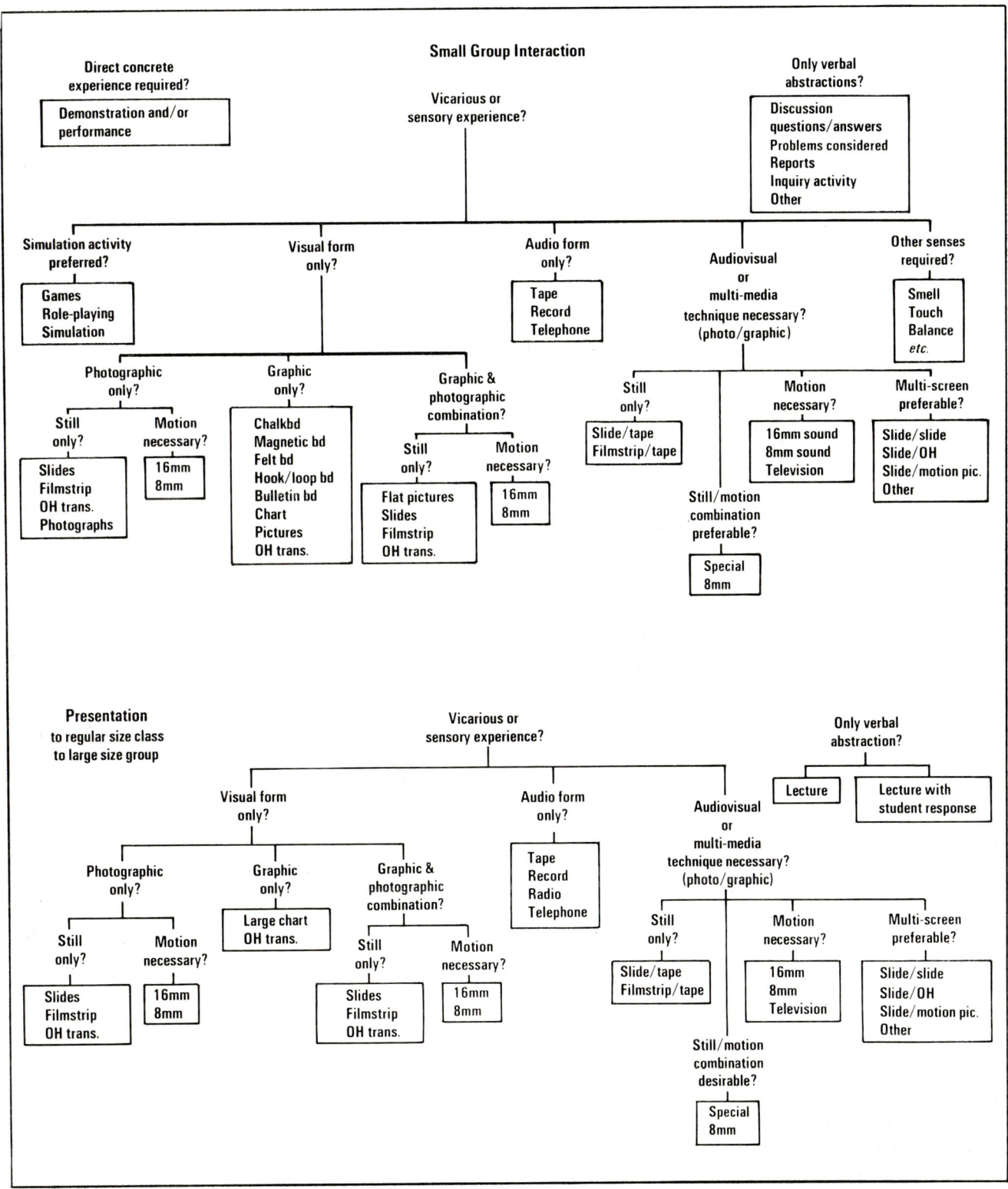

Fig. 5. Decision-point diagrams for matching media to subject content (from 'Which medium?' by J. E. Kemp, *Audiovisual instruction*, vol. 16, p. 32-6, December 1971, reproduced by permission of the author and publisher).

(c) Identifying the type of learning represented for each objective.

(d) For each objective listing the instructional sequence of events that would provide the appropriate learning conditions.

(e) Identifying the nature of the stimuli related to each instructional event (in terms of kinetic versus static, duration, etc.).

(f) Tentatively identifying the optimal medium that would present each stimulus.

(g) Reviewing overall objective sequences in order to make choices of media that would permit a presentation for a reasonable time interval before making a change to other media.

(h) Writing specifications in terms of instructions to the specialist who will prepare the media content (e.g., the programmer, script writer, etc.).

In his recent article 'Which medium?', Jerrold Kemp worked out a set of media decision diagrams where questions at decision points are asked based on the specific learning objectives intended, for example: is motion required; graphic only; photographic only, etc.? [149]. Two of these are reproduced in Figure 5.

The media taxonomy may provide some aid in working out appropriate media strategies, but a major constraint upon the teacher will always be availability. In science education PSSC study kits and the tremendous potential of related do-it-yourself media, from bicycle-wheel gyroscopes to plastic-comb electrostatic rods, the catalogue of single concept films, the drawing-pen-on-tracing-paper for overhead transparency masters, are all starting points to guide the teacher toward a rich source of display and communication.

IV. Basic learning systems

A. *Defining terms*

1. *System.* The term 'system' is generally defined as a set of interrelated and interacting units having some common properties. The components of a learning system, or the subsystem elements, all have the specified objective of working together in order that an individual who interacts with the system eventuates in learning. A slide projector, for example, could be a subsystem component in a learning system that includes the teacher, printed materials, a chalkboard, etc. Or the instruction can be systematized into a model composed of less tangible elements such as [150]: (a) specification of objectives, (b) selection of content, followed by (c) an assessment of entry behaviours, that feed into a five-element matrix: (d) determination of strategy, (e) organization of groups, (f) allocation of time, (g) allocation of space, (h) selection of resources. When these are accomplished there is finally, (i) an evaluation of performance, followed by (j) an analysis of feedback, to return to 'a' and 'b'. This would be a systems approach appropriate for a classroom. But at the other extreme Jantsch has envisaged a multilevel multigoal hierarchical education/innovation system for all of society in which the university would develop inter- and trans-disciplinary links between various levels [151].

2. *Basic learning system.* When we speak of a *basic* learning system we are restricting ourselves to a system that essentially is teacher-managed, a 'low-technology' system in the sense previously defined. The teacher is an essential element of the set which might consist not only of hardware components, but of selections, allocations, strategies, evaluations, all working together to accomplish the selected end. To illustrate the workings of such a system we might construct communication models, incorporate feedback loops, indicate how output correction signals are applied to regulate the input, etc.

3. *Intermedia/multimedia.* Meisler [152] draws a distinction between the similar terms, multimedia and intermedia as follows: a multimedia configuration tends to use a great deal of equipment to immerse the participants in a complex media-created environment. In one extreme case it would be the form of the Cinerama tri-screen multichannel sound configuration, the motive of which is to deeply immerse the individual. The Hayden planetarium has been doing this for many decades.

An intermedia configuration is described as 'a combination of common audiovisual media ... to present tightly programmed units of instruction'. In this sense one might consider it a device to provide redundant information cues at minimal interference, conveying multiple facets of an object or subject that itself may be identified with several sense stimuli.

The multimedia configuration is not as concerned with information, the message, or the level of redundancy, as it is with the medium. Indeed, interference might well be used as a source of stimulation. It might have educational values—the day will be remembered a year from now, perhaps also the chemistry experiment —but they are not nearly as well defined as in the intermedia combination.

B. *Linear intermedia systems*

1. *Static-projection/audio.* A learning system in which the essential element is a multi-channel media device with fixed sequential display is a linear intermedia learning system. An example is the filmstrip visual display and cassette audiotape combination. It is linear by the fixed sequence, although the frames in the visual projections are, of course, nonlinear, unless they were digitally encoded, as, for example, a representation of printed pages to be read by a group on a screen. It is customary for such a system to be in locked synchronization, where an inaudible pulse is magnetically encoded in the alternate channel of the tape to control the visual sequence.

A similar combination is the slide-projector/audio-cassette which can also be set in locked-synchro mode. Its inflexibility is only slightly better than that of the filmstrip combination in that, conceivably, a similar but updated slide can be interchanged with a former one without disturbing the system provided the audio information is relevant. This added capability is purchased at higher cost relative to the filmstrip/audio-cassette. A still more flexible arrangement is the newly developed 3 M slide projection system in which a complementary audio track is included on the border of each slide. But again flexibility is purchased at high cost. Of course the operation of these combinations can be modified by the inclusion of the teacher in the system to provide a cybernetic loop.

2. *Kinetic-projection/audio.* The widely used 8 mm and 16 mm sound film projection systems and video projection are almost identical examples of non-cybernetic teaching systems which present linear information. In the case of the video, however, if the tape is teacher-produced there is some degree of flexibility, as noted above in IIE, in that insertions may be made on feedback advice, but again at high cost. EVR and other video cassette systems: the RCA holographic vinyl-tape unit, and the Telefunken-Decca plastic disc, are all, of course, non-cybernetic. However the Sony video tape cassette, like other magnetic tape units, has revisability capability. In still-frame random operation mode EVR is essentially a filmstrip presentation without sound, and would perform the same function as a subsystem element, such as a slide projector, in an ordinary teacher-managed system. The single-concept 8 mm film with sound track is basically non-cybernetic; however, its short duration and random selection out of a catalogue makes it an ideal subsystem element in any cybernetic learning system. Short duration sequences in the above-mentioned media combinations could have similar properties.

3. *Other linear intermedia systems.* Of the various combinations and possibilities in linear intermedia systems, one of the most successful is the type of media display to be observed in the Royal Danish College of Dentistry (Århus), where a student examines in detail the specimen collections aided by a portable audio cassette playback unit [153]. The system is not so rigid as to preclude the possibility of revision of either the specimens or tape.

The concept of the use of real objects in audio combinations is basic to the immensely successful audio-tutorial carrel-based system developed by Postelthwait [154] and extended to virtually all the other sciences [155]. The great advantage of tactility pertains here and the programming includes the projection of slides, time-lapse single-concept films, etc. within the carrel framework. The extent to which the system responds to feedback can be made as flexible as desired, especially as the system extends beyond the privacy of the carrel into the tutorial or seminar modes.

C. *Cybernetic systems*

Of the three non-computerized cybernetic learning systems that have been of some interest in scientific education and have gone through several years of trial runs—the audio-tutorial, the Keller plan, and Seminar Autolecture—only the last two will be discussed below.

1. *The Keller plan.* This learning system was initially developed by Fred Keller and J. G. Sherman in psychology [156], extended to physics by Green of MIT in

1969 [157], and it is reported in use in several scientific fields [158]. It is described by Green as a 'self-paced, student-tutored, mastery-oriented' system. Material is divided into units equivalent to about a week's course work. Each consists of instructor-generated reading materials: a study guide, with an approach plan based on stated operational objectives, and four short equivalent examinations. One of the four will be given to the student randomly at his request and evaluated on a pass/fail basis by an undergraduate tutor, subject to teacher review. The tutor is responsible for individual dialogue with ten students. The study guide will lead the student to original articles, take-home kits, single-concept films, etc. in a highly structured manner. Attendance at the occasionally given lectures is not obligatory.

The freedom of pace enables mastery to be an expressed goal of the system. It offers a teacher a range of experimentation and of educational creativity despite the greater demands placed on him in the form of writing the guides, as well as the inevitable self- and public evaluation to which he is exposed. Assessment appears to offer no positive confirmation of the efficacy of the system in terms of its cognitive value on the basis of examinations. Nevertheless, there seems to be a positive attitude developed by most students, beyond the range of what might be construed as placebo (or Hawthorne) effect.

2. *Seminar/autolecture.* Seminar/autolecture is a media- and socially-oriented self- and group-paced learning system. Its initial conception was by Berman who conducted courses in physics with it at Rensselaer Polytechnic Institute in 1966-69 and the Goddard and Kennedy Space Flight Centers during the same period in short courses in space physics and mechanics [159]. It underwent a major development during the summers of 1968-71 at Harvard University by Baez [134], who conducted a year comprehensive physics course for over one hundred students each summer within a seven-week period.

The system consists of essentially two phases of equal duration (approximately 45 minutes each), preferably, but not necessarily, contiguous in time: the autolecture and seminar. The autolecture (defined in IIIC4 above) is not viewed as a substitute lecture but basically a tutorial or seminar, partly structured in the sense that the audiovisual medium performs a controlled activation function.

The medium content is teacher-generated and the format is an overhead-projector/audio-cassette combination channelled through a high-fidelity sound system and projected on a large wall, possessing both multi- and intermedia attributes (see IVA above). This combination is chosen because: (a) it has high-involvement properties provided the teacher adheres to specified criteria regarding (i) noise content relative to signal in both information channels, (ii) maintenance of a certain 'personal posture' in microphonic delivery and transparency composition; (b) the teacher can generate media information with minimal expenditure of time and effort because: (i) channel is generated at separate times under excellent control conditions and without fidelity loss, (ii) format is stipulated as unrigid so long as clarity and neatness are not sacrificed; (c) the medium has high cybernetic properties; (d) the medium can be duplicated rapidly and inexpensively, the visual component by reducing reprographic devices.

The production of the tape follows that of the transparency masters. After some experience, it has been found that these can be put into final form in a relatively short time—to tape one side of a C-60 cassette can take less time than is required to operate it in class—provided the subject content is outlined in advance. The concentration of the teacher on a structured, error-free development planned in advance is considered a positive factor, although accomplished at the expense of his spontaneity which is transferred to the seminar. There is no lack of spontaneity within the autolecture group, however. Revisions in the transparency can be entered at once by means of a felt pen or in several seconds with the aid of a thermal copier. Tape insertions are more difficult to accomplish because of the linearity of tape content. The teacher is therefore advised to maintain an accurate record of tape-transparency information correlations by means of the tape digital counter. Line-operated cassette units are, of course, preferred and it is considered to be important that the signal be fed into a wide-range amplifier-speaker system (see VB). The medium content is an expository/programmed mix. The group leader, called a moderator (a communication model of the S/A system appears in Figure 4, page 123), controls the synchronization on feedback signals from the group. Responses to programmed learning frames, replies to examination questions, as well as other commentary, are submitted to the teacher in the form of carbonless paper duplicates as data to be used in revising the autolecture content. Thus there is an external cybernetic loop, as well as internal loops within the semi-structured autolecture, led by an assistant or student, and also within the unstructured seminar, led by the teacher, or, if this is impractical, on occasion by his assistant.

The autolecture/tutorial is a student- rather than subject-oriented operating mode, wherein students in groups of four or less may interchange tutor roles and interact with the medium and programmes in a manner that more closely resembles self-pacing. It can supplement the normal autolecture by providing a remedial function, similar to that of the autolecture in self-study, or carrel, mode. Even more than in the Keller plan, teacher-generated media content exposes the teacher to self- and public evaluation.

No definitive cognitive assessments under controlled conditions exist; however, there is a clear indication that students learn by this system, and the large majority of the many hundreds of students who have been taught in this manner have shown a favourable attitude which seems to increase with exposure. Of significance are the independent findings of Baez and Berman that students exposed to an autolecture medium indicate a strong preference for it over any of the forms of video display. Although the assessment may have an element of subjectivity that casts some doubt on its validity, the data is so consistently in favour of the autolecture medium, by factors often of ten to one, that one can safely assume this medium to be preferred. There have been no thorough objective formative or summative attitudinal evaluations made. The research direction is toward obtaining data in this area [160]. Wilhelmsen of Bergen University considers the S/A system as significant in the area of social-dynamic studies [161]. It has already proved to be of value by Baez [162] in teacher training by gaining group leadership experience with a media stimulus as activator, as well as for trainees who generate their own autolecture media content and obtain data for evaluation studies.

Probably the most significant feature of seminar/autolecture is its validation programme combined with its capability of responding to validation. The requirement that the medium be shaped to the student imposes an informal flavour on it—indeed, in theory, making it more acceptable to students and teachers at the outset —and sets forth procedures to enable the teacher to edit it frequently and conveniently. The categories of validation are: (1) direct responding, (a) orally in class discussions, and (b) by writing verbal replies, calculated solutions, etc. when requested, as in responding to linearly sequenced frames; and (2) completing a validation form at the close of each autolecture of the type illustrated in Figure 6, below. The teacher-producer of the autolecture medium is advised of the needed revision by analysis of the written material and communications from the moderator.

```
AUTOLECTURE VALIDATION FORM

Please read before beginning autolecture      ++ = excellent
Try to complete this form within three minutes  + = good
Add comments if you wish                        0 = neutral
Indicate symbols on right                       - = fair
                                               -- = poor

THE AUTOLECTURE MEDIUM
Aural format
    sound level uniformity
    background noise absence
    sound reproduction
    speaker's voice quality
    aural presentation in general

Visual format
    hand lettering
    diagrams
    color usage
    reinforcement without distractions
    overall visual style

Subject content
    general organization
    ease of comprehension
    aural & visual content well related
    speed of presentation
    illustrative examples

THE GROUP ENVIRONMENT
Moderator
    delayed spoken commentary where necessary
    avoided distractions and irrelevancies
    allowed sufficient time for responding
    encouraged relevant discussions
    maintained an efficient pace

Fellow students
    helped me understand
    did not slow my progress
    did not make me feel inferior

THE AUTOLECTURE
    the medium-activated seminar overall
    the medium alone
    learning of important topics
    attitude change toward subject at close
--------
    topics least important to me are _____
    topics most important to me are _____
    topics requiring expansion are _____
    I spent ____ ____ ____ ____ in prestudy of micronotes
            40 min 20 min 10 min 0 min
```

Fig. 6.

The requirement of logical structure, based on prestated objectives, and the combination of stimulus-response-confirmation-validation makes the autolecture as a whole a unique type of programmed-learning experience, whether or not the classical format of linear-frame sequencing is adhered to.

V. Practical utilization of learning media

A. Low-cost visual displays

1. Criteria. The recurrent theme that appears throughout this report is that the selection of appropriate media depends upon two general criteria:

(a) *The format should be matched to the subject matter.* The details of a rabbit esophagus requires a real-picture representation, which can be shown most conveniently as a paper photograph for individual study, or 35 mm slide projection for group or individual study. A blackboard drawing simply won't do. The impact of a bullet and light-bulb is not to be described orally by the teacher when a slow-motion film can describe it far better. Indeed, science education demands not only precision of thought and operation but also precision of display.

(b) *The content should be shaped to the student.* The educational designers are hard at work at the boundaries of technology to apply the computer to the task of shaping subject content to the needs of the individual. It may well be that developing countries may jump the gap directly into computer-based learning by means of the most economical road possible—a massive central computer feeding tens if not hundreds of thousands of separate terminals. Lack of money has always been the primary problem in relation to educational and other needs of developing countries. But the political-sociological-psychological problems created by computer-based learning may well be as great.

When one thinks of low-cost media that adjust themselves to student needs, such obvious devices as the blackboard, flannel board, and magnetic board come to mind. The teacher with chalk in one hand and eraser in the other, covering the four walls of the classroom with equations, which his students diligently copy into their notebooks, has always been a source of ridicule. Yet, the mere fact that the eraser is there to correct, or otherwise edit the display, partially on feedback from the students, places it far ahead of forms of instruction that depend upon stimuli alone, set in an unyielding format.

On applying educational technology to the problem of developing low-cost media that are responsive to feedback—that is, cybernetic media—two categories stand out: (i) printed materials; (ii) overhead projections.

The widespread and low-cost availability of the photocopier in recent years has made it possible to distribute expendable printed matter to individual students who otherwise must depend upon bound textbooks for their personal study. These distributed materials are responsive to validation through classroom experience and updated by the teacher one page at a time. Photo-offset printing techniques have improved so rapidly that it is possible to produce many copies at a few minute's notice, either fullsized or reduced, and at very low cost.

2. Practical features of a modern overhead projector. Probably the most important application of technology to education from the point of view of a personal low-cost device that is readily available and inexpensive is the overhead projector. Its distinctive advantages have been described in IIB. It must be borne in mind that only a very small amount of research has confirmed the efficacy of the overhead projector in classroom teaching [163].

In purchasing an overhead projector one should consider the following features listed approximately in order of importance.

(a) *Even illumination and sharp focus throughout.* This presumes that the screen is set normally to the axis of projection of the image to avoid the so-called 'keystone' distortion. There is no optical way of otherwise eliminating this distortion although it is possible to adjust the positioning of the transparency to maintain a focused image on a screen or wall that lies at an incorrect angle. If the screen is set at the correct angle, however, there is no excuse for lack of sharpness in any part of the image. Several manufacturers eliminate the plastic Fresnel lens entirely, which may be damaged by heat and humidity. Although the image of the Fresnel lens appears as a universal background to the projected transparency, it is symmetric and generally not objectionable.

(b) *Low cooling-fan noise level.* Plastic fans and mountings are noticeably noisier than the more costly metal and rubber-mounted units. The overhead projector model with convection-cooled light source mounted above the transparency (noted in IIB) is an excellent choice where absolute silence is essential, but will not provide as brilliant an image as the conventionally designed overhead projector.

(c) *Detachable stage for cleaning optical elements.*

(d) *Elimination of glare.* This can be achieved optically without the need for intervening colour filters placed on the stage or in front of the operator's eyes.

(e) *Adjustable distance between the light source and optical elements.* This will eliminate the colour-fringing that appears on the screen when the distance from the projector is extreme.

(f) *Low silhouette of projector.* To prevent obstruction of the screen in ordinary classroom operation.

(g) *Provision for rapid insertion of new bulbs.* Some manufacturers arrange for premounting of a spare bulb for immediate replacement.

(h) *Transportable.* Many units are mounted on rolling tables or are otherwise light enough to be carried by hand.

(i) *Large flat working area at stage level.* Some projectors are desk-mounted. In any event, it should always be convenient to instal a frame on the stage that will guide unmounted transparencies as well as support a paper mask to reveal only sections of a transparency at any one time.

(j) *Front accessible noiseless light switch.* Some units have a front-mounted switching bar that may be pressed with very little effort to permit rapid and convenient elimination of the projected image particularly when changing transparencies.

(k) *Two light-intensity levels provided.* This is particularly important if projection in a large room is desired.

Most overhead projectors make use of a halogen light source. The glass surface of the bulb should never be touched with bare hands. If it is touched, it should be wiped with a cleaning solvent such as alcohol. These bulbs have a lifetime of approximately one hundred hours, making it especially important to hold a spare bulb in readiness. Some manufacturers, however, claim to extend the bulb lifetime several times by proper design of the on-off switch.

3. *Overhead-projection technique.* It is no secret that many students and teachers, accustomed to the informal spontaneity of the blackboard, dislike the overhead projector. Others are enthusiastic and appreciative of the device. The distinction between the opposing groups can be summarized by a single word: technique.

Anyone who has been to a professional meeting in any field, sitting in an auditorium, while the speaker reviews his five-year research project in a ten-minute talk illustrated by twelve slides displaying apparatus, equations, graphs, curves, etc., will appreciate the problems of visual communication, or lack of it.

A student, placed in the analogous position in a classroom, will yearn for the slow and easy way of the blackboard drawing, one line at a time, as the hour draws on. A teacher, uninterested in sharing the light, or limelight, with projected images, will pride himself on the mark of his profession, chalkdust on his jacket.

What has the overhead projector really to offer the teacher and student? Most immediate is a display that is carefully prepared in advance, anticipating all the problems and pifalls that the student may encounter. The anticipation is based on the wide experience of the teacher with students of similar learning backgrounds, whom he presumably had later tested to confirm his belief in his teaching effectiveness. Of equal importance is the convenience of duplicating the display in its entirety, except for colour differentiations, for personal use by the students. But not to be overlooked is a third aspect, perhaps of greatest importance, realized by the teacher but given less attention by the student. It is the direct and convenient way in which revisions may be made to shape the subject matter to the students— merely by writing on the transparency with a fine-pointed marker.

Of course, one may prepare an overhead transparency completely by fine-pointed marker directly on plastic sheets, which are manufactured in A4 continental standard size, ten-inch square size, or in long rolls that may be attached to the stage of the projector. Far cleaner and more permanent is the preparation of transparencies by thermal copier from ink drawings on tracing paper. Corrections to these masters may be made with the same ease as erasing pencil drawings (which also may serve as transparency masters). This assumes that the ink is scraped away with a surgical knife available at most shops that sell drawing supplies. The drawing paper should have a glazed surface to permit fine-line writing without absorption of the ink, and to permit the erasures. Drawing pens of 0.8 mm point diameter give excellent results. These masters, in preference to the transparencies themselves, are more suited to the making of photocopies for distribution.

The third aspect, shaping the subject matter to the student by revision of content, emphasizes the overhead projector's basic superiority to the slide and filmstrip projectors, or even the video display. Of course, one may argue that here the overhead projector is no better than a blackboard. To compete with a blackboard, the overhead projector must rely on its two first mentioned features. Moreover, we must direct our attention to the practical possibilities of overcoming the immediate objections to overhead projection. If a student is indeed more receptive to the gradual display of a complicated drawing, how can this be simulated by an overhead projection? If a teacher prefers that the displayed material not compete with his own presence in the classroom, how can this problem be solved?

The latter problem can be taken care of merely by temporarily switching the projector off—a procedure that hardly can be duplicated with a blackboard display. As the room lighting level is generally kept high during overhead projections, the procedure offers no practical difficulties. To solve the former problem, however, requires some talent in communication; some of the procedures described in IIB are recommended. The reprographic capability of transparencies is an important aid in teaching. Not only is the detailed taking of notes, with its attendant distracting influence, alleviated, but the pre-distribution of transparency copies permits precueing of the subject matter to the extent where complex formulae and diagrams become less confusing when projected for the first time.

Below are some of the more important guidelines that are suggested concerning the composition and projection of transparencies.

(a) Concentrate on the essence or structure of the subject with interesting examples to illustrate the points. The details should be left to the student's textbooks or other printed matter.

(b) Lettering should be hand-drawn and unmistakably clear. Diagrams and verbal material should be balanced attractively. Drawings should be simple and somewhat abstract to clearly point out the essential ideas at once.

(c) Avoid machine-printed lettering or engineering drawings unless they are essential for clarity.

(d) Avoid the use of cardboard or plastic frames for each transparency. Instead, fit a transparency guide, made of rulers, for example, on the stage of the projector.

(e) Colour may be inserted directly on the transparencies with permanent fine-markers in rough shading, for discriminating features and adding esthetic interest. Avoid plastic colour sheeting unless it can be attached quickly and effortlessly.

(f) Separate a transparency into sections expressing separate ideas, and reveal only one relevant section at a time beginning at the top. Arrows drawn on the left can cue the operator.

(g) Produce a format and content that invites continual responding either by stimulating the discussions in class, or by requiring direct replies to questions or the completion of unfinished work such as labelling an unlabelled diagram.

These guidelines emphasize two points of view: (i) a student will identify with and be more interested in a personally made informal drawing accompanied by carefully but personally drawn lettering, than he would rigid machine-made ones. This is an unproved hypothesis for which there is some evidence and quite in line with McLuhan's conception of the casual and undetailed medium as sensitizer of involvement feelings. (ii) By avoiding a formal structure it is not only easier to devote a greater effort to the planned display, and to course planning in general, but equally important, it is easier to rework the material in accordance with student validations.

4. *Examples of overhead-projection transparencies.* If one were to attempt to develop a scientific point, for example, the observed level of radioactivity at a monitoring station following a nuclear explosion, how *not* to make a transparency is illustrated by Figure 7, page 133. The copying of curves and legends directly out of a journal may be acceptable for distributed printed materials, but not for projections. Printed materials are studied personally *at one's own pace*. Projected materials are studied together with others *at the pace of the group*. This is one reason why many lectures fail to teach when a speaker goes into detail and a large proportion of students cannot follow, that is, maintain the pace. There are, of course, sound reasons for this lack of comprehension, such as the failure to provide the students with a precued structure (note Ausabel's advance organizer in IIIA). It takes but a few minutes for the teacher to trace out the essential points on a transparency and produce a format that tests the comprehension of his students and encourages their involvement (see Fig. 8, page 133).

Figure 9 (page 134) illustrates a simple, easily drawn transparency, that makes clear, at a glance, exactly what an 'autolecture' medium is, in terms of its apparatus. Figure 10 (page 134) is far more complicated, but easily drawn, or in this case traced. It definitely requires precueing to save time in class. In his oral discussion the teacher would point out that although the planes of Mars's orbit and that of the earth are inclined to each other, still the Mariner spacecraft can

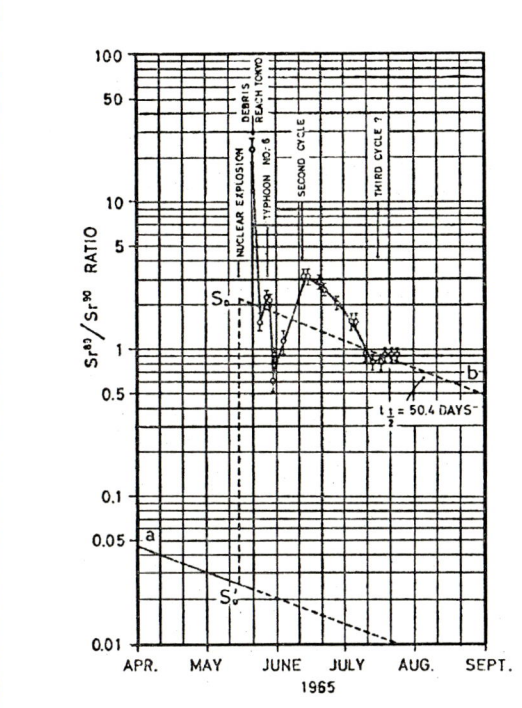

Fig. 1. Variation of the Sr^{89}/Sr^{90} ratio in rain after the Chinese nuclear explosion of 14 May 1965. The error limit indicated is due primarily to the counting statistics. The straight line aS_0' is the Sr^{89}/Sr^{90} ratio in the stratosphere prior to 14 May. The straight line S_0b indicates the estimated Sr^{89}/Sr^{90} ratio in the stratosphere after the second injection of strontium isotopes by the Chinese.

Fig. 7. A non-acceptable transparency. (From 'Global circulation of strontium after the Chinese nuclear explosion of 14 May 1965', by P. K. Kroda *et al., Science*, vol. 150, p. 1289-90, 3 December 1965. Copyright 1965 by the American Association for the Advancement of Science.)

be released from the earth in the direction of the earth's orbital plane, and reach Mars when the planet crosses the earth's plane.

In Figure 11 (page 134) a transparency is shown at two different levels of written commentary, depending on the level of comprehension of the students. If the material were precued it would probably be safe to make use of the second one. In any case, when projecting, the written verbal material might initially be entirely covered over while the teacher gives his own

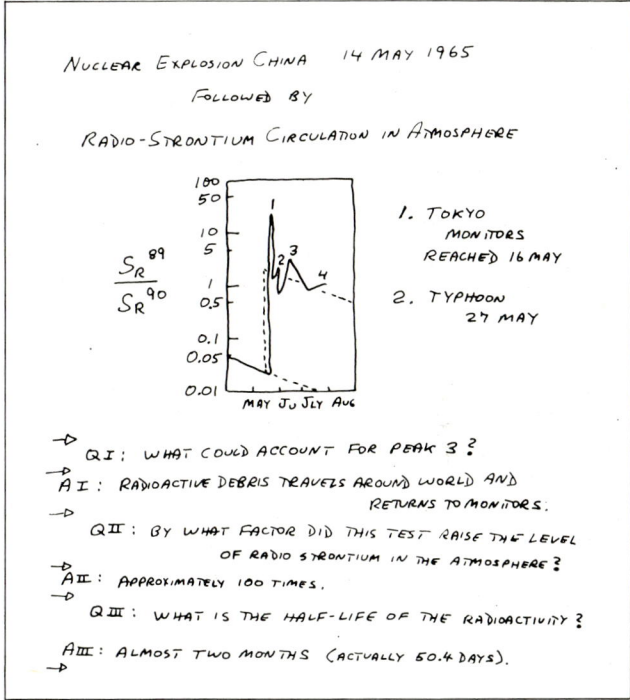

Fig. 8. A transparency made from Fig. 7. The master is traced with drawing pen on glazed translucent paper to outline the essential parts. Arrows on the left guide the operator in uncovering a paper mask to reveal questions and answers.

spoken commentary. For example, he might comment on how the horizontally moving body would have continued along the dotted path were there no gravity force, but that gravity actually causes it to fall down toward the earth's centre. He would then point out that it doesn't strike the earth because it is moving so fast horizontally, and that after a time it is evident that the curving earth has kept a fixed distance below the object which has now become a satellite in circular orbit. At this point he reveals the rest of the transparency making sure that now his students have grasped the complete idea. The micronote in Figure 2 (page 116) illustrates a transparency that projects a sequence of linear programmed frames and confirmations one at a time.

B. *The personalized audiocassette*

1. *General use.* There are several occasions in science education where an audiotape is particularly useful, for example, in listening to commentary while observing bio-

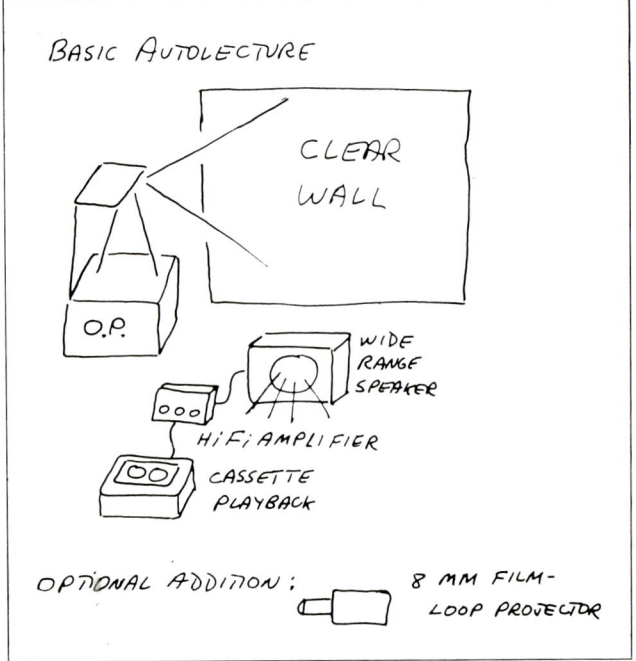

Fig. 9. A hand-drawn and hand-lettered transparency that concentrates on the essential elements. Colour may be added on the final copy by shading with fine markers.

Fig. 10. A hand-drawn transparency that requires precueing by the students. Note the corresponding micronote in Fig. 3.

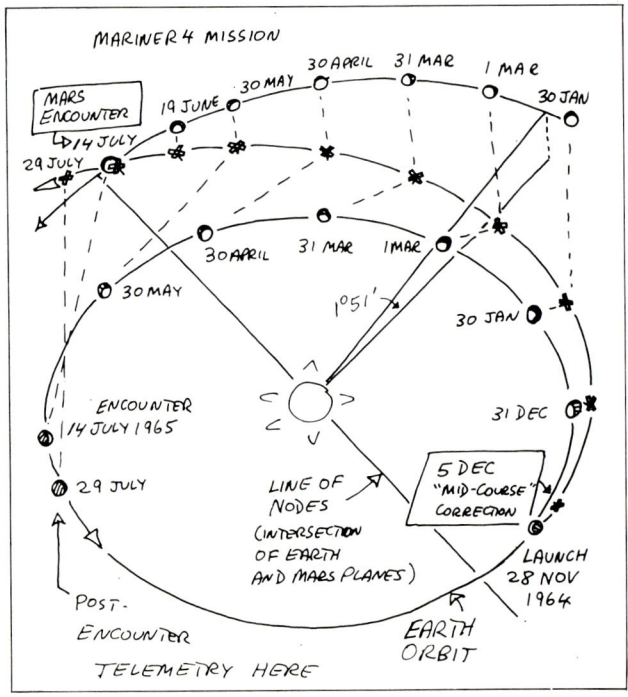

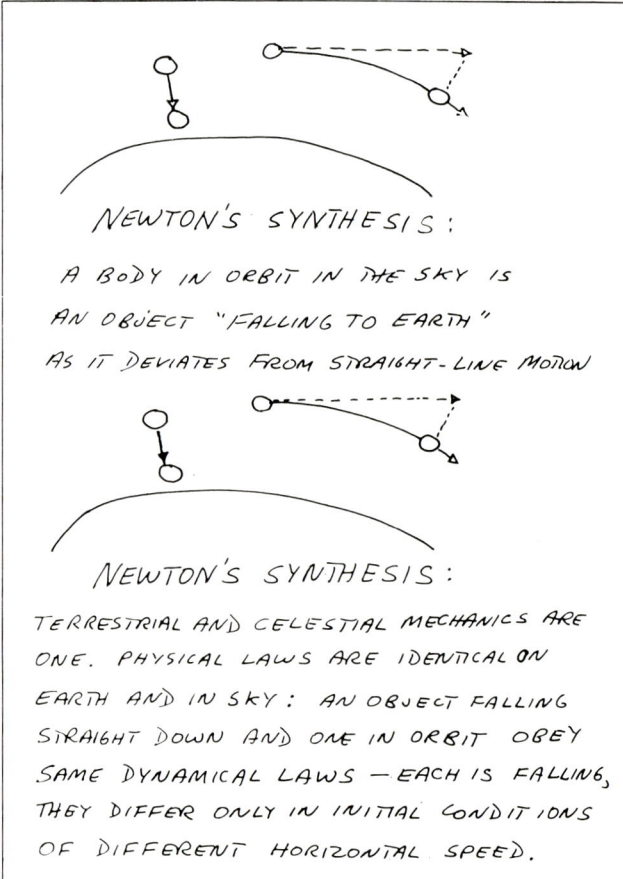

Fig. 11. Transparencies at two levels of difficulty, the lower one providing a deeper insight, but requiring precueing.

logical specimens, as described in IVB. In many parts of the world, particularly in developing countries, a lesson is sent by radio while simultaneously the students work with predistributed printed or projected materials. This teaching arrangement is called 'radiovison' [164]. The school often copies the radio programme on an audiotape and runs the lesson for classes that meet at other times. Furthermore, it can be re-run, whole or in part, and presumably can be distributed to students who wish to borrow it. This procedure has been extended even to universities where a small group will listen to a lecture while reading printed notes (presumably in an uncontrolled manner), and the tape may be stopped at any time by an individual for the discussion of a point. Alternatively, the tape can be run by an individual in a self-paced mode [165].

2. *Practical features of a modern audiocassette recorder.* If a teacher wishes to record his own tape for his students it is recommended that he consider the audiocassette recorder as the most convenient device to work with. In purchasing a unit, one should consider the following features, listed in approximate order of importance.

(a) *The recorder should be of semi-professional or professional quality.* It should be powered by electric line, never by batteries unless expressly designed for professional use in the field. Noticeable 'wow' and 'flutter' are absent and the frequency range is wide enough to record music faithfully. A cheap recorder will not only be fatiguing to listen to for long periods but will not stand up to heavy-duty operation.

(b) *The recorder must provide several inputs and output plugs.* There must be a direct microphone input, and output to an external amplifier even though the player may contain its own amplifier. There must also be a headphone plug. The amplifier to be coupled to the recorder may be medium-powered but it should have a reasonably flat frequency range and low hum level. A so-called acoustical-suspension, or pressurized-enclosure, loudspeaker provides excellent frequency response in a small volume but requires slightly higher power to operate because of its reduced efficiency. For this reason distortion may result if it is connected directly to the internal amplifier of a cassette recorder.

(c) *Recordings should be made with a professional microphone.* The standard microphone that may come with a non-professional recorder is likely to be unsatisfactory. This would be apparent at once when making comparisons on a given tape. Cheap crystal microphones are to be avoided; however, an expensive condenser microphone is not essential. A satisfactory directional dynamic microphone generally costs $ 50-75. At least one manufacturer (AKG) provides feedback control on one model to prevent howling when recording with a loudspeaker nearby. Otherwise, headphones must be used for monitoring the sound.

(d) *A pause lever is essential.* The pause function should not be assigned to the standard tape-stop lever.

(e) *The recorder should possess a digital counter.*

(f) *The recorder need not incorporate an automatic volume control.* If it does, be sure that it does not shift the volume level on the tape if the recording mode is interrupted.

(g) *Ferrite or crystal fine-gap recording and playback heads are desirable.* These are incorporated in all professional recorders and will stand up to heavy use.

(h) *A Dolby background noise silencing circuit is recommended.* Tape recordings have always been plagued by background hiss which now can be virtually eliminated.

(i) *The power should turn off at the end of a tape playback.* This feature is especially useful if many people are expected to use the recorder.

(j) *A 'bias' switch should be provided to permit the use of chrome-oxide tapes.* These tapes produce superior recordings; however, high-density ferrous-oxide tapes provide quite excellent-quality results in a professional recorder.

(k) *A slow-speed rewind lever would be a convenience.* This feature is yet to be found in the general tape-recorder market apart from language-laboratory units. It would enable one to more conveniently return to a passage that has just been heard than by means of the fast-speed rewind lever.

The most frequently used cassette is the C-60, with playing time thirty minutes per side. One manufacturer (Agfa) adds three minutes reserve time per side to its tapes. Slightly more economical to use in terms of playing time and storage are the C-90 and C-120 cassettes, which run 45 and 60 minutes per side respectively. There is, however, a somewhat greater risk that the thin mylar-based tape in these longer rolls might be broken, especially during rewinding. Moreover, the rewind and fast-forward cycles may be very long for these tapes.

As there are two tracks per side for stereo operation, one may double the useful recording time for each tape by confining the signal to a single track. On the other hand, if both tracks are recorded simultaneously from a single input microphone the background noise will be lowered.

3. *Tape-recording technique.* Some guidelines for recording audiocassette tapes are suggested below.

(a) Rather than present a formal lecture, the teacher should talk into the microphone as if a single listener were standing behind it. Slight errors that occur naturally should be retained; however, gross errors of fact should, of course, be eliminated, as should excessive redundancy.

(b) If material is read into the microphone, this reading should never be detected as such by the listener. An extemporaneous style should be developed through practice. Eventually it should be possible to record a tape from notes (as would be on a transparency) in barely more time than it would take to present the same subject matter in a classroom; the playing time generally is much less.

(c) Tape recordings should not be made in a dead room or in one with full sound absorption but in a room with normal drapery and carpeting, away from chiming clocks, moving automobiles, blower-fans, air conditioners, crying children, ringing phones, etc.

(d) Changes in volume level, as might occur with recorders with volume-level limiters when changing from play to record mode, should be avoided. Audible clicks resulting from inserting new material can be prevented by making use of the pause control: set the recording level to zero, run the tape back slightly, release the pause control, return the recording level to its normal position, and begin talking.

(e) Recorded inserts should be made only at the end of a sentence or phrase that is followed by almost a full second of silence.

(f) If a tape should break, carefully separate the plastic sections of the cassette, place 'Scotch' tape temporarily across the cylinder of wound tape to prevent it from unwinding. Handle the loose ends of tape with tweezers and fasten the broken ends together with scotch tape on the shiny (non-coated) side. Trim the 'Scotch' tape to the magnetic tape on the edges, reinsert the tape in the holder, keeping it outside the guide wheels, and snap the cassette together. If no screws are available hold the cassette together by wrapping tape tightly around it lengthwise.

(g) The location of key audio passages will be facilitated by recording the digital number of the counter.

(h) Because of the difficulty of deleting passages and inserting others of unequal length it is recommended that each seven or eight minutes of recording time be followed by three minutes of silence. Thus, if revisions are to be made, only one subsection would be affected. In playing back the tape, the silent zones would be overridden by the fast forward lever, noting the position of the digital counter.

(i) The intervals that require pausing in class, replying to programmed frames for example, can be produced with the aid of the pause lever rather than by inserting a silent interval directly on the tape.

(j) Some editing can be accomplished by making the master tape on a professional reel-to-reel recorder and transferring the finished recording to a working cassette.

VI. Conclusions

A. *Uniqueness of science education*

1. *Media and the student.* The various modern media of educational technology have been analysed to some extent in this report. The taxonomy sought to identify the formats required to match a given subject—whether 3D real form display is needed (as in crystallography), or real motion pictures (as in a physics demonstration), or 3D diagrammatic animation (as in representation of orbits), or plane diagrams (as in circuit analysis).

Then, the various media devices, or subsystems, were described in terms of their ability, from the teacher or administrator's point of view, to produce the various formats efficiently. Finally, the question of shaping the media to the student was considered, of designing media that really respond to students, beyond the mere requirement that students respond to the media. In short, media and media systems were looked at from their ability to be sensitive to student validation and to adjust themselves accordingly.

This last theme was continued in Section V in focusing on media that the teacher of very limited means, as would be found in a developing country, could produce for himself. These media could have the potential of working into small groups of students, or a single student in self study, produce the means of validation and teacher self-evaluation, and continue reworking until the result desired by student and teacher alike, in mutual co-operation, is attained.

It is clear that ordinary unvalidated classroom lecturing may not even reach the average of the class, and the teacher may not even realize this. Examinations may yield data that relate only to home study. If a teacher wishes to employ media to reach individual students he could well draw on the successful techniques of programmed instruction, while avoiding some of its pitfalls. That is, he must invite some response by the student who, in turn, is expected to evaluate this response and demonstrate at once whether the entire procedure has been successful in meeting its objectives. And if not, it must be adjusted for another attempt.

In practice, the analysis of the evaluation is basically statistical and depends on a group-average response, for it is the average of a group of students at which the adjustment is generally aimed. But, provision must be made for the student at the lower end of the curve, or for that matter, at the higher end. Hence the effort in modern educational circles on defining the input behaviour of the entering student, as well as the objectives of the unit of instruction. For the student must find his right group, and the teacher must lead him through other groups to find it. Then the medium, shaped to the group, will be shaped as well to him.

2. *The needs of science.* A great deal of the information in this report can be applied to non-scientific fields. There are, however, aspects of science education that are quite unique; it would be well to comment on the particular role that science plays here. In science education an emphasis is placed on: (i) understanding concepts and principles, (ii) applying concepts and principles to deduce new information, (ii), arranging the environment and applying the principles to generate new information, and (iv) applying deduced and generated information to induce general principles. Basically, science education involves the clear understanding of the statement of a problem and the methodology of arriving at solutions.

All that any educational medium can do is provide an efficient environment for learning. For a scientist, learning goes far beyond the acquisition of factual information. It is concerned primarily with the development of positive attitudes toward scientific methodology: defining objectives, thinking critically and without bias, designing experiments, measuring with precision, analysing errors, and drawing precise conclusions.

The most effective way that media can be utilized in the education of a scientist—aside from the accurate and efficient display of factual and conceptual information—is to set its own example. Suppose I should 'teach' quantum chemistry by projecting a few pages from an advanced text on a screen, repeat the words to the students, and then proceed to examine them. If I then collect the grades and conclude that I have a very dull class, on what scientific basis have I drawn my judgement? Have I accounted for all the variables? Yet such judgements are rendered daily in and out of science education. Only the rise of programmed learning and similar education movements requiring validation as a basic premise, and the protests of students, have brought out in education a universal conscience: if a student hasn't learned, the fault lies in the medium.

B. *Recommendations*

1. *Identification of problems in the employment of learning media world-wide.* A partial listing follows.

(a) The relative place of structured vs. non-structured education, or of shaped behaviour vs. free expression. Clearly, the precise specification of end objectives and the guidance of the student toward the attainment of these objectives in a well-determined way strikes at the very core of the concepts of free expression and experimentation. From a pragmatic viewpoint the forms of extrinsic programmed instruction do work well, at least in the communication of facts and principles. But if we extend this point of view to all science education, how will new principles be discovered?

(b) A disaffection of students with a hardware emphasis related to feelings that education is becoming depersonalized. For some students there is little distinction between the isolation in an auditorium filled with others like him and in a carrel in simulated dialogue with a computer.

(c) An insensitivity of some media systems to a feedback signal from the students and an inability to update content.

(d) An overcentralization of some media systems and the possibility of generating an undesirable authoritarian stance.

(e) An inertia of teachers in applying innovations and a search by them for a role beyond that of equipment manager.

(f) A shortage of reliable data on the effectiveness of media and media systems and an uncertainty in developing foolproof research strategies. No matter what media system is being tested, we must always cope with the sum total of a student's previous experience. Instructional television in the 1970s will not be received in the same way by students as it was in the 1950s for various interacting reasons. On the one hand, some students will be excited with the novelty of a new approach; on the other, some will be irritated with any deviation from the 'time-tested and true'.

2. *Some possible solutions.* One might consider the following list of suggestions.

(a) Provide formal education at the 2-5 year age level through a prepared environmental approach, and incorporate proved innovations at an early age. By exposing young people, particularly future teachers, to new ideas, there will be, later on, little resistance to innovation without real cause, either by students or teachers.

(b) De-emphasize competition among students and make greater use of co-operative group study. In other words, create an environment where students will learn from each other, guided by the media.

(c) Concentrate attention on software (or, as it sometimes is called, teachware) and provide the guidance to teachers on how to use media effectively, and how to develop media content and novel approaches that would encourage student motivation and originality. This guidance should include the study of information theory regarding the reinforcement between otherwise competing channels, the effects of noise level, and overloading of the student's decoding mechanism. But it should also include learning theory regarding advance organizers, transference, overt responding, etc.

(d) Establish a personal posture in media format and display and in student-teacher student-student contacts.

(e) Emphasize mastery learning for all with challenging problems for the highly motivated or talented.

(f) Balance the informational or structured level with the innovational or unstructured level. The *informational* level would tend to employ centralized media in a modular prepackaged framework, clearly delineating operational objectives in the cognitive and motor-sensory domains, classical programmed learning techniques, and open-ended completion times with remedial routings. The *innovational* level would employ highly flexible and responsive (cybernetic) media systems and small groupings emphasizing social awareness, scientific curiosity, historical and philosophical issues, and critical analysis.

3. *The overriding world problems.* Baez has set down the 'three Cs' of science education—Compassion, Competence, and Curiosity [166]. They could apply to all of education, taking its lead from science, as indeed Baez recommends. As he points out (especially if one adds a fourth C: Creativity) they could lead a frontal attack on the four 'P-problems' that drain the world—Population, Pollution, Poverty and the Pursuit of Peace.

Appendix

LEARNING MEDIA PERIODICALS

Audio-visual communications, United Business Publications, Inc., 200 Madison Avenue, New York 10016.
AV communications review, National Education Association, 1201 Sixteenth Street, Washington, D.C., 20036.
Audiovisual instruction, National Education Association, 1201 Sixteenth Street, Washington, D.C., 20036.
British journal of educational technology, National Council for Educational Technology, 10 Queen Anne Street, London, W1M 9LD.
Educational broadcasting international, Tavistock Square, London, XCI H9JP.
Educational/Instructional broadcasting, Acolyte Publications, Inc., 647 N. Sepulveda Boulevard, Los Angeles, Calif. 90049.
Educational media, 1015 Florence Street, Ft. Worth, Texas, 76102.
Educational screen and audiovisual guide, Chicago, Illinois.
Educational technology magazine, 140 Silvan Avenue, Englewood Cliffs, New Jersey 07632.
Media and methods, North American Publishing Company, 134 North 13th Street, Philadelphia, Pa. 19107.
NSPI journal, National Society for Programmed Instruction, Trinity University, 715 Stadium Drive, San Antonio, Texas, 78212.
Programmed learning and educational technology, 27 Torrington Square, London WC1.

NOTES

($AVCR$ = AV Communications Review)

1. James D. FINN, A possible model for considering the use of media in higher education, *AVCR*, 15, 1967, p. 153-7.
2. Arthur I. BERMAN, The seminar/autolecture: a case study of a new teaching systems, in: S. J. Eggleston, (ed.), *Paedagogica Europaea*, Braunschweig, Germany, Georg Westermann, 1972, p. 86-104.
3. Arthur I. BERMAN, *Balanced learning,* New York, Harper and Row, 1973. Rudy Bretz has introduced an alternate media classification scheme that emphasizes the carrier and an intermedia approach in *A taxonomy of communication media,* Englewood Cliffs, New Jersey, Educational Technology Publications, 1971.
3a. Denis GABOR, Holography, 1948-1971, Nobel lecture, 11 December 1971, *Prix Nobel,* Stockholm, 1972, p. 169-201.
4. Marshall MCLUHAN, *Understanding media,* New York, McGraw-Hill, 1964, 365 p.
5. Donald E. BROADBENT, *Perception and communication,* London, Pergamon, 1958.
6. Joseph H. KANNER, Teaching by television in the army. *AVCR,* vol. 16, 1968, p. 178-87, points out that in the pioneer studies with instructional television in the United States Army, the most significant aspect noted was the ability of an instructor to observe his mannerisms.
7. Niels MEYER (Technical University of Denmark), private communication; Evan Herbert, Technology for education, *International Science and Technology,* Aug. 1967, p. 28-97.
8. *Educational media,* vol. 1, March 1970, p. 26.
9. Richard I. EVANS and Peter K. LEPPMANN, *Resistance to innovation in higher education,* San Francisco, Jossey-Bass, 1967; Philip H. Coombs, *The World Educational Crisis,* Oxford, London, 1968; Richard Hooper, Educational technology in the USA—a diagnosis of failure, *AVCR,* vol. 17, 1969, p. 245-64.

10. Robert W. Locke, Has the education industry lost its nerve?, *Saturday review*, 16 January 1971.
11. Charles F. Hoban, Communication in education in a revolutionary age, *AVCR*, vol. 18, 1970, p. 363.
12. Vernon S. Gerlach and Donald P. Ely, *Teaching and media*, Englewood Cliffs, New Jersey, Prentice-Hall, 1971.
13. Cecil I. Garrison, *1001 media ideas for teachers*, McCutchan, Berkeley, Calif., 1970.
13a. Arthur I. Berman and Albert V. Baez, Seminar/Auto-lecture Experiences, *American journal of physics*, vol. 38, 1970, p. 313-9.
13b. Alfred M. Bork, The Harvard Project Physics Film Programme, *Physics teacher*, vol. 8, 1970, p. 163-8.
13c. Leon D. Harmon and Kenneth C. Knowlton, Picture processing by computer, *Science*, vol. 164, 1969, p. 19-20.
13d. R. Cotterill, private communication.
14. Vannevar Bush, *Science is not enough*, New York, Morrow, 1967.
15. Willy Agtby, College of Commerce, Århus, private communication.
16. Charles M. Rossiter, Jr., Rate of presentation effects on recall of facts and of ideas and on generation of inferences, *AVCR*, vol. 14, 1971, p. 318-24.
17. E. Foulke, (ed.), *Center for rate controlled recordings newsletter*, University of Louisville, Kentucky.
18. Godwin C. Chu and Wilbur Schramm, *Learning from television: what the research says*, Institute for Communication Research, Stanford University, Stanford, Cailfornia, 1968, 222 p.; L. P. Greenhill, *Research in instructional television and film*, Department of Health, Education and Welfare, Washington, D.C., 1967.
19. R. Dubin and T. C. Taveggia, *The teaching-learning paradox*, Center for the Advanced Study of Educational Administration, University of Oregon, Eugene, 1968. In his recent survey, Frank Costin (Lecturing versus other methods of teaching: a review of research, *British journal of educational technology*, vol. 3, 1972, p. 4-31) concludes that popular derogation of the lecture method is unwarranted.
20. The reader is referred to recent sociological studies such as Margaret Mead, *Culture and commitment*, Doubleday, New York, 1970, or Lewis Feuer, *The conflict of generations: the character and significance of student movements*, New York, Basic Books, 1968.
21. The Genesys programme of the University of Florida is a closed-circuit video system that attempts to solve the feedback problem by telephonic linkage to the instructor. See M. E. Forsman, 'Graduate engineering education via television', *IEEE International Convention Digest*, IEEE, New York, 1968.
22. I. Keith Tyler, Educational implications of the TV medium, *AVCR*, vol. 12, 1964, p. 61-74.
23. Chu and Schramm, *op. cit.*
24. W. J. McKeachie, Research on teaching at the college and university level in: N. L. Gage, (ed.), *Handbook of research on teaching*, Chicago, Rand McNally, 1963, p. 1153.
25. Stanley K. Gryde, The feasibility of 'Programed' Television instruction, *AVCR*, vol. 14, 1966, p. 71-89.
26. G. L. Gropper and A. A. Lumsdaine *Studies in televised instruction*, Pittsburgh, American Institute for Research, 1961.
27. C. R. Carpenter and C. P. Greenhill, *Comparative research on methods and media for presenting programmed courses in mathematics and English*, Pennsylvania State University, University Park, 1963.
28. L. J. Briggs, D. Plashinski, and D. L. Jones, Self pacing versus automatic pacing on the subject-matter trainer, in A. A. Lumsdaine and Robert Glaser, eds., *Teaching machines and programmed learning: a source book*, Washington, D. C., National Educational Association, 1960, p. 591-2.
29. Gryde, *op. cit.*, p. 84.
30. Robert L. Crist, Group use of programed instruction as a means of generating homogeneous study groups, *AVCR*, vol. 17, 1969, p. 201-9; G. O. Leith, *Second thoughts on programmed learning*, London, Councils and Education Press, 1969.
31. Charles F. Hoban, New media and the school, *AVCR*, vol. 10, 1962, p. 353.
32. F. McLean, EVR, *Endeavour*, vol. 29, 1970, p. 18-23; Peter C. Goldmark, *IEEE spectrum*, vol. 7, 1970, p. 22.
33. For example, W. T. Lippincott, D. W. Meek and F. H. Verhoek, *Experimental general chemistry*, Philadelphia, W. B. Saunders, 1970.
34. *New York Times*, 28 November 1970.
35. *International Herald Tribune*, 24 March 1972.
36. D. W. Allen and K. A. Ryan, *Microteaching*, Reading, mass., Addison-Wesley, 1969; Arye P. Erlberg, Microteaching, *Journal of educational technology*, January 1970; W. R. McAleese and D. Unwin, A selective survey of microteaching, *Programmed learning and educational technology*, vol. 8, 1971, p. 10-21; Philip C. McKnight and David P. Baral, *Microteaching and the technical skills of teaching: a bibliography of research and development at Stanford University, 1963-9*, Stanford, Calif., 1969; E. Perrott and J. H. Duthie, TV as a feedback device: microteaching, *Educational TV international*, vol. 4, 1970, p. 258-61.
36a. Dennis Gabor and George W. Stroke, Holography and its applications, *Endeavour*, vol. 28, 1969, p. 40-7.
37. Dennis Gabor, Winston E. Kock, and George W. Stroke, Holography, *Science*, vol. 173, 1971, p. 11-23.
38. Emmett N. Leith and Juris Upatnieks, Progress in holography, *Physics today*, vol. 23, March 1972, p. 28-34.
39. Dennis Gabor, *Nature*, vol. 161, 1948, p. 177.
40. E. N. Leith and J. Upatnieks, *Journal of the Optical Society of America*, vol. 53, 1963, p. 1377; George W. Stroke, *An introduction to coherent optics and holography*, 2nd ed., New York, Academic Press, 1969.
41. Robert J. Collier, Christoph B. Burkhardt, and Lawrence H. Lin, *Optical holography*, New York, Academic Press, 1971.
42. Robert J. Collier, Holography and integral photography, *Physics today*, vol. 21, July 1968, p. 54-63.
43. B. F. Skinner, The science of learning and the art of teaching, *Harvard educational review*, vol. 24, 1954, p. 86-97.
44. David A. Gilman The origin and development of intrinsic and adaptive programing, *AVCR*, vol. 20, 1972, p. 64-76.
45. Arthur I. Berman, Programed learning and the post-sputnik crisis in education, in: *Space flight physics*, New York, Doubleday, 1973.

46. Jean PIAGET, *Science of education and the psychology of the child*, New York, Viking Press, 1971, p. 78.
47. Derek ROUNTREE, *Basically branching*, London, MacDonald, 1966.
48. D. J. KLAUS, An analysis of programing techniques, in: R. GLASER, *Teaching machines and programed learning I: Data and directions*, Washington, D.C., National Education Association, 1965, p. 118-61.
49. George L. GEIS, Variety and programed instruction or what can't be programed? *AVCR*, vol. 14, 1966, p. 109-15.
50. Maurice BALSON, The effect of sequence presentation and operant size on rate and amount of learning, *Programmed learning and educational technology*, vol. 8, 1971, p. 202-5.
51. Sigmund TOBIAS, Effect of attitudes to programed instruction and other media on achievement from programed materials, *AVCR*, vol. 17, 1969, p. 299-306; Sherman H. FREY, Shinkichi SHIMABUKURO and A. Bond WOODRUFF, Attitude change in programed instruction related to achievement and performance, *AVCR*, vol. 15, 1967, p. 199-205.
52. T. F. GILBERT, Mathetics: the technology of education, *Journal of mathetics*, vol. 1, 1962, p. 7-73.
53. Lawrence M. STORUROW, Computer-Aided Instruction, in: *The schools and the challenge of innovation*, New York, Committee for Economic Development, 1969, p. 270-319.
54. L. M. LACKNER, Current research on programed texts and self-instructional learning in mathematics and related areas, *AVCR*, vol. 15, 1967, p. 181-99.
55. R. M. GAGNÉ and L. BROWN, Some factors in the programming of conceptual learning, *Journal of experimental psychology*, vol. 62, 1961, p. 313.
56. Robert T. FILEP, Individualized instruction and the computer: potential for mass education, *AVCR*, vol. 15, 1967, p. 102-12.
57. Hannah ARENDT, *The Human Condition*, New York, Anchor, 1959.
58. O. K. MOORE, From tools to interactional machines, in F. G. KNIRK and J. W. CHILDS, (ed.), *Instructional technology*, Holt, Rinehard and Winston, 1968, p. 99-105.
59. William Clark TROW *Teacher and technology*, New York, Appleton-Century-Crofts, 1963, p. 5.
60. Allen L. HAMMOND, Computer-assisted instruction: many efforts, mixed results, *Science*, vol. 176, 1972, p. 1005-6.
61. A subject dealt with extensively in Anthony G. OETTINGER's controversial book, *Run, computer, run*, Cambridge, Mass., Harvard University, 1969.
62. For a good analysis of computer applications, especially applicable to science education, see G. SCHWARZ, O. M. KROMHOUT, and S. EDWARDS, Computers in physics instruction, *Physics Today*, vol. 22, September 1969, and Computers in review by the editors of *Scientific Research*, 27 October 1969, p. 21-49.
63. *Physics today*, vol. 23, February 1970, p. 52.
64. Alvin TOFFLER, *Future shock*, New York, Random House, 1970, p. 202.
65. C. P. SNOW, *The search*, New York, Scribners, 1960, p. 20.
66. Samuel DEVONS and Lillian HARTMANN, A history-of-physics laboratory, *Physics today*, vol. 23, Feb. 1970, p. 44-9.
67. Gerald HOLTON, The relevance of physics, *Physics today*, vol. 23, 1970 (Nov.), 23; Albert V. BAEZ, The spirit of science—Worldwide, *The science teacher*, 1969, p. 36; F. E. DART, Science and the world view, *Physics Today*, vol. 25, June 1972, p. 48-54. See also John W. GARDNER, *Self-renewal: the individual and the innovative society*, New York, Harper and Row, 1963; W. D. HITT, A self-renewing educational system, *Battelle technical review*, July 1968.
68. P. J. TANSEY and D. UNWIN, *Simulation and gaming in education*, London, Methuen, 1969.
69. James A. ROBINSON, Simulation and games in: P. H. ROSSI and B. J. BIDDLE, (eds.), *The new media and education*, New York, Doubleday, 1966, p. 39-135.
70. Alvin TOFFLER, *op. cit.*, p. 377.
71. Brian SUTTON-SMITH, Child's play, *Psychology today*, vol. 6, December 1971, p. 67.
72. Jean PIAGET, Physical world of the child, *Physics today*, vol. 25, June 1972, p. 23-7.
73. *Time*, 5 May 1972, p. 35.
74. John L. MARTINSON and David C. MILLER, Educational technology and the future of the book, in: Sidney G. TICKTON, (ed.), *To improve learning*, vol. 11, New York, Bowker, 1971.
75. A. I. BERMAN and A. V. BAEZ, *op. cit.*, p. 316.
76. *Astronautics and aeronautics*, editorial, July 1971, p. 29.
77. In BERMAN's experiments with groups observing transparencies made from masters printed by IBM selectric orator type compared with hand lettering of equivalent material, the hand lettering seemed to be preferred provided it was at the same level of legibility and unaccompanied by distractive stimuli.
78. Arthur I. BERMAN, *Balanced learning, op. cit.* Part I.
79. Margaret MEAD, *Science*, vol. 173, 11 April 1969 editorial, and *Culture and commitment, op. cit.*
80. William GLASSER, *The identity society*, New York, Harper and Row, 1972.
81. Emmanuel G. MESTHENE, *Technological change*, Cambridge, Mass., Harvard University, 1970, p. 33.
82. On this theme, see Charles REICH, *The greening of America*, New York, Random House, 1970.
83. For interesting interpretations of this film, see Jerome AGEL, (ed.), *The making of Kubrick's 2001*, New American Library, 1970.
84. *Time*, 23 November 1970, p. 49.
85. *Time*, 13 October 1967, p. 32.
86. C. R. CARPENTER, A constructive critique of educational technology, *AVCR*, vol. 16, 1968, p. 16-23.
87. A. H. MASLOW, A theory of human motivation, *Psychological review*, vol. 50, 1943, p. 370-96.
88. A. W. MELTON, (ed.), *Categories of human learning*, New York, Academic Press, 1964.
89. Robert M. GAGNÉ, Learning theory, educational media and individualized instruction, *Educational broadcasting review*, vol. 4, 1970, p. 49-62. Reprinted in R. HOOPER, (ed.), *The curriculum*, Bletchley, Open University Press, and Edinburgh, Oliver & Boyd, 1971, and S. G. TICKTON, *op. cit.*
90. N. E. MILLER, *et al.*, Graphic communication and the crisis in education, *AVCR*, vol. 5, 1957.
91. The reader is advised to consult SKINNER's controversial *Beyond freedom and dignity*, Knopf, New York, 1971 for his current views on behavioural control, derived from operant conditioning theory, and compare his *Walden II*,

New York, MacMillan, 1948, describing the 'utopia' created when his theory is put in effect in educational, governmental, and other institutions, with A. S. Neill's *Summerhill: a radical approach to child rearing*, New York, Hart, 1960, based upon love, approval, and minimal constraint.

92. Robert M. Gagné, *The conditions of learning*, New York, Holt, Rinehart and Winston, 1965, and Instruction and the conditions of learning, in: L. Siegal, (ed.), *Instruction: some contemporary viewpoints*, San Francisco, Chandler, 1967, p. 291-313.
93. David P. Ausabel, *The psychology of meaningful verbal learning*, New York, Grune and Stratton, 1963.
94. Subsume is defined: to consider an idea as part of a more comprehensive one, to bring under a rule, or take up into a more inclusive classification from *The Random House dictionary of the English language*, unabr., New York, Random House, 1967.
95. Jean Piaget, Development and learning, and the development of mental imagery, in: R. E. Ripple, (ed.), *Readings in learning and human abilities*, 2nd ed., New York, Harper and Row, 1971.
96. J. S. Bruner, R. Olver, and P. Greenfield, *Studies in cognitive growth*, New York, Willey, 196; see also his interview in *Psychology today*, December 1970.
97. Jerome S. Bruner, *The process of education*, Cambridge Harvard University, 1960, p. 12.
98. *ibid.*, P. 51.
99. *ibid.*, p. 63.
100. Arthur Koestler *The act of creation*, London, Hutchinson, 1964.
101. M. Wertheimer, *Productive thinking*, London, Tavistock, 1961.
102. K. Lewin, *Principles and topological psychology*, New York, McGraw-Hill, 1936.
103. E. C. Tolman, A psychological model, in: T. Parsons and E. A. Shils, (eds.), *Toward a general theory of action*, Cambridge, Mass., Harvard University, 1951.
104. The reader is referred to the 'Physics for Children' issue of *Physics today*, vol. 25, June 1972, particularly the article by Jean Piaget, (previously noted in reference 72), R. Karplus, Physics for beginners, p. 36-47, and J. Griffith and P. Morrison, 'Reflections on a decade of grade-school science', p. 29-34.
105. Curtis Wilson, How did Kepler discover his First and Second Laws?, *Scientific American*, vol. 223, March 1972, P. 93-106, quoted Pierce in his article. See also Charles S. Pierce, *Values in a universe of chance*, P. P. Wiener, (ed.), New York, Doubleday, 1958.
106. K. Popper *The logic of scientific discovery*, New York, Harper and Row, 1965 (original German edition, 1934).
107. Peter Caws, The structure of discovery, *Science*, vol. 166, 12 December 1969, p. 1375-80.
108. Ruth Beard, *Teaching and learning in higher education*, Middlesex, Harmondsworth, Penguin, 1970, p. 61.
109. Benjamin S. Bloom, (ed.), *Taxonomy of educational objectives, Handbook I: Cognitive domain*, New York, McKay, 1956; B. S. Bloom, D. R. Krathwohl, and B. B. Masia, *Taxonomy of educational objectives, Handbook II: Affective domain*, New York, McKay, 1956.
110. Herbert Klausmeier and Richard E. Ripple, *Learning and human abilities*, 3rd ed., New York, Harper and Row, 1971, p. 117.
111. Benjamin S. Bloom, Some theoretical issues related to evaluation, in: R. W. Tyler, (ed.), *Educational evaluation: new roles, new means*, University of Chicago, 1969.
112. Norman MacKenzie, Michael Eraut, & Hywel C. Jones, The clarification of objectives, in: *Teaching and learning*, Unesco, Paris, 1970, p. 102.
113. J. Bruner, *op. cit.*, and *Toward a theory of instruction*. Cambridge, Mass., Harvard University, 1966, as noted in MacKenzie, *et al., op. cit.*
114. Robert F. Mager, *Preparing instructional objectives*, Palo Alto, Calif., Fearon, 1962.
115. N. MacKenzie, *et al., op. cit.*, p. 113.
116. Edwin Fenton, *The new social studies*, New York, Holt, Rinehart, and Winston, 1967.
117. Göte Klingberg, *A scheme for the classification of educational objectives*, Gothenberg School of Education, 1970.
118. Paul Douglass, The contract as an academic tool, *Improving college and university teaching*, vol. 19, 1971, p. 110.
119. Alexander Schure, Science education and instructional systems, in: Robert A. Weisgerber, (ed.), *Instructional process and media innovation*, Chicago, Rand McNally, 1968. For additional guidance on writing behavioural objectives see H. H. McAshan, *Writing behavioural objectives*, New York, Harper and Row, 1970.
120. Richard L. Turner, Conceptual foundations of research in teacher education, in: B. Othanel Smith, (ed.), *Research in teacher education*, Englewood Cliffs, New Jersey, Prentice-Hall, 1971, p. 12.
121. W. J. McKeatchie, Research in teaching: the gap between theory and practice, in: C. B. T. Lee, (ed.), *Improving college teaching*, Washington, D.C., American Council on Education, 1967, p. 211-39, and *Research on college teaching: a review*, Washington D.C., November 1970. Costin, in his review (see note 19), suggested that Dubin and Taveggia's study could be subject to further analysis based on a differentiation of achievement for specific course objectives.
122. R. M. Beard, *Research into teaching methods in higher education*, London, Society for Research into Higher Education, 1967.
123. J. McLeish, *The lecture method*, Cambridge, England, Cambridge Institute of Education, 1968. See also L. A. McManaway, Teaching methods in higher education: innovation and research, *Universities quarterly*, vol. 24, 1970, p. 321-9.
124. B. Othanel Smith, *op. cit.*, (note 120), p. 4.
125. N. MacKenzie, *et al., op. cit.*, p. 135.
126. As A. Laing put it (The art of lecturing, in: David Layton, (ed.), *University teaching in transition*, Edinburgh, Oliver & Boyd, 1968, p. 68): '...teaching is not purely an intellectual affair. The teacher cannot teach a subject without at the same time also inducing some sort of attitude toward that subject.'
127. Although studies with young children show a more pronounced learning effectiveness for stimuli in the auditory channel rather than the visual (E.C. Carterette and M. H. Jones, Visual and auditory information processing in chil-

dren and adults, *Science,* vol. 156, 1967, p. 986-8), when dealing with older literate subjects the visual channel is far superior (F. R. HARTMAN, Single and multiple channel communication, *AVCR,* vol. 9, 1961, p. 25-62).

128. J. R. HARTLEY, Programmed learning, in David Layton, *op. cit.* (note 126), p. 103.
129. H. PETERSON, *Great teachers,* New Brunswick, New Jersey, 1946, p. 175.
130. A. K. C. OTTOWAY, Teaching small groups, in David Layton, *op. cit.* (note 126), p. 53.
131. Maurice BROADY, The conduct of seminars, *Universities Quarterly,* vol. 24, 1970, p. 273-84.
132. Frank COSTIN, *op. cit.* (note 19), p. 13.
133. W. J. MCKEACHIE, in: N. L. Gage, (ed.), *op. cit.* (note 24), p. 1126.
134. Albert V. BAEZ, Innovative experiences in physics teaching, *American Journal of Physics,* 1973.
135. Arthur I. BERMAN, The media-activated seminar, *Educational technology,* 1973 (in press).
136. William JAMES, *The principles of psychology,* vol. 1, New York, Henry Holt, 1890, p. 403.
137. Donald E. BROADBENT, *op. cit.* (note 5), p. 297-300.
138. Anne M. TREISMAN, Verbal cues in selective attention, *American journal of psychology,* vol. 77, 1964, p. 215.
139. Murray G. PHILLIPS, Learning materials and their implementation, in: F. G. KNIRK and J. W. CHILDS, *op. cit.* (note 58), p. 94.
140. Wayne OTTO, Hierarchical responses elicited by verbal and pictorial stimuli, *American educational research journal,* vol. 1, 1964, p. 241-8.
141. Witfield BOURISSEAU, O. L. DAVIS, Jr., and Kaoru YAMAMOTO. Sense impression responses to differing pictorial and verbal stimuli, *AVCR,* vol. 13, 1965, p. 249-58.
142. Adrian P. VAN MONDFRANS and Robert M. W. TRAVERS, Paired-associate learning within and across sense modalities and involving simultaneous and sequential presentations, *American educational research journal,* vol. 2, 1965, p. 88-99.
143. Robert M. W. TRAVERS, The transmission of information to human receivers, *AVCR,* vol. 12, 1964, p. 373-85.
144. Werner SEVEREN, The effectiveness of relevant pictures in multiple-channel communication, *AVCR,* vol. 15, 1967, p. 386-401.
145. BERMAN made an attitudinal pilot study to determine the extent to which auditory reinforces visual verbal content that accompanies relevant diagrammatic material in university students. It was clear that outlined verbal visual subject matter that paralleled auditory information was more acceptable than either minimal visual verbal content essential to diagram comprehension or much larger amounts of visual verbal content not entirely relevant to auditory content.
146. Hower J. HSIA, On channel effectiveness, *AVCR,* vol. 16, 1968, p. 257-67. See also Intelligence in auditory, visual, and audiovisual information processing, *AVCR,* vol. 17, 1969, p. 272-82.
147. S. A. MUDD and E. J. MCCORMICK, The use of auditory cues in a visual search task, *J. applied psychology,* vol. 44, 1960, p. 184-8.
148. Leslie J. BRIGGS, The design of multimedia instruction, in: F. G. KNIRK and J. W. CHILDS, *op. cit.* (note 58), p. 62.
149. Jerrold E. KEMP, Which medium?, *Audiovisual instruction,* vol. 16, December 1971, p. 32-6.
150. Vernon S. GERLACH and Donald P. ELY, *op. cit.* (note 12), p. 12.
151. Erich JANTSCH, Inter- and transdisciplinary university: a systems approach to education and innovation, *Higher education,* vol. 1, 1972, p. 7.
152. Richard A. MEISLER, Technologies for learning, in: S. G. Tickton, (ed.), *op. cit.* (note 74), p. 223-632.
153. Professor P. KNUDSEN, private communication.
154. S. N. POSTLETHWAIT, J. NOVAR, and H. T. MURRAY, Jr., *The audio-tutorial approach to learning* (2nd Ed.), Minneapolis, Burgess, 1969.
155. Heather ADAMSON and F. V. MERCER, A new approach to undergraduate biology, *Journal of biological education,* vol. 4, 1970, p. 155-76. S. C. DRIVER, Programmed individual study methods in the learning of science, *Australian science teachers journal,* vol. 16, May 1970, p. 27-33. Jack V. EDLING, Individualized instruction the way it is: 1970, *Audio-Visual Media,* vol. 4, 1970, p. 12-6. L. R. B. ELTON, *et al.,* Teaching and learning systems in a university physics course, *Physics education,* vol. 6, 1971, p. 95-100. Peter FENNER and Ted F. ANDREWS, *Audio-tutorial instruction: a strategy for teaching introductory college geology,* CEGS Programmes Publication, no. 4, 1970. R. J. GILLESPIE and David A. HUMPHREYS, The application of a learning resource system in teaching undergraduate chemistry, in: D. G. CHISMAN, (ed.), *University chemical education,* London, Butterworths, 1970.
156. Fred S. KELLER, Goodby teacher, *Journal of applied behaviour analysis,* vol. 1, 1968, p. 79-89.
157. Ben A. GREEN, Jr., Physics teaching by the Keller Plan at MIT, *American journal of physics,* vol. 39, 1971, p. 764-75.
158. B. V. KOEN, *Engineering education,* vol. 60, 1970, p. 735; J. S. MCMICHAEL and J. R. COREY, *Journal of applied behaviour analysis,* vol. 2, 1969, p. 79. H. J. SANDERS, *Chemical and engineering news,* 9 October 1972, vol. 13, p. 18-41.
159. Arthur I. BERMAN, Questions/answers about seminar/autolecture, *Audivisual instruction,* vol. 13, 1968, p. 847-57; Seminar/autolecture, *Today's education,* vol. 57, December 1968, p. 33-6.
160. see Agnes G. REZLER, *The assessment of attitudes,* Geneva, World Health Organization, 1972.
161. Lars WILHELMSEN, private communication.
162. Albert V. BAEZ, *op. cit.* (note 134). In the Harvard courses Baez held complementary lecture-demonstrations live; he also coupled single-concept films into the autolecture medium whenever the need arose.
163. James CABECEIRAS, Observed differences in teacher verbal behaviour when using and not using the overhead projector, *AVCR,* vol. 20, 1972, p. 271-80.
164. Anna FOXALL, Television and radiovision in the teaching of modern mathematics: a comparative study, *British journal of educational technology,* vol. 3, 1972, p. 236-44.
165. G. W. ACKERS and J. K. OOSTHOEK, The evaluation of an audio-tape mediated course. *British journal of educational technology,* vol. 2, Part I, May, Part II, November, 1972.
166. Albert V. BAEZ, Integrated science teaching as part of general education, in: P. E. RICHMOND, (ed.), *New trends in integrated science,* vol. II, Paris, Unesco, 1973.

Integrated multi-media systems for science education which achieve a wide territorial coverage

by A. R. Kaye
Institute of Educational Technology at the Open University, United Kingdom
and M. J. Pentz
Faculty of Science at the Open University, United Kingdom

Summary

The main emphasis in this paper is on the special problems which arise in designing and implementing integrated multi-media science courses for students working at a distance, i.e., in their own homes, and with little face-to-face contact with their teachers. The discussion is restricted to the situation in which open-circuit broadcasting of television and radio material is included as part of the teaching system.

The first section outlines briefly the particular problems posed by teaching science-based subjects at a distance. It stresses the need for the use of a wide range of different media and modes of teaching and the production of course materials by multi-skilled teams working together within the same organization. The second section of the paper discusses examples of different types of distant teaching systems (correspondence colleges, educational divisions of broadcasting organizations, university and high school extension departments). Then follows in the third section a detailed description of science teaching at the Open University in the United Kingdom, which at the time of writing is probably the most advanced multi-media distant teaching organization in operation. As such it represents the most sophisticated development of the 'state-of-the-art' in the application of educational technology to the large-scale teaching of science. The problems of designing self-instructional printed materials linked to broadcasts and home experiment 'kits' are discussed with reference to recent research on the analysis of learning materials. The role of face-to-face teaching by part-time tutorial staff, and that of residential schools based on intensive laboratory work, is also described. A brief description of the course production process is included, in the hope that this may be relevant to proposals in other countries for setting up similar teaching systems. The procedures adopted for formal assessment of student achievement (a combination of continuous assessment and final examination), as well as the procedures used for course evaluation, are also mentioned.

The final section discusses the future potential of such teaching systems in the light of predicted technological developments in dissemination of information, in rapid access to source material, and in the establishment of adaptive two-way links between students and teaching centres. Such developments suggest the possibility of including a greater degree of flexibility in the 'teaching at a distance' situation.

In conclusion, the paper stresses the importance for society of including wide-scale teaching of science at a generalist level within a framework of permanent education.

Introduction

Integrated multi-media systems for distant study

There is really nothing new about 'multi-media' teaching, nor is teaching at a distance (e.g., in correspondence courses and external degree programmes) a recent development. However, it is only in the last few years that systematic attempts have been made to integrate the use of different media within a framework of distant teaching. It is this combination which has led to the really novel developments in instructional design, in the roles of teachers and learners within such systems, and even in the nature of the material taught.

The recent impetus to develop such systems appears to have arisen from a combination of socio-economic factors and developments in educational technology. Amongst the most important of the former factors in the developed countries is an increasing demand for out-of-school educational facilities, in connexion with both leisure and career interests. This, looked at in the context of many of the current problems associated with conventional secondary and tertiary education (vastly increased student numbers, the criticisms levelled by the proponents of the 'de-schooling' movement [1][1], the increasing pressure on governments to provide more cost-effective education) is related to the development of the concept of *'éducation permanente'* [2]—the provision of a flexible pattern of educational facilities available at any time in a person's life. In the developing countries, with their extreme shortages of qualified teachers and a much greater need for massive educational provision, the problems are of course very much greater. Recent relevant developments in educational technology include techniques for analysing teaching materials, especially for presentation in a self-

[1]. For references, see Bibliography, page 175.

instructional mode (e.g., programmed instruction, CAI), associated with an increased stress on the so-called 'individualization of instruction' [3] (the production of teaching materials sufficiently flexible to cope with individual differences in learning styles). At the same time, there has been a welcome change away from a succession of 'single-medium fixations' (note the rise and fall of instructional radio, instructional television, and programmed instruction, each originally seen as almost universal panaceas) [4] to a more comprehensive systems approach which sees each medium, and each teaching mode, as playing a valuable role within an integrated structure. In recent years, of course, we have also seen the beginnings of new developments in the hardware aspects of communicating teaching material (cable-TV, dial-access systems, satellite communications, etc.) which are probably going to play an increasingly important role.

The combined effect of these and other factors has produced a climate in which flexible, open-entry [2], educational systems which can provide resources, teaching, and contacts in people's homes or neighbourhoods at any stage in their lives, will become increasingly available.

To be fully effective, such systems need to combine the centralized production of basic teaching materials (which may be distributed via postal and broadcast services) with some sort of local facility for developing personal contacts and face-to-face teaching; they need to include most, if not all, of the following main channels of presentation and evaluation:

Channels for presentation of instruction

written materials (set books, specially prepared texts and notes)
film and television
visual materials (diagrams, slides, film-strips etc.)
audio materials (tapes, disks, radio)
equipment (laboratory kits, models)
face-to-face teaching

Channels for evaluation of, and feedback on, student learning

self-assessment materials (e.g., programmed texts, structured exercises)
peer group contacts
teacher contacts (tutorials: face-to-face/recorded/by telephone)
some form of continuous assessment of progress (e.g., by correspondence with tutors, or by machine marked tests)

Although all of the items in these two lists may be found in a school-based (intramural) teaching system, it is probably true to say that only in the context of comprehensive distant study teaching systems have these items actually been carefully integrated into a total package. Of necessity, opportunities for face-to-face teaching are minimal in such systems, and this is one of the main factors which has led to the need for careful analysis of the material to be taught, coupled with a close examination of the potential of each of the media deployed. This implies that instructional materials for such distant study systems need to be developed along very different lines from those normally adopted in a conventional school or university teaching situation (although there are parallels with some of the large curriculum development projects, such as Nuffield or PSSC).

Most importantly, courses which are to be delivered via a multi-media distant study system need to be prepared by multi-skilled *teams* including subject matter experts, media experts, educational technologists, and editorial and administrative staff; and the teams need the freedom to operate in such a way that the nature of the material being taught (and *not* a particular broadcasting organization or the particular educational ministry or college concerned) dictates the way in which the different media within the systems are used. This really implies the creation of independent institutions to develop and operate such systems if their maximum potential is to be realized.

The teaching of science-based subjects in a multi-media distant study system

Science and mathematics provide a special challenge as subjects for teaching at a distance. Everything mentioned in the preceding section on the analysis of teaching materials, appropriate use of media, and the need for a team operation, takes on a special significance in the context of science education. At the risk of stating the obvious, it is worth listing some of the more important characteristics of science and mathematics as

---

2. That is, entry in which there is no selection based on age, educational background, occupation, etc.

disciplines which create particular difficulties for a distant-teaching system.

Language. Scientific and mathematical disciplines have languages of their own which pose special problems in communication. Imagine a person with no specialist knowledge in either history or physics, say, confronted with a research paper on some aspect of the French Revolution and a research paper on high energy nuclear physics. It is not difficult to imagine which paper he would most easily be able to understand. And this is not just a distinction between 'arts-based' and 'science-based' disciplines. Communication between *scientists* in *different* disciplines can be notoriously difficult even at relatively non-specialist levels; this is more than just a question of terminology—there are conceptual and methodological differences between, for example, the biological sciences and the physical sciences.

Structure. Any body of concepts and knowledge which makes up any particular 'area' of science as it is taught at, say, university level, takes for granted a set of 'lower-level' concepts, terms and operations without which understanding is greatly impaired, if not impossible. This reflects the complex nature of the structure of knowledge within any scientific or mathematical discipline.

Quantification and precision. Most of science is essentially concerned with parameters that can be measured and quantified. This emphasis on precision and on manipulation of numerical data combined with a respect for objectivity in observing and interpreting natural phenomena again differentiates 'science-based' subjects from the 'arts-based' subjects, in many areas of which much greater stress is put on the value of subjective interpretations and criticism.

Laboratory work. Much of science as it is practised involves the testing of hypotheses by setting up experimental situations or observing natural phenomena 'in the field'. Science as it is taught must reflect this, so there is a need to put students in laboratory situations and let them experience the 'raw materials' and techniques of the disciplines being taught.

The 'spirit of science'. Science teaching needs to reflect what has been called 'the spirit of science'—'the particular approach to rational inquiry exemplified by scientists, driven by a belief in its efficacy and by a restless curiosity...' [5].

Information. There is the problem of the 'knowledge explosion' in the sciences—the exponential increase in the amount of new research findings in science and technology.

Obviously, the relative importance attached to these various factors will depend on the underlying purpose and philosophy of the science course in question (an undergraduate course aimed predominantly at producing more scientists will differ in some of these dimensions from one aimed predominantly at educating people in the social and political implications of scientific decision-making).

The first three of these factors (language, structure, precision and measurement) taken together imply that, in the preparation of science materials for distant study, special care must be taken to produce models of clear exposition. This is just as important for communication amongst the members of the team producing the material as it is for the finished products and the students who will be using them. There are techniques available for summarizing the structure of teaching materials—for example, the preparation of specific learning objectives, the designing of 'maps' or networks showing the logical dependencies between different concepts, and the development of summaries of assumed 'entry' and 'exit' behaviours in terms of required and taught terminology and concepts. The materials resulting from these techniques can be related to specific assessment questions and problems with which students can monitor their learning through answers and commentary, provided directly or indirectly.

For students, one of the potential advantages of studying at a distance is the chance to adopt individual study patterns (which may be precluded in a group situation). So it follows that course materials for distant study should be as *flexible* as possible, without sacrificing any of the clarity and underlying structure which is so essential if confusion and misunderstanding is to be avoided. Given the complex structure of much scientific knowledge, it may seem difficult to combine the teaching of this with an adequate degree of flexibility, but it should be stressed that a high degree of interrelatedness between concepts does not necessarily imply one best sequence for learning. It is fair to say that many science teachers (and teaching materials) prescribe learning sequences which are unnecessarily rigid, and fail to recognize that students may learn more effectively by adopting idiosyncratic strategies for finding their way around a particular subject. An interesting approach here (in an on-campus context) is that described by Epstein for presenting specialized research topics to first-year college students with only a minimal scientific background [6].

Work that may eventually prove of value in designing individual (self-instructional) systems for distant study with sufficient flexibility to allow for any number of different learning strategies within the context of a thoroughly specified knowledge structure, is that currently being undertaken by Pask [7]. A general description of the implications of this and similar work in the context of recent thinking on the value of behavioural objectives is discussed by MacDonald Ross [8].

At the present time, however, it would be fair to say that no distant study systems have satisfactorily solved the problem of providing optimally flexible study materials. Of course, imaginative use of the different media can go some way to remedy this. Briggs [9] points out that the major advantages of multi-media systems are concerned with the need for '... different conditions of learning for different teaching objectives' and that '... a greater variety of media produces a greater chance of accommodating individual differences in learning styles'.

The problem of providing students with adequate laboratory experience seems at first sight almost insuperable for students working at a distance, even when short-term residential schools are included as part of the system. Obviously, the magnitude of this problem depends on the subject being taught, the level of specialization, and the underlying rationale for the teaching. However, this also is an area where a detailed analysis can result in different objectives and functions being carried by different media and modes of teaching (e.g., home kits), and the time spent under 'real' laboratory conditions being used to maximum effectiveness. The pioneering work by Postlethwait in this area is of obvious relevance to teaching at a distance [10].

It is somtimes said that the degree of planning and organization required to produce integrated self-study courses is incompatible with the need to convey the 'spirit' of science, the sense of excitement and discovery that characterizes so much of science as it is practised. There is certainly a danger, in the nature of the teaching system, that unless these factors are regarded as of primary importance throughout all stages of development of a science course, the course will suffer seriously. But broadcasts, archive film material, and face-to-face contact with practising scientists at tutorials and residential schools can be used very effectively here. And 'open-ended' investigations using relatively simple equipment and readily available materials can be built into the work carried out by students on their own at home (see, for example, Adamson and Mercer [11]).

The problem of the increasing amount of new scientific information becoming available, linked with the need for access to a wide variety of specialized journals, is at the moment a very serious one for distant study courses at university level. Hopefully, the implementation of new developments in the hardware of information retrieval and presentation systems (dial-access, cheap microfiche readers) may go some way towards reducing this problem in the not-too-distant future. At the moment, the best solution seems to be to build sufficient flexibility into higher-level courses (which may have to run for five or six years to recoup development costs) so as to allow for the dissemination of new research information on a year-to-year basis (e.g., via mimeographed reprints, audio tapes, or live broadcasts).

The main purpose of this rather lengthy introduction [1] to the particular problems posed by teaching science at a distance is to stress the need for:

flexible multi-media systems which contain a wide range of different components;

the production of courses by 'multi-skilled' *teams,* all members of which (not just the subject-matter experts) have a background in the subject being taught (thus the radio and television producers, the educational technologists and the editors must be scientifically 'literate'). This has implications for the organizational structure of the institution which is providing the courses—to be maximally effective it requires a new type of organization which is independent of traditional teaching institutions, and broadcasting bodies. Currently the Open University is probably the best available model for such institutions, and it is interesting to trace the trends in distant study systems over the last decade so as to see why this is the case. This is the purpose of the next section of this paper.

---

1. Some of the points raised are developed further below, pages 154 f.

Trends in the development of distant study systems

Patterns of organization and development

It is not possible in a paper of this size to describe in detail all the distant study systems offering science-based courses. It is, however, feasible and relevant to examine the context in which such systems have developed and for this purpose we have looked at some existing systems under five main headings, [1] depending on their organizational structure:

systems that have developed within the context of correspondence schools and institutes, or within extension departments of schools and universities (e.g., the North-Western Polytechnic in Leningrad, Wisconsin University Extension Division, the University of New South Wales, the NHK Senior High School);

systems that have been set up specifically to use broadcasting to bring courses to off-campus students (e.g., Chicago City College, Télé-CNAM) and which are run by an established conventional teaching institution;

systems that are primarily dependent on broadcasting organizations (e.g., the Telekolleg);

experimental systems which are more or less responsible to Ministries or Ministerial bodies (e.g., the Warsaw TV Polytechnic, the Canadian TEVEC Project);

independent organizations which have been established specifically to design and provide integrated multimedia courses for distant study (e.g., the Open University).

This order corresponds to some extent to a trend in the last few decades from the use of the multi-media approach within correspondence education systems, through the development of systems in which broadcasting was initially conceived as having a dominant role, to independent systems which use many media and modes of teaching within the framework of an integrated distant study system. Examples of systems in each of the above categories are very briefly described in the following pages. The references given indicate sources of more detailed information.

The science-based subjects taught by the institutions mentioned are listed in Table 1 below.

The importance and effectiveness of much traditional correspondence education tends to be ignored by many people working in conventional schools and universities in Europe. However, discussion of the development of multi-media systems cannot afford to ignore the way in which traditional correspondence study has prepared the ground.

For example, in the Soviet Union there are currently about three million students studying by correspondence for qualifications at both specialized secondary level and in higher education, mainly in science-based subjects and mathematics. These courses have nearly always been multi-media, since the correspondence component is integrated with residential schools, local consultation centres where students can meet each other and their tutors, and practical work experience in specially equipped laboratories at many major factories. More recently, many courses have been modified to include open-circuit television and radio broadcasts. The television programme listings in *Pravda,* for example, show one to one-and-a-half hours of broadcasting each evening on the educational TV channel specifically for students taking correspondence courses. The courses are run either by special correspondence institutes (of which there were eighteen in 1964, including one for deaf students and another for blind students) or by correspondence departments of conventional universities. A flourishing example of the former is the North-Western Polytechnic in Leningrad. In a recent paper about the work of this institution, the author lists the basic premises on which effective distant teaching in science and technology must be based [1]:

'that the student must not be left to work in total isolation, and that some face-to-face teaching and television teaching must be included in the system to overcome this problem;

'that there must be adequate provision for laboratory teaching;

'that there must be an adequate network of local centres where students can meet each other and part-time teachers;

'that residential accommodation for short-term periods must be provided;

---

1. The range of examples under some of the headings (especially the first two), is by no means comprehensive; furthermore, for the purposes of this paper, we have made no reference to the widespread use of distant teaching methods in developing countries.

Table I. Science-based courses produced by the institutions listed.

Institution	Science, mathematics and technology subjects	Level
North Western Polytechnic, Leningrad, U.S.S.R.	Most disciplines within the three areas; vocationally oriented	University
Wisconsin University Extension Division, United States	Astronomy; chemistry, engineering; geology; mathematics; pharmacy, physics; veterinary science	University and high school
The University of New South Wales, Australia	Physics; chemistry; biology; mechanics; algebra; calculus Refresher courses for dentists, engineers, doctors	Bridging courses High school university
NHK Senior High School, Japan	Mathematics, science	High school
Chicago City TV College, United States	Physical sciences (general); biological sciences (general); astronomy, geology; physics; mathematics; data processing	University
Télé-CNAM, France	Higher mathematics; electronics; data processing; new maths	Post-secondary
Telekolleg, Federal Republic of Germany	Physics (+ chemistry); mathematics; biology; technical drawing; electrical engineering; chemistry	Secondary
Warsaw TV Polytechnic, Poland	Mathematics; descriptive geometry; chemistry; physics; strength of materials; fundamentals of electronics	Bridging courses between secondary and university level
TEVEC Project, Canada	Mathematics	Grades 5-9 of elementary school
Open University, United Kingdom	A wide range of courses in most disciplines in these three areas (see Appendix 1 to this paper)	University (No prerequisite of academic qualifications)

'that the subject(s) studied must be related to the student's job;
'that written teaching material should be specially prepared by the teaching staff of the correspondence university.'

Education by correspondence in the U.S.S.R. is closely integrated with more conventional educational institutions (full-time day schools and evening schools) in terms of syllabuses, examinations and the sharing of facilities. Working adults are actively encouraged to participate in adult education. Not only do many factories have special teaching laboratory facilities, but at most places of work there are viewing rooms set aside for the reception of educational broadcasts. Students on higher education courses may be given thirty to forty days' extra paid holidays each for studying, together with four months' paid leave for preparation of their final year thesis.

In other European countries [2], correspondence education has never achieved the status or comprehensiveness that it has done in the Soviet Union. This is to some extent associated with different economic factors and the wider availability of conventional educational institutions. It is also associated with a more rigid approach to study at a distance, in that systems which combine correspondence components with other teaching modes (tutorials, residential sessions, laboratory experience) and with other media, are less widespread ('... the text book and the printed duplicated course remain the basic instruments of correspondence education ...') [2]. However, in recent years many correspondence institutions have started combining their activities with such audio-visual aids as films and tapes (Sweden), open-circuit radio and television (e.g., the experiments conducted by the National Extension College in the United Kingdom), with face-to-face teaching (e.g., mobile teaching laboratories in the Netherlands), and with group study (e.g., the People's Schools in Sweden).

In the United States, correspondence education developed primarily within the framework of extension courses given by many state and city universities. Notable amongst these were Wisconsin and Chicago, which ran extension departments as early as the 1890s. After World War II, there was a marked increase in enrolments in university extension courses as a result of programmes of studies for veterans sponsored by the 'GI Bill'. At Wisconsin, an additional impetus to extension course development was the establishment of the USAFI Centre near the university campus in Maddison (1942), for which the university developed a new range of distant teaching programmes. During the last fifteen years, Wisconsin has been running extension courses using television and radio components. The most sophisticated of these is the AIM (Articulated Instructional Media) Project, started in 1964, which is built around the following principles [3]:

'The method of instruction should be appropriate to the subject matter, the student, and the learning situation.

'Learning derives from active involvement in appropriate activities; hence the instructional methods used should stress student-centred learning activities.

'Any particular method should be used only for those instructional purposes in which it has proven successful, or in connexion with a careful evaluation of its success.

'Economy and efficiency of instructional method should be constantly assessed.

'Periods for advisement must be included regularly to help students anticipate and adjust to changes in method, to help them see the course or sequence as a whole, and to help them branch off (or out) of a program, course or sequence.

'It is important to evaluate student achievement per se, and not to confuse the methods used for teaching/learning as an evaluative criterion in place of student achievement.'

The AIM courses combine tutoral functions, counselling, laboratory sessions, residential stays, local study facilities, mobile 'learning laboratories', and audio-visual resources (TV, radio, telephone hookups), as appropriate for each subject matter.

At the same time as the development of this project, attempts were made to increase the professional quality of course materials by providing special units on correspondence study in graduate courses in the adult education curriculum of the school of education.

In Australia [4], where the low population density in many regions has made it impossible for all children and students to attend day schools and universities, there has been a long tradition of correspondence education. Each state correspondence school now broadcasts regular radio and television programmes for the primary and secondary schoolchildren who are working on their own or in small groups under adult supervision. Written work is sent in to the school headquarters to be marked and commented on, and regular work sheets and leaflets are sent to the children. Some schools have made two-way radio links available between centrally based teachers and pupils. At the post-secondary level, the University of New South Wales in Sydney offers correspondence courses combined with radio and television broadcasts, and seminars and laboratory work at the University itself.

In Japan [5, 6], senior high school correspondence courses were started shortly after the Second World War (mathematics and science being added to the senior high school curriculum in 1949); by 1963 there were nearly 100,000 students taking high school courses by correspondence and it was these courses which to a large extent provided the impetus for the development of the NHK Junior High School in 1963. This is an organization independently established from the NHK Broadcasting Corporation, and courses are jointly prepared by the High School staff and NHK production staff. The curriculum offered is the same as that given in local schools but is divided into four years instead of three. Correspondence students use textbooks selected from those authorized by the Ministry of Education. Work books or study guides and the accompanying broadcast materials are planned by a group made up of representatives from the National High School Correspondence Research Association, the textbook authors, the broadcasting lecturers, and members of the NHK Correspondence School Broadcasting Division. The broadcast lecturers actually write the study guides.

In its early years, the NHK High School played a valuable part within the Japanese educational system, but today the function of the school is under close scutiny because of decreases in enrolments, in the quality of students applying, and in the numbers graduating. This is associated with the great increase in recent years in the proportion of students attending full-time high schools.

Organizations established to bring courses to off-campus students by using broadcasting and other media

Two well-known examples of institutions in this category are the Chicago TV College (which began teaching in 1956) and Télé-CNAM in France (started in 1963).

The Chicago TV College [5] was established to bring the entire junior college syllabus to people studying in their homes in the Chicago Metropolitan area. The television college is administered by the Dean of Television Instruction. Each course is in general produced by a single teacher recruited from the intra-mural teaching staff and he is required to present the television programmes, draft the course study guide and prepare assessment material. Each course is reviewed and updated or modified as necessary every five or six years. Television is used primarily for demonstrations (of experiments, materials, specimens, etc.) and presentations; the main part of the students' learning being through test materials, printed study materials and practical work. For practical work, students can sign up for laboratory sessions held in campus laboratories on Saturday mornings.

Some of the major problems experienced by the college ('... academics who cling to lectures and blackboards and cannot see that the medium can shape and improve instruction; directors who find academics dull and who yearn for Hollywood productions ...')[1] are probably associated with the fact that each course is prepared by one teacher taken off full-time conventional teaching and that the directors are on contract to the college from the local educational TV station—in other words, there is no real team approach. A limited team approach was tried initially, but soon abandoned, as the college teachers found it 'uncongenial' [7].

Télé-CNAM [5] in France was initially set up to provide centrally produced television broadcasts to twelve viewing centres in Paris via a point-to-point microwave system. Now the programmes are also broadcast on the second television channel of ORTF to the north of France and the Paris region. The broadcast material is integrated with correspondence work and with practical work under the guidance of '*animateurs*'. Courses are aimed at providing refresher training in scientific and technical subjects for adults already engaged in a profession (see Table 1). Students whose work is deemed satisfactory can take the CNAM examination in the same way as the intra-mural full-time students.

Systems primarily dependent on broadcasting organizations

The German Telekolleg [5] started its operations in 1967; all of the course materials (TV programmes, lesson sheets, information documents, self-test materials) are prepared by Bayerischer Rundfunk, and Saturday tutorial sessions (Telekolleg days) are organized by the Bavarian Ministry of Education which provides the necessary teachers. The same teacher who looks after a group of fifteen to twenty students in these Saturday meetings also marks and comments on their written work. The curriculum corresponds to that of the Berufsaufbauschule (a school intended for young people in employment with only an elementary school background).

At the moment, plans are being developed for setting up a Telekolleg II, for the Abitur examination, and also for a Teleberuf—an out-of-school teaching system for vocational training for apprentices, working people, and industrial training instructors.

1. Private communication.

Experimental systems responsible to ministries or ministerial bodies

The Polish 'TV Polytechnic' [5] was established in 1966 to provide a basic course in technology for adults and school leavers in employment, to allow them to reach a sufficient standard to enter a conventional university for higher studies as full-time students. The project is sponsored by the Polish Government with the assistance of Unesco. Televised lectures are used to cover the most conceptually difficult material, and students can obtain course materials from bookshops. Exercise books for self-assessment material and programme scripts are also available. Some students attend evening classes related to the course, others may be able to use 'consultation centres' set up in some factories; isolated students can make use of correspondence facilities provided by the universities. The production of the course has had a salutary effect on reorganizing and updating the curriculum in technology in the seventeen universities which teach the subject in Poland. It has also stimulated suggestions for further television courses—notably post-secondary courses in mathematics and physics (which started in 1968), and a course in electronic data processing for engineers (currently being considered).

The main emphasis in these Polish experiments seems to be the provision of broadcast television material to complement and strengthen the existing facilities available for out-of-school education, rather than the production of an entirely new integrated system.

The Canadian TEVEC project [5] was run from 1968 to 1969 in north-eastern Quebec to bring 9th grade education in French, English, mathematics and social science to adults in an area of high unemployment and with poor schooling facilities. Television and radio programmes were used in conjunction with correspondence material (explanatory brochures and computer marked tests), repetition centres where teachers could be contacted, and counsellors who visited students in their homes. 34,000 adults enrolled in the programme when it started and it has been judged so successful that a similar multi-media scheme on a larger scale is planned to start this year.

Independent organizations specifically established as multi-media teaching institutions

The Open University of the United Kingdom [8] is the only large-scale institution which has been established with sufficient funding to attempt to fully exploit the potential of integrated multi-media distant teaching. Its teaching activities cover the entire country. Like all British universities, it is an autonomous institution with the freedom to design its own curricula. Unlike any other university in the country, students need no academic entry requirements at all to enrol on a degree course. It started its teaching activities in 1971 and now has a *full-time* staff of 1,320, of which 264 are centrally-based academic (teaching and research) staff, and 128 are regionally-based academic (tutorial and counselling) staff. The number of *part-time* regional tutorial and counselling staff is currently about 4,200. The main academic divisions in the university are the six faculties (arts, educational studies, mathematics, science, social science, and technology) and the Institute of Educational Technology (all located at the central university campus) and the senior tutorial staff (based in the thirteen regional offices throughout the country). The television and radio production staff are recruited by the BBC, and work in partnership with the faculty staff in preparing programmes. The producers are almost invariably graduates in the discipline areas within which they work.

The course materials, and the way in which they are produced and evaluated, are discussed in more detail in Section 3 of this paper. Basically, they consist of specially prepared correspondence texts, radio and TV broadcasts, home experiment kits (in science and technology) and various other items such as audio tapes, models, film strips, etc., where appropriate. Students can attend regular evening or weekend class tutorials at the 250 or so study centres throughout the country, and, on science-based courses at least, are obliged to attend one-week residential summer schools held at various host universities. A large proportion of the study centres are equipped with computer terminals which can be used by students on those science-based and mathematics courses which involve computing work.

A student attains an ordinary degree by accumulating six course credits; eight course credits are required for an honours degree. Each full course consists of 34 weekly 'Course Units' of work (about ten to fifteen hours' work a week), and to obtain a credit for a course a student must perform satisfactorily both on continuous assessment (regular tutor and computer marked tests) and on a final examination. At the moment the equivalent of 16 full-credit courses are available; by 1975, this will have increased to at least 45.[1] There are currently about 35,000 students taking Open University courses as part of their degree programme. There are also some students taking single courses for particular vocational or personal reasons without the intention of obtaining a degree.

General conclusions

There are several general points worth making at this stage. Firstly, that many existing multi-media distant teaching systems may have only a temporary or stop-gap role to play. This is perhaps especially the case in countries like the Soviet Union, Poland and Japan, where effective combinations of correspondence tuition, broadcasting, and face-to-face teaching have played (and still do play) a very important role in providing badly-needed educated and technically skilled manpower. But as conventional intra-mural facilities and trained teachers become more widely available, and as the backlog of adults who missed much of their schooling dries up, so these systems become less widely used.

This appears to be the case, for example, in the Soviet Union, where the rate of enrolment on correspondence courses for secondary specialized and higher education courses has now decreased, while intake into conventional day schools is rising.

In Japan, as we have already mentioned, the role of the NHK Senior High School is now being re-examined. There is a parallel here with the Italian Telescuola, which was designed for school children in areas where formal schooling was not available. During this time it was fairly successful, but then became unnecessary as more and more local schools were built.

Systems set up for these reasons were generally seen as being second-best alternatives to conventional intramural schooling. The danger is that they may become too inflexible and be unable to adapt to a change role when universal basic intra-mural education is achieved in the countries concerned. This changed role, which is beginning to make itself felt in some of the richer

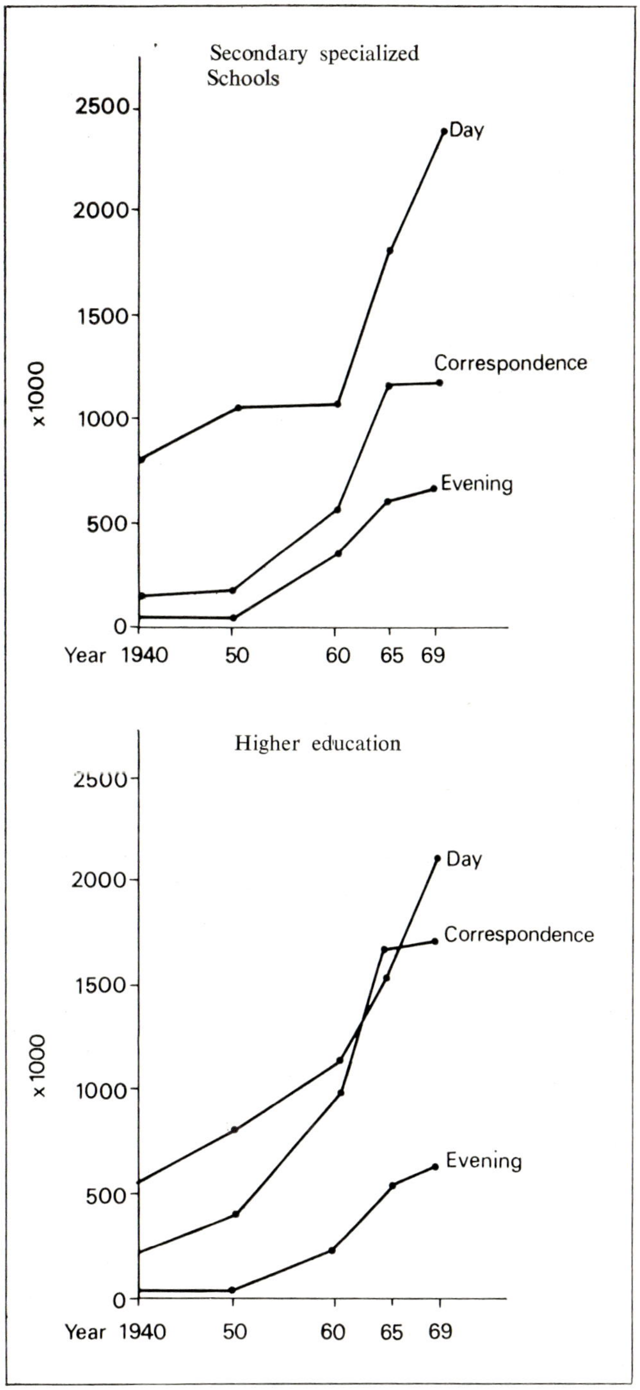

FIG. 1. U.S.S.R.: Numbers of students in all disciplines in secondary specialized schools and in higher education, in thousands. Source: *Narodnoe khoziaistvo SSSR v 1969 g.* Moscow, 1970, p. 675.

---

1. See Appendix 1 for list of current and planned course titles in sciences, technology and mathematics.

Western countries, is one of providing educational facilities which will be chosen in preference to conventional schooling—where study at home, at one's own convenience, becomes a virtue rather than a necessity. This implies the provision of a variety of courses and materials designed to satisfy a wide range of needs (professional updating, vocational training, cultural enrichment, etc.) with the minimum of barriers to access. The success of home study systems in terms of student achievement is relevant in this context.

For example, in the first years of operation of the Open University's Science Foundation Course, over 70 per cent of finally registered students obtained a course credit (i.e., passed on the basis of examination and continuous assessment).

Reports on many of the systems mentioned above stress the high motivation of students; where comparative studies have been carried out, they demonstrate that student achievement can be as good as, if no better than, that obtained in intra-mural institutions teaching the same curricula (e.g., Chicago TV College, NHK Senior High School, and Télé-CNAM) [5]. In many cases, even the amount of personal attention received by students on distant study systems is considered of a better quality than that given in intra-mural institutions (see, e.g., Lefranc on distant study systems in France) [9]. This is a direct outcome of an analysis of the learning process which has concentrated face-to-face tuition where it is most needed, and not dissipated the time of teaching staff in presenting information in crowded lecture halls.

The second main point we wish to make concerns the role of teachers within integrated multi-media systems. It is unlikely that the full potential of such systems will be realized when they are doing no more than providing duplicates of existing intra-mural school and university curricula; this is because such curricula, developed specifically within an intra-mural context, place certain constraints on both content and teaching method—it is disturbing, for example, to see standard school and university textbooks carrying much of the teaching load in some distant study systems, and television being used to give 'lectures'. Even more importantly, it is because course preparation in such systems is carried out predominantly by intra-mural teaching staff who find it difficult to realize the extent to which the media can shape and improve their teaching, and who find it difficult to adopt an integrated team approach to course preparation.

The Open University was fortunate in this respect because in its design it profited from the experience of existing institutions—the necessary educational technology expertise, and a novel team approach to producing courses could be built into the system from the beginning; (the nature of the analysis which led to the particular design of the Open University is described by Neil [10]). An interesting outcome of the nature of such a system is that the term 'teacher' becomes inappropriate '... it could equally be applied to the subject specialist working centrally, the subject specialist as an intermediary, the television and radio producers, and the educational technologist...' [1].

The Open University is not, of course, the only example of an approach to preparation of teaching materials which has brought about a change in the traditional concept of the role of teachers. The Deutsches Institut für Fernstudien (DIFF) in Tubingen, established in 1967 on a grant from the Volkswagen foundation, is working along similar lines in the preparation of multi-media packages, including broadcasts, for delivery in intra-mural universities, and for in-service training for teachers [12]. It is not unreasonable to suppose that these new developments in preparing teaching materials will soon start having a valuable impact on teaching in conventional institutions. After all, students in many large intra-mural universities and colleges have little more personal contact with their teachers than students studying at a distance. For example, after only eighteen months of operation, the widespread availability of Open University materials in the United Kingdom is arousing a great deal of interest in traditional educational circles.

The influence of educational technology on the design of a multi-media distant study system—the Open University

A general description of science-based courses

Course components

The main components currently being used in science-based courses at the Open University are shown in

Figure 2. They have been divided into those used by students, those used by part- and full-time tutorial staff, and those needed for the administration of courses.

The list is more or less self-explanatory—it is obvious that the wide range of course components, and the fact that most of them are specially prepared by the University staff, necessitates both a complicated production schedule and a team operation. The broad outline of the pattern of components used is determined to some extent by fixed budgetary constraints. Each course is allotted a budget, the amount depending on the credit rating of the course (courses are either full- or half-credit—34 or 17 weekly 'Course Units' in length). Each Course Unit invariably (at the moment) includes one television broadcast; radio broadcasts are allocated either to each Unit or to every two Units. The part of the budget allotted to broadcast production is fixed, and cannot be used for funding other components; the broadcasts, for this reason, and because of their fixed transmission times, represent the least flexible, but possibly one of the most valuable components of each course (see section below on the use of television and radio in science courses).

There are two other fixed budget heads—one for home kit equipment, the other for tutorial and assessment functions. Within these budgetary constraints, which may well be changed in the light of experience over the next few years, each team is free to use whatever components seem most suitable for the subject matter in question. Not all the components shown in Figure 2 are thus necessarily included in every course (with the exception of the broadcasts), and Course Teams are free to suggest further components within the limits of the budget provided. In most courses, the main teaching load is carried by specially written cor-

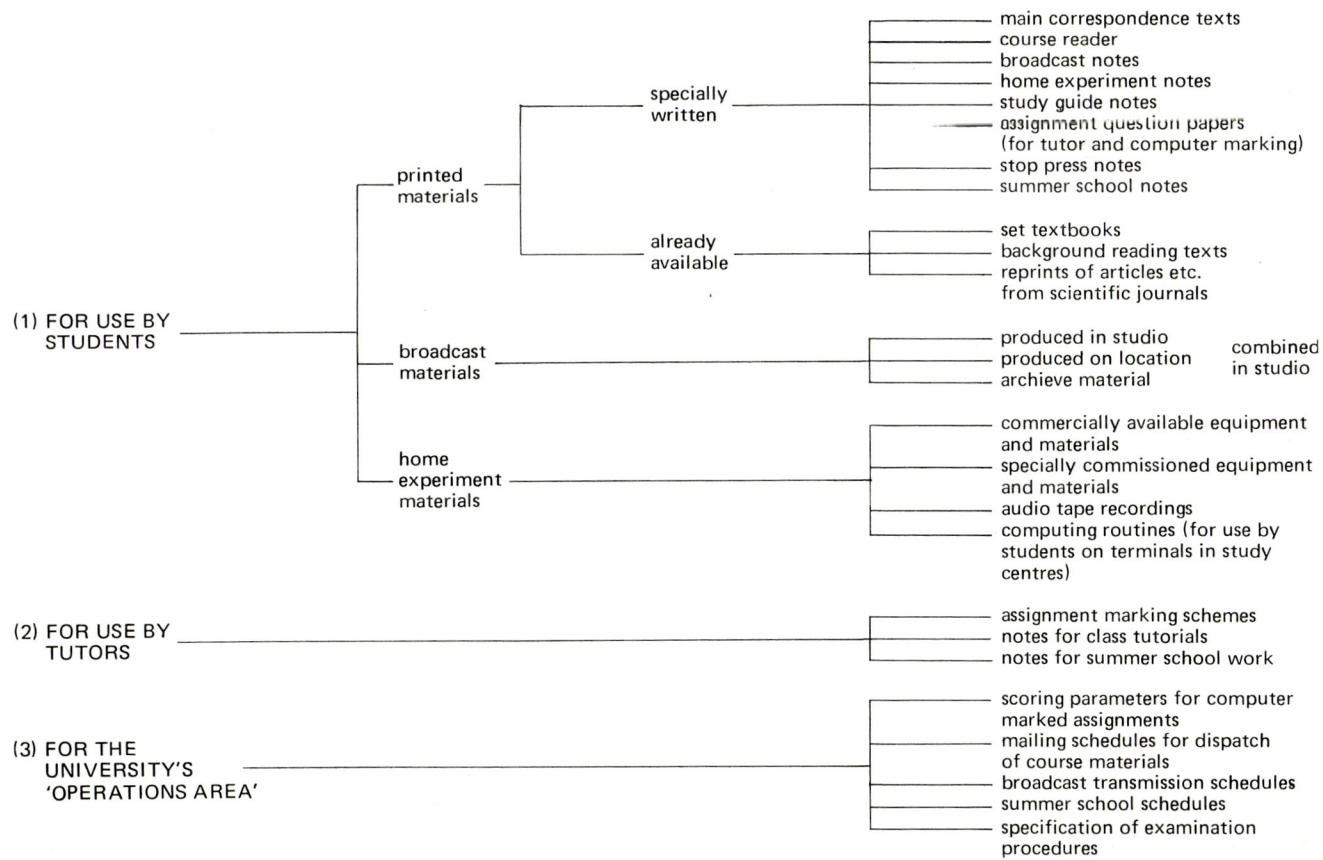

FIG. 2. Main components of Open University science courses

respondence texts, which are specifically designed for self-instruction, and differ in several important features from conventional textbooks. These texts are designed to remain basically unchanged throughout the life of the course (four years). The other printed components can be changed from year to year, either in the light of student feedback, or because of new developments in the discipline in question, or as a matter of policy (the tutor- and computer-marked assignments). Furthermore, there is a provision for remaking a proportion of the television and radio broadcasts for the second year of transmission. On several current courses, students are provided with cheap cassette tape recorders, and the audio-tape components of such courses can be changed from year to year fairly inexpensively.

We have already pointed out that one of the features which distinguish the Open University from the majority of the other distant study systems discussed in the previous section is the very high proportion of materials that are created by the university, and the much lesser use of existing textbooks, etc. This is partly related, of course, to the fact that we are not teaching within the constraints of a fixed syllabus; but there is a second, much more important factor. The first courses produced in the university were four interdisciplinary foundation courses (in humanities, mathematics, social science and science), and these courses have to some extent set the pattern for the second- and third-level courses currently being given and produced. It was soon made apparent that if these foundation courses were to achieve their objectives, and actually teach people with a wide range of backgrounds, many of whom had not studied for a long time, then we could not rely on existing books to carry the main teaching load. Furthermore, in most of the areas covered, certainly in science, there were no existing texts that would have been suitable. Accordingly, a lot of effort was put into writing new materials, and writing them in such a way that they contained ample opportunities for students to monitor their progress as they worked through them.

Even for higher level (more specialized) courses, there are problems in using set books. Very few university-level texts in science are suitable for fitting into a multi-media system that relies almost exclusively on self-instruction, and it is difficult in many areas to find texts which cover the material which a team wishes to teach and which are also up-to-date. Furthermore, there is a limit on the amount of money which students can be asked to to spend on books, and local lending libraries do not hold a wide range of specialized science texts. So for these and other reasons, existing textbooks, although used on nearly all courses, cannot form the main 'spine' of a course, and the correspondence texts, specially written by the Course Team members, continue to be the main components even on the more specialized courses. In some cases, they may be supplemented by course readers, containing papers and articles reprinted from other sources, or specially commissioned from scientists outside the university. Correspondence texts and course readers (but not the other printed materials) are on sale in bookshops for people who wish to obtain them without enrolling as students. An interesting 'spin-off' effect of the university's operations is that many standard texts, prescribed as set books for use on courses, have been reprinted in cheap paperback editions because of the vastly increased sales market opened up by the university.

Students spend between about eight and eighteen hours each week studying on each science-based course. The way this time is divided up between different components depends on the nature of the material, the level of the course, and the students' individual interests and backgrounds. In the science foundation course, notional times on different components were given in the course study guide as:

Studying the correspondence text	6 hours
Answering self-assessment questions	1 hour
Receiving TV and radio programmes	1 hour
Doing home experiments	2 hours
Answering computer-marked assignment material	½ hour
Answering tutor-marked assignment material	1 ½ hours
Attending a study centre	1 hour
Total	13 hours

On second-level courses, the main difference from the foundation course is in the time spent in group tutorial work and its organization. Current second-level courses have three-hour Saturday morning tutorials at two-monthly intervals (i.e., three such sessions for each half-credit course), whereas study centre tutorial classes are held at fortnightly intervals on the foundation courses. However, these three-hour sessions are generally held in laboratories, so it is possible to give students much greater opportunities for practical work experience than at study centres. The laboratories used are those at technical colleges and universities in the various regions, generally in institutions where the tutors teaching Open University students work full-time. Despite the fact that course-based tutorials for higher level courses may not be given at study centres, students on such courses still use the centres for other purposes —to meet students, to play back copies of the television and radio broadcasts (on film and on audiotapes respectively) or to talk to their counsellors. Counsellors are part-time members of staff, recruited from local institutions in the same way as tutors. They provide advice on general study problems and choice of courses, and, in many cases, general subject-matter problems (the more specialist, discipline-based tutor handles the more complex subject-matter problems, either in tutorials, or by correspondence, tape, or telephone).

Students on science-based courses are obliged to attend a one-week residential summer school for each course (or full course equivalent) in which they are enrolled. These schools are each staffed by a small nucleus of central faculty staff, together with tutors from other universities specifically hired for each one-week period. These tutors are nearly always already 'in the system' in that the great majority of them also act as part-time correspondence tutors. Time at the schools is predominantly devoted to intensive laboratory work sessions (from 09.00 to 17.00 each day) supplemented with seminars, talks, and problem-solving sessions in the evenings. In biology and earth science schools, the laboratory work can be integrated with field work (in fact, one of the critical factors in choosing university locations for some science summer schools is the fauna, flora and geological characteristics of the area).

The nature of the laboratory work carried out at the summer schools is discussed in more detail in the section below on the teaching of science practical work.

Designing materials for distant study

The particular design of the course components used by the students was influenced by three major factors:
the learning mode is essentially a self-instructional one, with only limited facilities for questioning tutors;
the students have a wide range of backgrounds (educational, occupational, etc.) and interests; they are all adults, many of whom have not done any serious studying since leaving school;
the system had to be as flexible as possible within the given constraints.

With the exception of the variability in educational background associated with an open-entry policy, these are all factors which are of major importance for any multi-media distant study system.

The first of these factors has implications both for the design of the components (correspondence texts, broadcasts, etc.) and for the design of the system (constant monitoring and evaluation must be built in as a matter of routine). The matter of evaluation, together with formal assessment procedures, is covered in the section on assessment and evaluation. At this stage, we intend to consider the ways in which the components, especially the written ones, have been particularly designed for the self-instructional mode.

Appendix 2 shows examples of pages from various science course correspondence texts; in order, these show:
1. A unit study guide sequence;
2. A table of the main terms, concepts, and principles used or mentioned in the course unit (Table A);
3. A set of unit learning objectives;
4. A conceptual diagram;

TABLE 2. Functions of Open University science correspondence text items

	(1) Study guide	(2) Table A	(3) Unit objectives	(4) Conceptual diagram	(5) Student-active components of corr. text	(6) Self-assessment questions
Indicates unit 'entry requirements'	√	√				
Suggests learning strategies	√				√	
Specifies a basic 'mastery level'			√		√	√
Provides feedback					√	√
Gives overall picture of structure of material covered		√	√	√		

5. An example of a page from a unit text, showing an 'activities-centred' approach to presenting the material;
6. A page from a block of self-assessment questions (SAQs)—such questions are included at the end of each main unit section.

Table 2 shows the main functions of each of these items, most of which appear in all the specially written Course Unit correspondence texts.

Taken together, these items provide a fairly comprehensive structure within which the student can monitor his own learning, and acquire an overall picture of the Course Unit material. The structure variously specified in the objectives, in the table of terms and concepts, and in the conceptual diagram, refers of course to the whole unit (the written materials, broadcast materials, and experimental work). The questions and problems posed in the tutor- and computer-marked assignments are generally cross-referenced to the specific unit objectives, and the resulting grades and comments provide a valuable, if somewhat delayed, further channel of feedback.

A few comments on the items shown in Appendix 2 are apposite at this stage.

The study guide sequence shown from the foundation course in science, was produced after the first year of operation of the course (1971), as a result of some of the study problems that arose.

Feedback received during 1971 indicated that many students were having problems in identifying the most important parts of a given unit, despite the indications given in the unit objectives, in the table of main terms, and in the self-assessment and assignment material. One of the main problems seems to be to convince students that, for any given unit of material, there are different levels at which they are free to study it, and that the primary objective of the course is not to get them to remember every single piece of information presented. This is why we are now including such 'study guide sequences' in many of the Course Units.

Similarly, one of the major reasons for including lists of learning objectives in each unit was to put across a related point—that there is a distinction between learning based essentially on recall of definitions, facts, etc., and learning of higher-level skills and abilities associated with the interpretation of facts and the production of novel rearrangements of given items of information. This is one of the greatest problems in trying to teach at a distance—that of convicting students that a course is not trying primarily to fill them with bits of pre-digested information which they regurgitate on demand. In this context, the pre-specification of learning objectives has both advantages and disadvantages. Certainly, one can give these objectives to students to stress that those based primarily on recall of information are of much less importance than objectives concerned with interpretive and synthetic skills. But the very process of specifying such objectives and giving them to students reinforces the impression of a model of the educational process in which learning is completely predetermined, and in which there is no place for originality. For this reason, we stress to students that the objectives given represent a basic level of mastery of the unit materials, sufficient for them to achieve an adequate level of competence on the formal assessment criteria, but that they do not preclude the production of novel interpretations and analyses. Indeed, a proportion of the tutor-marked assignment materials are specifically designed to identify such work, increasingly so in higher-level courses.

There are other reasons why listing such objectives does not provide a total specification of all the intended outcomes of a given course. An obvious problem is that of defining the level of specificity or generality of a set of objectives, and the ways in which illustrative test items can be derived from them. More fundamentally, it is questionable whether the inter-relationships between complex concepts in subject matter of this level can be described adequately by a list structure such as a set of objectives.

But, given these reservations, the value of such objectives in defining and communicating a basic level of mastery within a subject area cannot be denied. The extensive sets of such objectives [1] being produced for science courses at the Open University play an important role both in communication within the course teams and with the students, and also in the rational design of test materials.

The conceptual diagrams included in many of the Course Units represent an attempt to show the relationships between the main concepts discussed in the unit in a way that is not possible with a list structure. It is apparent that students find these diagrams most useful when they have completed their study of a unit, as an aid to consolidation. They can also act as a useful stimulus to students to attempt to produce their own conceptual diagrams, which may quite justifiably differ from those given in the text.

The table of terms, etc. used in each unit (Table A) was originally envisaged purely as an aid to communication within the course team. As can be seen from the example in Appendix 2, each term is listed as either being taken as a prerequisite (from general knowledge, from an earlier unit, or from another course), or as one that is introduced in the unit itself (to be developed in that particular unit or in a later one).

These tables are valuable in course preparation, as they enable authors to see quickly how their own unit subject-matter fits in with that of the other units being written by their colleagues, and it was decided in finalizing the science foundation course to include them in the units as they were printed for the students.

The presentation of the teaching material in the correspondence texts generally differs from a straightforward textbook style in that attempts have been made to actively engage the student in manipulating the terms and concepts being discussed. Some correspondence texts have been written as classical programmed texts. Others, like the example shown in the appendix, involve the student in plotting graphs and carrying out observations with instruments from the home kit. Still others contain long structured exercises involving the students in analysing, for example, the data from a series of related experiments. Even where texts are predominantly written in a normal prose style, there are always problems and questions built in at intervals to allow students to monitor their progress.

Blocks of self-assessment questions (cross-referenced to the list of unit objectives) are included at the end of each major section of text, and especially after any excursions into sections of a set textbook or course reader. Model answers and comments on these questions are provided at the end of each unit. Such questions play a particularly important role in identifying the relevant points in parts of set texts, and students are generally asked to do such reading with a prior series of questions and problems in mind. In designing these questions, and deriving them from the unit objectives, care is taken to see that they are representative of the sorts of questions which are asked in the regular computer- and tutor-marked tests.

At the beginning of this section we mentioned the need to make the unit materials as flexible as possible, and this means both from the point of view of coping with a wide range of previous educational backgrounds amongst our students, as well as catering for different 'styles' of learning. We have approached the first problem, in the foundation course at least, by structuring the text into three 'channels'. The middle channel constitutes the main text, the lower channel contains additional remedial material for students experiencing difficulty, as gauged by performance on the self-assessment questions, and the upper channel elaborates the theme of the main text for students who wish to follow up particular points. In this way, we hope to avoid boring the student who already knows some science without frightening off the beginner. Material in the upper channel is not assessed in the assignments or examinations, but students who wish to go on to a higher level course in the particular area are advised to read it. The concept of a lower channel of remedial and support material has proved particularly valuable for giving students who are poor in mathematics the necessary mathematical background to the foundation course. The second problem, that of designing the courses in such a way that they communicate effectively to large numbers of students without imposing too rigid a pattern of study, is not so easy to deal with. For a start, the teaching programme is paced to a large extent by the regularity of broadcast transmission dates and the due dates for submission of assignments. At the moment, nothing can be done about the broadcast 'pacers' without entirely decoupling the television and radio components from the correspondence texts; and in any case, 'pacing' is held by some to be an *advantage* for the independent learner. As for the second 'pacer'—the assignment due dates—it is difficult to see how one can operate a system of continuous assessment which supplies feedback to students without having fixed due dates. And we felt it would be quite wrong to base the entire assessment of student achievement on an end-of-course examination alone.

It has been suggested that a valuable function of the different media in a multi-media system is to provide a degree of redundancy (by presenting the same materials on different media) on the assumption that some students learn most effectively from television or film, and others more effectively from printed materials [2]. As far as the television component goes, we could not afford to adopt this policy, as so much of the course materials in science require a visual, moving medium for their presentation. The best we can do in this respect is, for example, on the courses on which students are loaned tape-recorders, to present some of the material already covered in the text in a slightly different manner (e.g., as a discussion between the author and a colleague). Radio can also be used in this way, of course, but audio-tapes represent a more flexible approach as they are not tied down to transmission times and can be replayed at will. Another possible use being considered for tapes is for communication between students and tutors, even as an optional alternative way of submitting assignment work. On some science courses, tapes are being used in connection with home experiments. As far as the printed materials are concerned, there is the possibility of including with each unit some sort of 'study order choice guide' which would help the student to select different paths through the Course Unit depending on which parts interest him most. To do this properly would imply quite a radical restructuring of the way in which the texts are planned, however. But, even as they are designed at the moment, it is far easier for students to adopt different strategies and sequences of learning by judicious use of the objectives, self-assessment questions and tables of main terms, etc., than it would be if they were studying from a conventional textbook.

The use of television and radio in science courses

In most of the science courses, the pattern of use of the television component of each unit follows that established on the foundation course. Each Course Unit contains television material which is more or less closely integrated with the correspondence text, and in many cases, with the home experiments and assignment based on the unit in question. Television is used to cover objectives which would be difficult or impossible to achieve using the other media (see, for example, the objectives associated with dynamic, process or sequence meanings listed by Kaye [1]). It also, of course, plays a general motivational role in helping students to identify with the course team, and to overcome their feelings of isolation.

The great majority of the programmes are made up of two or more items, often unrelated, which play specific roles *vis-à-vis* certain parts of the correspondence text or home experiment or assessment material. Items for inclusion in a broadcast are generally selected on one or more of the following criteria:
to enable the student to see and often to 'participate' in the experimental work necessary for a thorough grasp of the material taught in the correspondence text, and which he cannot carry out with his own home kit. In some cases, such experiments might be impossible for students at a conventional university to share in (e.g., bubble chamber experiments with large particle acceleration) because the equipment is too large or complex or expensive. In some programmes, students actively participate in the experiment being shown by, for example, taking readings off the screen and then using the data as the basis for part of a tutor-marked assignment;
to provide, where necessary, guidance and assistance for the student in his own experimental work;
to teach conceptually difficult material where visualization of dynamic structures is essential—e.g., in explaining many *processes*. Here devices such as time-lapse photography and animation are of obvious value;
to give the student a vicarious experience of, e.g., field work and laboratory work through the medium of film, either specially taken for the Open University, or from existing film archives.

It is generally the case that there is far more material which merits inclusion in television programmes on these criteria than there is transmission time available.

The programmes are linked in to the remainder of the unit components by the pre- and post-broadcast notes which respectively inform the student which material he should have studied before the programme derive maximum benefit from it, and what activities or to problems he should now try after having seen the broadcast. The post-broadcast notes invariably include a summary of the programme, to aid later recall, and often contain self-assessment questions and comments; or they might refer the student to a home experiment which involves him in analysing data he has seen produced in an experiment on the programme. One example of such an approach is from a unit on high energy physics in the foundation course, where the television programme is devoted to showing the various processes involved in obtaining bubble chamber pictures of nuclear interactions. The student is then supplied, in his 'home experiment' material, with a set of real bubble chamber pictures, templates, and energy-loss curves and has to analyse the interactions, then summarize his findings and send them to his correspondence tutor as part of a tutor-marked assignment.

There are certain problems associated with such close integration of the television component with the remainder of the unit material. We have already touched on one of these—the way in which the transmission times pace the student's work on the unit as a whole. Each programme is transmitted twice, once on a weekday evening, and once on a Saturday or Sunday morning, and this provides a limited amount of flexibility. (Open University TV transmissions are on the BBC 2 network between 17.30-19.30 on weekday evenings, and between 09.30-13.00 on weekend mornings.) The regular broadcasts can be seen as a virtue in that they provide a stimulus for students to keep 'on schedule' with their studies, but they nevertheless militate against students working at their own pace. There are also production problems associated with tight integration of the broadcast components—such integration implies that the programmes can only be finalized when the rest of the unit materials are nearing completion. Yet increasing pressure on our studio facilities means that a certain number of programmes inevitably have to be made when the remainder of the unit is only in draft form. For these and other reasons, some course teams are experimenting with the possibility of designing tracks—one essentially based on printed materials, one on the television broadcasts, and one on the radio broadcasts. The three separate tracks would then be synthesized toether in the final few units of the course.

The material of the radio programmes is often much less closely integrated with the remainder of the unit teaching material than that of the TV programmes and they tend to be built around a definite theme and form self-contained programmes. The radio programmes fulfil, in the main, valuable 'enrichment' functions, such as:

bringing well-known scientists and other outside speakers into contact with the students, through discussions with Open University staff;

bringing the students a flavour of the excitement and fascination of science, 'to show why scientists find science fascinating'; this can be done supremely well in the intimate atmosphere created by a discussion amongst a small group of speakers, in a way that is almost impossible using the written word;

placing the material taught in the correspondence text into its social, philosophical and historical background.

However, radio is also used extensively for reinforcing the teaching of more difficult points, often with the aid of charts, diagrams, etc., which the student is asked to refer to during or before the programme and which are essential to a full understanding of what is being discussed.

Each radio programme is about 25 minutes long, but we have found it valuable to restrict the actual 'programme' material (made well in advance) to 20-22 minutes, and to keep 3-5 minutes spare for last-minute announcements which can be recorded up to a few hours before the programme goes on the air.

Various other uses which have been suggested for radio, and which several course teams are trying this year, include:

giving recorded discussions and 'model answers' on computer-masked assignment questions, based on the analysis of student responses provided by the computer, or on other feedback data such as special questionnaires, or information from part-time tutors;

holding 'teach-in' sessions on points of particular conceptual difficulty in the texts, based on the learning difficulties experienced by the *previous year's* intake of students.

The teaching of science practical work

We have already mentioned the two main channels of practical work—the home kits and the summer schools. The activities and experiments associated with the home kits are generally closely integrated into particular Course Units, whereas the summer school work is more or less decoupled from any given unit. The fact that our students must do the great majority of their work at home was a major stimulus for us to provide integrated practical work right through each course. This contrasts strongly with traditional university teaching where courses of 'lab. sessions' often seem to be given more or less in isolation from the material being taught in a parallel course of lectures.

But teaching of practical work is not restricted to use of the home kit materials and work at summer schools; we have looked at the objectives of our teaching of experimental work under several main headings. Briefly, these consist of objectives associated with: (i) development of manipulative skills; (ii) accurate observation, measurement, and collection of data; (iii) analysis and presentation of data; (iv) interpretation of data and critical evaluation of experimental design; (v) production of novel suggestions, e.g., for designing further experiments.

Most people would agree that these are ranked in order of increasing importance. And if it is categories (iii)-(v) above in which we are particularly interested, then there is no reason why the operation of the experiment and the associated data collection cannot be shown on the television, and the students given the data to analyse, interpret and evaluate. In fact, we mentioned just such an experiment in the preceding section of this paper.

It would, however, be unwise to rely entirely on vicarious exposure to experimental work. This is where the summer school is so important, for here we can supply students with a much wider variety of materials and instruments than in the home kits, and they have the chance to work on open-ended experiments with the opportunity to test alternative explanations of the phenomena they are investigating.

However, the use of open-ended experiments is not restricted to the summer school. It is not impossible to set up true problem-solving experiments using home kit materials. The first experiment in the foundation course in science involved students in trying to work out the relationship between the time taken for a ball to roll down an inclined plane, the distance rolled, and the inclination of the plane, as in the original Galileo experiment. The second part of the experiment required them to critically examine the reason for the inconsistency which Galileo found in extrapolating his results to the case of free fall. In one of the more advanced second-level courses ('The Biological Bases of Behaviour'), one experiment involves students in investigating whether goldfish can distinguish between different colours. The experiment itself runs over a course of 16 days, involving initial training of the animals, and a series of experimental trials under different conditions. Care must be taken to devise an experimental design with a kit on a second-level course on electromagnetics in the same course is designed to test the implications of a particular theory of selective attention by using a tachistoscope to present short unfinished sentences to an experimental subject. Analysis of the data from the experiment introduces the student to the use of simple statistical methods. A cathode-ray oscilloscope (supplied with a kit on a second level course on electromagnetics and electronics) is used to examine the characteristics of different electronic circuits which the student can design. These, and many other home kit experiments, provide examples of 'open-ended' work in which the main responsibility for designing the experiment, and interpreting the results in the light of the design, rests with the student.

Many of the home kit projects deliberately exploit the opportunity for using students to collect mass data, which can then be analysed centrally, and the results 'fed back' to the students via a television or radio programme. In the science foundation course, students are each asked to take measurements of atmospheric contamination by sulphur dioxide. The measurements are taken under standard conditions, at fixed times throughout the country, using a sampling apparatus provided with the kit. Students return their results on forms which can be fed directly into the university's computer; the results of last year's experiment aroused a great deal of interest [3] and the intention is to repeat the experiment each year over the next three years.

In the second-level course on 'The Biological Bases of Behaviour', one of the projects involves students in carrying out a habit reversal experiment, using a tank and various cue materials provided in the kit, and goldfish which students buy. Each student will send his results to the university, again on a computer-read form, and the statistical analysis of all the data will be made available to the students.

Many of the home kit materials are used, not for carrying out *experiments* in the strict sense of the word, but for providing relevant 'experiences' which illustrate particular points being made in the text. Examples are the use of a hand spectroscope to examine different flame spectra in conjunction with study of spectral lines in the correspondence text, observation of the study of co-acervate drops under the microscope as an illustration of one theory of the origin of life, microscopic examination of prepared slides of tissues, and examination of a pickled sheep's brain in conjunction with study of a text on brain anatomy.

An important result of the production of home kits on such a large and economically feasible scale is the design and mass production of new instruments which otherwise might never have got past the prototype stage. Some examples are the Open University McArthur microscope (a compact hand microscope manufactured in plastic), a ball-operated binary calculator and tutor (BOBCAT) which is used to demonstrate mechanical binary counting, and a sound level indicator (Noisemeter). The BOBCAT and Noisemeter are both supplied to students on the technology foundation course—'The Man-made World'. Furthermore, several of the kits contain cheaper, specially designed versions of standard laboratory instruments. Examples of these are the Open University Foxall Colorimeter (which is battery operated), a tachistoscope, and an oscilloscope.[1]

On second-level courses, the laboratory-based three-hour tutorial sessions provide an opportunity for students to handle materials and instruments which cannot be supplied with kits. This is particularly important for biology students, for example, who can observe a much wider range of living and preserved specimens, with a tutor at hand to deal with problems and queries. These short sessions also act as a valuable chance to familiarize students with a 'real' laboratory situation, and to practise some basic manipulative skills (e.g., preparation of slides, use of sophisticated instruments) before they start the intensive summer school week.

It is too early yet for us to know how successful the combination of home kits, laboratory based tutorials and summer schools will be in giving students the necessary background in practical work. Indications so far are encouraging, but there are problems, especially for students who are completely new to science, and who may develop inappropriate techniques in using the home kit materials. This is one area in which the local turorial sessions are so important.

Course production methods

The total number of specially written correspondence texts and television and radio programmes produced, or in production, since the university started designing courses in 1970 is at the time of writing approaching 500. An equivalent number of the other course components listed above (with the exception of home kit materials, specific only to science based courses) must be added to these figures to give some idea of the enormous output that has been achieved in such a short time. This section of the paper is devoted to a very brief discussion of the procedures used in the planning, co-ordination and production of these materials.

In the science faculty alone, there are currently thirteen course teams in operation. Each team includes up to twelve academic subject-matter experts (one of whom is elected chairman), three or four television and radio producers, an educational technologist, an editor, and one or more course assistants. Some of the members may be working simultaneously on two or more courses—although this is more a matter of expediency than desirability.

The first task of the team is to decide on the broad outline and aims of the course they are going to produce. In the foundation course this was a particularly difficult procedure, as we knew very little about our potential students, and plans for the development of future courses, for which the foundation course would have to form a sound basis, were only very provisional. There was general agreement that the course should treat material from the four main disciplines in the faculty (biology, chemistry, earth sciences and physics) in an integrated way, that the course should make students aware of the social implications of science, and that no prior knowledge of science would be needed by students before they started the course. On the basis of these considerations, and the particular skills and interests of the scientists in the course team, it was possible to conceive a broad working outline on which to proceed.

For the second- and third-level courses, the nature of the initial decisions about course subjects can be refined to some extent by polling students already in the system about what courses they would like to see provided, and by building on and developing concepts introduced in the foundation course. Of relevance here is the fact that the University has adopted a policy of allowing students total freedom in selecting the courses they wish to take, the only proviso being that students must take two foundation courses, one of which must be passed before any higher level courses can be started. Students are advised, however, to take the science foundation course before embarking on higher-level science courses.

In their initial planning, as well as in considering content outlines, course teams need to take into account the given budgetary constraints, the projected completion dates (texts for printing must be ready at least twenty weeks before the date on which they are mailed to students) and the overall university studio recording, printing, publishing, mailing and data processing schedules. A special project control office co-ordinates the demands of all the different course teams on such central university operations areas.

The chart in Figure 3 is a simplified version of a more detailed planning network (see Lewis [4]) which shows the main steps in the development of a course from an initial brief outline to its completion. This process is essentially divided into three main phases.

---

1. Lists of home kit components included in various science and technology courses are included in Appendix 3.

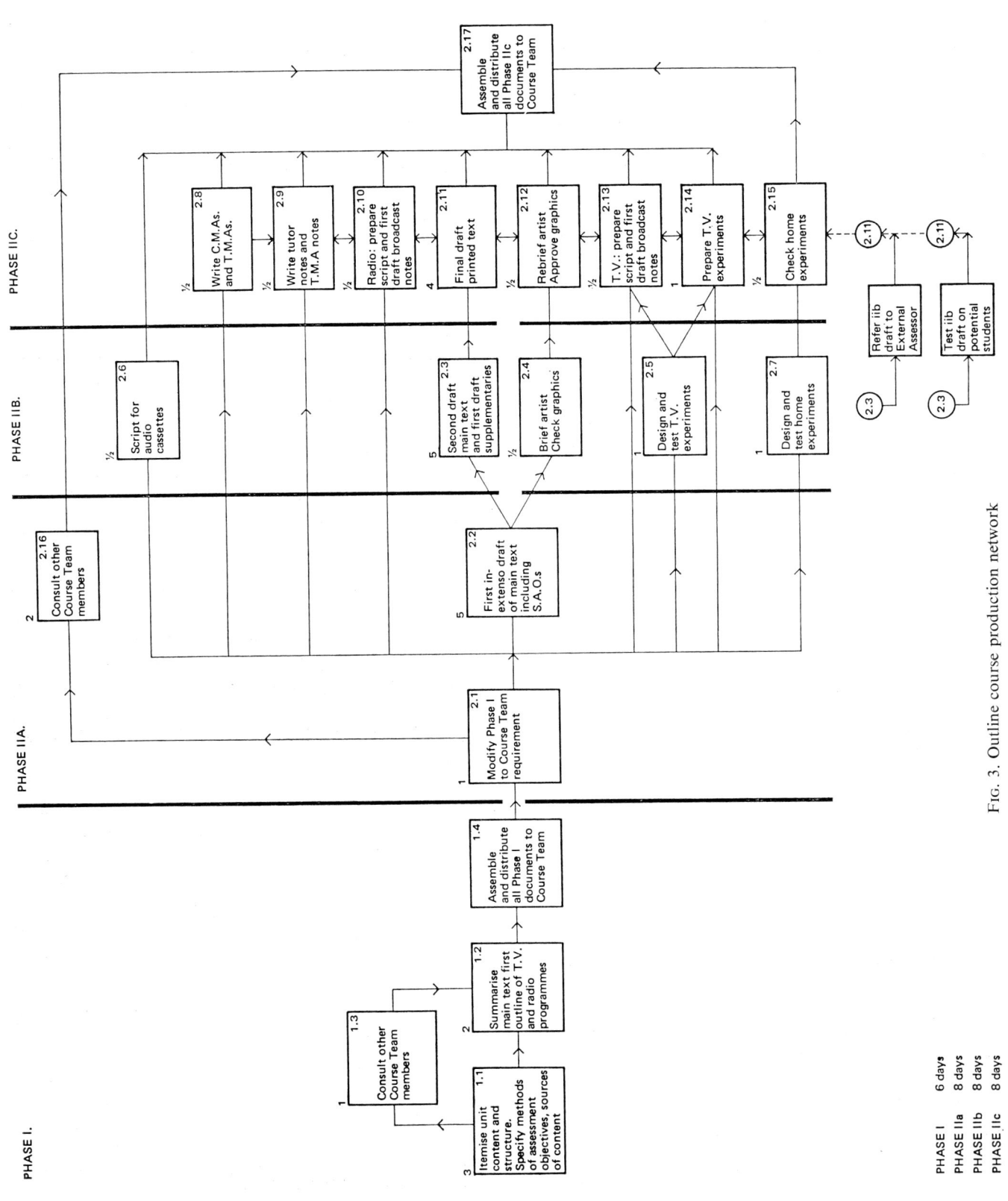

Fig. 3. Outline course production network

Phase I starts from the allocation of a number of course units (generally 1 to 3) to each 'course unit author' on the team. He is then required to produce a 'Phase I draft' of each unit by a specified date. The time allowed to prepare this draft is 6-8 days, usually spread over a fortnight. The Phase I draft includes the following items:

1. A statement of general aims—a short paragraph is usually sufficient.
2. A list of unit objectives—these are the objectives to be achieved through study of *all* components of the unit (correspondence text, set text, broadcasts, home experiments). The objectives are ranked to show their relative importance.
3. An indication of the methods by which the unit objectives are to be assessed—i.e., by self-assessment questions (SAQs), computer-marked assignments (CMAs) or tutor-marked assignments (TMAs).
4. A preliminary list of the things (concepts, important terms, laws, principles, etc.) that are
   (a) assumed to be known from earlier units of the course, or from prerequisite courses or from 'general knowledge';
   (b) introduced and explained in the unit.
   (This list has been found to be extremely useful, in avoiding mismatches between units within the course and between the course and prerequisite courses.)
5. A synopsis of the unit—written in sufficient detail to give other course team members a clear idea of the author's intentions, since they need to take these into account in planning their own units.
   The synopsis is accompanied by a topic network/conceptual model study chart. This is either a network showing how the various parts of the unit are related conceptually to each other, and to other units in the course, or it is simply a chart showing the route, or routes, the student should follow through the unit. Frequently both types of diagram are included.
6. An indication of set-book reading to be prescribed and background reading to be recommended.
7. Preliminary suggestions for the television and (where applicable) radio components of the unit.
8. An outline of proposed home experiments or an indication of home experiments under consideration.
9. A checklist of any special devices expected to be included in the unit, such as audio cassettes, film strips, maps or charts, offprints, etc.

During the preparation of the Phase I draft, the author consults the educational technologist about the formulation of the unit objectives, the forms of assessment to be used, and the design of suitable student activities. He also discusses with other course team members the inter-relationships between his unit and theirs, makes initial inquiries about possible home experiment kit components, and discusses ideas about the broadcast component of the unit with the BBC producers.

The Phase I draft is circulated to the course team and discussed at a course team meeting, where it is either approved or sent back for revision. Once the draft is approved, the author can start work on his first Phase II draft. Experience has shown that three distinct stages of preparation are required in this phase of the production of a course unit.

In Phase IIA, for which the time allowed is 8 to 10 working days, normally spread over a period of 3-4 weeks, the author produces his first *in extenso* draft of the main correspondence text, including other self-assessment questions, circulates them to the course team members and discusses their comments with them. During this phase of the operation, the initial objectives and content outline usually require some modification—the process is an iterative one.

In the light of course team comments, the author now proceeds to produce a second draft of the main text, and at this stage (Phase IIB) he prepares a first *in extenso* draft of certain *supplementary text components*. The supplementaries that can be drafted at this stage include:

1. *A Contents Checklist* (including non-print components, if any). Special items like maps, charts, offprints, filmstrips or audio cassettes, which may have been listed indicatively in Phase I, are specified definitely at this stage.
2. *A Unit Study Guide*. This may indicate to the student different 'routes' through the Unit corresponding to different levels of study commitment. It may tell the student which objectives and which sections of the Unit he may leave out if he is running short of time. And it can help him keep track of where he is in a Unit which involves study of a set textbook together with the main correspondence text.
3. Proposed test items for *Computer Marked and Tutor Marked Assignments*. The inclusion of these items, as with the SAQs, is essential if the Course Team, and the External Assessor (see below) is to

get a clear idea of the level at which the material is to be learnt.

4. Notes on the assignments for course tutors.

In addition to preparing the second *in extenso* draft on the main text and the first draft of supplementaries, the author carries out four other important activities in Phase IIB:

(a) He briefs the graphic artist in the design office on the preparation of diagrams and illustrations for the texts.
(b) He consults with the producers and with the laboratory technicians on the design and preparation of experiments, demonstrations, models, etc., intended for inclusion in the television component of the unit.
(c) If he intends to include an audio cassette in the unit, he prepares a script and makes a first recording. Both recording and script will be circulated to course team members.
(d) He completes the design and carries out the first tests on the home experiments. The main object of this phase of preparation of the home experiment is to check that the experiment will work under home conditions and that it can be done in a reasonable time.

The time allowed for Phase IIB is 8-10 days, normally spread over a period of 3-4 weeks.

The Phase IIB materials are circulated to course team members for comment, and at the same time are submitted to the external academic assessor and, whenever possible, subjected to some form of developmental testing. A period of 2-3 weeks must accordingly be allowed for these processes to take place.

The external academic assessor is chosen by the author, subject to course team approval, as an expert in the subject area covered by the unit, to comment on the accuracy, clarity and scientific validity of the material.

When funds and time permit, the complete set of Phase IIB materials is sent to a group of a dozen or so potential students for them to work through. This is of some value in identifying areas of special difficulty or ambiguity in the text. Students are asked to return the material with their comments and their attempts at the draft assessment material, and a meeting is then arranged between them and the author, once their comments have been analysed by the educational technologist.

There are various reasons why this development testing, although certainly useful, does not provide a fully satisfactory answer to trying out draft course material [5]. Firstly, it is not possible to try out the total 'package'—it is very rare that the television, radio and home experiment components reach a testable stage at the same time as the correspondence text. Secondly, the texts themselves are rough drafts, cheaply duplicated, without any of the professional design and finish of the final products—often without even the final diagrams and illustrations. Thirdly, the students find it difficult to put themselves in a situation where they are studying 'for real'. And then there are distinct differences in the ways in which different individuals feedback—some students cover the texts with useful comments, others ('low commenters') hardly make any written comments, and one has to tease responses out of them at a face-to-face meeting. There are also organizational problems in that it is not usually possible to send students such draft units to study in the same sequence as they will finally appear in the completed course, because each author is developing his units at the same time as the others; furthermore, production schedules often become so tight that there are always some units in a course for which there is insufficient time for testing. Finally, there is the problem that if an author does not like the comments made by the students, he can justify ignoring them on the grounds that the particular group of testers is unrepresentative, stupid, bloody-minded, etc. For these reasons, we feel obliged to mount our major evaluation and revision effort during the first year of operation of a course, and this aspect is discussed in more detail in the next section of the paper.

In Phase IIC, the author prepares a final draft of *all* printed text components of the unit. He will have had comments from the course team and from the external assessor on his IIB draft, and in some cases, the results of developmental testing.

The supplementary texts will now include first drafts of home experiment notes and of radio and television broadcast notes. These texts can only be finalized after the experiment has been tested with the equipment and materials in the form actually to be sent to students and after the radio and television programmes have been recorded.

In addition to finalization of all text materials (except home experiment and broadcast notes), the parallel activities in Phase IIC are:

1. Checking and approval of all graphics.
2. Preparation of draft scripts for the radio and tele-

vision components—usually done by the producer in consultation with the unit author.
3. Further preparation of television experiments, for studio recording or for pre-filming on location.
4. Checking of home experiments with at least mock-up or prototype of the actual apparatus and materials to be used by the students.

The Phase IIC materials are assembled and submitted to the course team for formal approval.

The time allocated for Phase IIC is, again, 8-10 days,[1] normally spread over a 3-4 week period.

The final phase—Phase III—of the production of a course unit includes the following activities:
1. Final preparations for the television programme, rehearsal and recording.
2. Rehearsal and recording of the radio programme.
3. Finalization of texts for broadcast notes.
4. Check of home experiments in the form in which they will be performed by students and finalization of the home experiment notes.
5. Approval by the course team of video and audio tape recordings of the television and radio programmes (which may, in the absence of such approval, have to be wholly or partially remade).

The text components of the unit now go to the faculty editors for preparation and marking-up for printing, and the unit author is required to check galley proofs and, finally, page proofs.

The time allocation for Phase III is 10-12 man-days.

The work of the course team is by no means completed when all the components of the course unit have been produced. The unit author is expected to produce draft questions for the end-of-year examination and to participate in the planning and preparation of the summer school programmes, including the design of experiments and the preparation of laboratory notes, summer school tutors' notes, etc.

Once the course is going out to students, a 'caretaker course team' is set up, combining a nucleus of members from the original team with one or more staff tutors.

In concluding this account of the course production methods developed in the faculty of science at the Open University, two general points are worth noting.

The first is that the minimum academic staff time allocated to the production of a single course unit is 40 man-days (or 400 man-hours). The average is certainly above 50 and in difficult cases (where more than the minimum number of drafting stages is required, or where a television programme may involve a lot of filming on location) the unit time may well exceed 60 days. If these figures are compared with the 8-18 hours spent by the student on studying the unit, the ratio of academic staff production time to student learning time is in the range 30-70 to one. At first sight this may seem excessive, until it is recalled that these course materials are necessarily of a highly structured character, in many ways comparable to programmed learning texts, for which such ratios of production to learning time are not uncommon.

A second general point which should be made is that, in giving this simplified account of the course production system, we have glossed over many problems which make the process far more complex than it might appear.

Firstly, it is notoriously difficult to make any accurate time estimates for completion of the various phases outlined in Figure 3. Different course teams operate in different ways within the system, and different individuals vary very much in their approach to writing and to making broadcasts.

Secondly, there is the problem of sequencing—ideally, it would make for closer integration between units if they were produced in series, one after the other, so that the authors of later units could know precisely what had been included in earlier units in the course. But such a 'serial production' method would add considerably to the total time required to produce a course, which already exceeds a year. So would the alternative device of 'stockpiling' units at the IIC stage and then having them revised and integrated by an editorial group from the course team.

Thirdly, there is the problem of different lead-in times for different components. If manufacturers are to have the necessary time to produce large numbers of home kit items, then they often have to be ordered before the final shape of the course has become clear. The same problem arises with provision of set books—publishers have to be alerted well in advance. Copyright clearance procedures for text and television programmes often have to be initiated at a very early stage.

For a more detailed discussion of these and other problems, and the reasons why they imply that

---

1. All the time allocations quoted in this section refer to academic staff or Course Unit author man-days and do *not* include the time of educational technologists, television or radio producers, laboratory technicians, graphic artists, academic 'course assistants', etc. A 'day' is reckoned to be a minimum of ten hours.

established production control methods such as PERT and CPM are not applicable to our situation, the interested reader is referred to Lewis [6].

Assessment and evaluation

Introductory remarks

The purpose of this section is to give a brief description of the related problems of student achievement and evaluation of teaching materials and procedures. Although the discussion will be within the context of the Open University, many of the points made will be relevant to any distant teaching situation in which courses are being given and designed by the same institution.

Any comprehensive assessment and evaluation system must fulfil at least three main functions:

formal assessment of student achievement: to determine which students, after a given course of instruction, have reached a satisfactory level of competence or mastery in the stated objectives for that course; this implies the careful integration of assessment material into the design of the course (hence our stress in the previous section on the simultaneous development of assessment material and teaching material);
continuous feedback: to give students the information and feedback which they need to judge their level of achievement and to modify their learning accordingly;
course evaluation: to measure the effectiveness of different teaching strategies and materials and to act in general as a 'quality control' mechanism on the basis of which effective improvements can be made.

To these might be added the function of identifying the excellent student who is performing at a qualitatively higher level than that indicated by the stated course objectives.

TABLE 3. Assessment and evaluation procedures

	Formal assessment of student achievement	Feedback to students	Course evaluation
Self-assessment exercises and problems		Model answers and comments supplied; (immediate)	
Regular computer- and tutor-marked assignments (CMAs and TMAs) Summer school assessment by tutors Terminal examination	by combining the various scores obtained	Absolute and relative grades obtained; model answers on CMAs and tutors' comments on TMAs; (3-6 weeks' delay)	Various analyses of score data Various analyses of score data

Letters and 'phone calls from students Regular reports from tutors University course unit reporting system		Replies provided can give rise to 'mid-course corrections' via stop-press mailings and radio announcements; used for modifications to course for the following year	Locate areas of difficulty; errors in texts
			Summarize comments from students and tutors in each region
			Information on:
			  study times,
			  viewing and listening patterns,
			  study centre visits,
			  interest and difficulty indices

'In depth' studies of learning with small panels of students			Systematically locate areas of difficulty and provide detailed information for modifying specific units

Obviously, these functions have to be deliberately built in to any distant teaching system from its inception, as the more informal channels through which at least a proportion of them are carried out in many conventional teaching situations just do not exist. Table 3 shows the various items used at the Open University to carry out each of these functions.

The various assessment and evaluation functions must be taken into account from the moment a course starts being written, and the necessary procedures for analysing and taking action on the information collected once a course is in operation must be the responsibility of the course team concerned. This, in fact, is one of the main functions of the caretaker course team which we have already mentioned. Figure 4 shows the various channels of communication between the caretaker course team, the students, the regional organization and the data processing division, without which we would be unable to monitor the effects of our course materials.

Formal assessment of student achievement

As we have already mentioned, the regular homework assignments and self-assessment materials are based on the objectives of the weekly Course Units. In ensuring that these questions representatively sample both the different objectives and different content areas of each unit, we found it useful to draw up simple matrices like the one shown below, which help authors to make explicit their assessment intentions.

Level of objectives based on	Course Unit section				
	1	2	3	4	etc.
1. Recall, recognition					
2. Comprehension					
3. Interpretation					
4. Evaluation, synthesis					

Notional times to be spent on different assessment components to be entered in matrix cells.

One particularly valuable result of this procedure is that it enables authors to maintain an appropriate balance between questions testing lower-level objectives (e.g., ones associated primarily with recall and recognition of facts), from higher-level ones concerned with interpretation and synthesis of material. Such a device is also useful in ensuring that the self-assessment tests are representative of those given in the formal assignments, and that objective questions and short-answer questions are used to test the particular skills appropriate to each testing mode.

Each science course contains a computer-marked assignment for every two units, and a tutor-marked one for every four units. Students return their answers to CMAs on special forms which are processed by a document reader linked to the university computer. They receive computer printed letters within a few weeks of submitting the assignment which inform them of the grade obtained. In addition, they can consult their counsellors for more detailed comments on each CMA, as the course teams provide counsellors with model answers. TMAs are submitted by students directly to their correspondence tutor, who marks and comments on the work on the basis of notes prepared by the course team, and returns the script to the student. He also sends in to the university the marks he has given to be recorded on each student's file.

Each student is awarded an overall grade for continuous assessment based on the best three-quarters of the assignments he has submitted during the year (this allows sufficient flexibility to cater for personal problems students may have in completing all the assignments). Each student also receives a grade for his work in the end-of-course examination. These examinations are held at centres throughout the country. Students who receive pass grades or better on *both* the examination and continuous assessment receive a credit for the course, whereas those who fail on both do not receive a credit at all. Students who pass on the one but fail on the other are considered as borderline students, and an assessment made by their tutors at the summer school (on a three-point scale) is taken into account in the arbitration procedure.

The fact that there are two elements in the assessment scheme (the examination and the summer school assessment) which are based on work done under supervised controlled conditions, is sufficient to act as a check on students who might consider playing the system by getting somebody else to do their homework

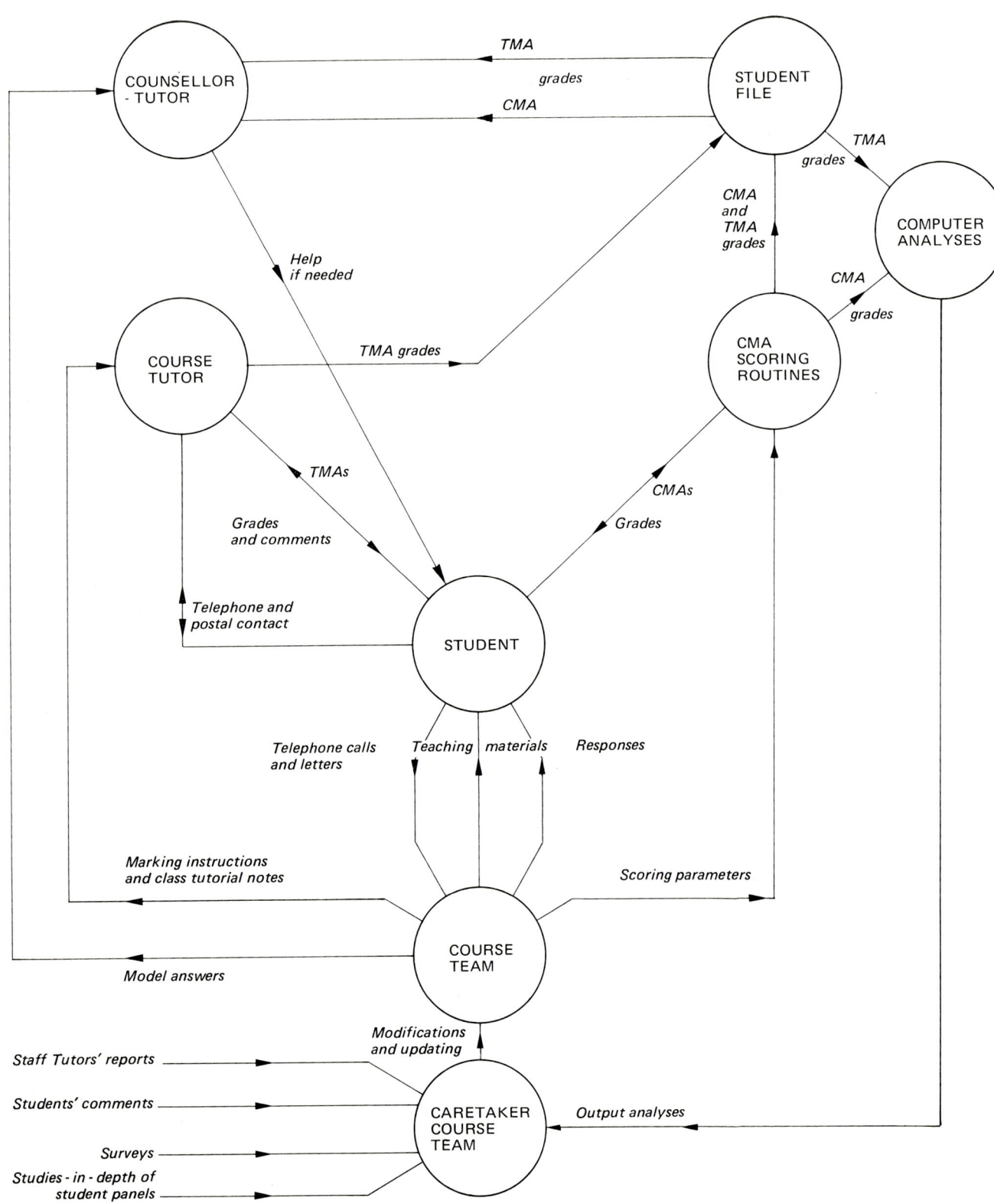

Fig. 4. Assessment and evaluation: communications within the Open University system

assignments for them. In fact, we have no evidence to suggest that this is happening—the numbers of students on the science foundation course in 1971 who did very well on continuous assessment and very badly in the examination were small.

Some interesting points have emerged from the analyses of data received on assessment. Firstly, despite the fact that in developing the questions for the CMAs, it was not possible to go through the normal procedure of pre-testing each of the items on a representative sample of students (because there were no students who had already taken the course), there are good correlations between scores on the tutor- and computer-marked tests—the same students who do well on the one type of test do equally well on the other.

These results are especially encouraging in view of the fact that we are using a wide range of different question types for the CMAs, some of which are very different from the traditional 'one from several' multiple choice question and have never been tried before. Secondly, our correspondence tutors on the whole are impressed by the quality of the work submitted by students, and compare it favourably with the work of the students in the conventional universities and colleges where they are employed. Thirdly, the educational background of Open University students does not seem to play as large a part as might have been expected in their achievement on course work. For example, in the science foundation course in 1971 there were about 200 students with no previous educational qualifications at all yet half of this group achieved a credit for the course.

Course evaluation

We have already made it clear that it is almost impossible, by the nature of the University's centralized production process, to produce courses in such a way that they are found totally satisfactory from the start. And we have referred to the reasons why the inclusion of developmental testing in the production cycle is only of limited value in identifying problems and faults in the material. Obviously the central course teams are no longer designing courses in a state of *total* ignorance about the characteristics of our students and the ways that they study, but the fact remains that there is little possibility of knowing what problems students are going to have with specific units until the first year of operation of a course.

Course teams have in their budgets a fixed sum for making alterations to each course from year to year: these can involve rewriting the note material, remaking a proportion of television and radio programmes, and completely changing the examination and assessment material. There is also the possibility of making some modifications to a proportion of the correspondence texts if they are due for reprinting. When this budgetary allocation was first made, it was assumed that much the same sum would be needed for each of the four years of a course's life. But it is now increasingly apparent that the major rewriting effort needs to be concentrated into the first year of operation of a course. For this reason, we have tried this year on two of the second-level courses to mount a major evaluative effort on what is a less expensively printed and produced version of the materials, on the basis of an agreed commitment from the course team to rewrite the courses entirely should the feedback from students suggest that this is necessary.

Students on these courses are sent lengthy feedback questionnaires and log-sheets for each unit, which ask for indices of understanding for each of the main concepts, procedures, principles, etc., discussed in the unit. They also ask students to provide information on which self-assessment questions they find difficult, and why. Also included are questions on time spent on the various components of each unit, and ratings of their usefulness. Data from these returns can be used as an immediate basis for action by the course team in producing stop-press mailing notes or radio announcements. More importantly, these data can be scanned at a later stage in conjunction with data from other sources (the analyses of assignment scores, the reports from tutors, and the general information from the weekly standard course unit reporting form). The multi-faceted picture of each course unit built up by looking at all these data together can provide fairly powerful indicators of what needs changing in a particular unit. For example, consider the information obtained from a computer print-out of the numbers of responses made to the different alternatives in a set of multiple-choice questions from a CMA, and suppose such an analysis showed a majority of students giving incorrect answers to a certain question. This information by itself is practically useless—students may have chosen an incorrect answer for lots of different reasons (misinterpretation of the question, ambiguity in the wording). But if it turns out that students in their questionnaire returns are reporting difficulties with the self-assessment materials based on complaints made in study centres about areas of difficulty, then a picture begins to emerge on which action can sensibly be taken.

One very interesting point has already arisen from the work being carried out in the in-depth studies with groups of students returning detailed unit-by-unit questionnaires. This concerns the problem of *sample size*. It is obviously very expensive and time consuming to monitor detailed questionnaires from all the students on a course; and if the information is required for immediate action, it would need something like a tenfold increase in the number of analysts processing the results, to get any information ready in time. In a recent report by Moss and Chapman [7] on their work in this area, recommendations to the course team were based on samples of thirty or so returns from panels of students (out of total course numbers, on different courses, ranging from 300 to 1,200). Various checks were then made to compare the panels of students with the corresponding overall student population of each course in terms of grade distributions and mean unit study times. In every case, the panels were found to be representative of the overall population on these parameters. This means it is possible to use such panels to *predict* fairly accurately difficulties which a majority of students might be experiencing, and to get remedial information out to tutors very rapidly. This work is still in its early stages, but it should result in the establishment of a set of routine course-unit specific procedures based on panels of students which can supply rapid accurate feedback to authors during the first year of operation of a course.

Other research being started or planned into student learning from the weekly course units includes the carrying out of 'in-depth' interviews with small groups of students and the development of techniques to identify different study patterns and learning styles amongst students. These are being backed up with detailed course-specific surveys of the ways in which tutors use the marking schemes which the teams produce for the TMAs, and the value of different experiments, teaching techniques, and procedures used at study centres and tutorials.

Within the science faculty, a small research group is currently being established to start work on a detailed analysis of science pratical teaching in multimedia systems. This will encompass studies at a theoretical level into the various rationales proposed for practical work, but will also investigate the use of a combination of television, filmstrips, notes, and home kit material in the teaching process. At yet another level, part of the project will include a wide-ranging search for teaching experiments and 'experiences' which can be incorporated into the home experiment component of our different courses.

In the university as a whole, overall research is being conducted into student characteristics (e.g., the influence of educational and occupational backgrounds on achievement) and comparative studies are being mounted into viewing and listening patterns, study centre attendance, etc. Other data currently being analysed concern the reasons which students give for dropping out of courses. Such data will provide a valuable framework in which to fit the more specific and detailed information obtained by course teams on comprehension of particular unit materials.

The future potential of multi-media distant study systems

This concluding section looks briefly at the future potential of integrated systems of the type we have described, in both developed and developing countries; it concludes with a discussion of the importance of *science* education within such systems.

During the next decade or so, future trends in multi-media distant study systems will probably be strongly influenced by the availability of various new technological developments. These can be looked at under three main headings—developments concerned primarily with channels of dissemination; those concerned primarily with rapid and efficient access to materials; and those concerned with establishing adaptive two-way links between student and teaching centre.

Under the first heading we might list the use of satellites [1] and at the other extreme, cable, for dissemination of video and audio components of courses. There is no reason, for example, why centrally produced TV and radio programmes broadcast by satellite should not be used in conjunction with locally produced ('study centre' type) programmes made available over cable. Despite its limitations, broadcasting is still the cheapest way of transmitting audio-visual material to dispersed students; and hopefully these limitations will be overcome when cheap and reliable video recording and playback devices become available. We should also mention the dissemination of written material in the form of microfiche (on the assumption that cheap portable microfiche readers do become widely available) [2], and new developments in printing technology [3], which will drastically reduce the time and cost involved in preparing textual materials. Developments such as these are of special relevance to 'higher level' science education, where the rate of generation of new information already outstrips the rate at which conventionally printed materials can be made available on an economic basis. This is already a problem for second- and third-level science courses at the Open University, where the main 'core' of the teaching material—the professionally printed and published correspondence text—must remain essentially unchanged for four years.

Our second heading—that concerned with easy *access* to a wide variety of materials by the student, is also particularly relevant to science education. A multi-media system which combined a facility for rapid access at will to reference material would overcome many of the difficulties associated with trying to teach university-level courses to students who cannot get to conventional libraries. Local study centres with terminals linked to computer-based 'libraries' of material, presented either on a TV screen or by a conventional teletype print-out device could go a long way to solving this problem. Such a system could provide the necessary flexibility for students to follow up particular points of interest by obtaining original source material.

Computers will play an important part in the developments covered in our third main heading—that of opening up further adaptive two-way links between the student and the teaching source. Although many of the claims made by proponents of computer-assisted instruction (CAI) may have been somewhat premature, there is no reason to suppose that eventually they will not be realized. Indeed, many relatively successful CAI systems are already in operation [4]. There is no need in this paper to point out the valuable role which a sophisticated CAI system with well-developed and tested software could play in helping students with very different backgrounds to study science at a distance. For example, in the science foundation course at the Open University, such a facility would be invaluable for taking students with little mathematical background through the material in the lower (remedial and introductory) channels in the course. A prerequisite for the widespread use of CAI in a distant study system is the development of cheap computer terminals, including ones with visual display facilities.

The use of many of the devices which we have listed is, of course, going to have an interesting effect on many of the other components of multi-media systems. For example, there will be the possibility of devising more flexible assessment procedures, where less reliance than usual will be placed on conventional written assignment materials, and where computers could monitor students' progress through a course on a much more individual basis than is possible at the moment.

The main point of our excursion into technological 'star-gazing' is to stress the way in which the devices mentioned could accentuate the existing trend towards more '*person-centred*' and *flexible* educational provision. We have already mentioned the increasing discussion in recent years in the more developed countries of the concept of '*éducation permanente*'. If this is to be realized, it implies an almost total restructuring of our traditional patterns of education. Alain Drouard [5] has listed four prerequisite conditions which need to be achieved at a political level before the concept of lifelong education could take on any reality. They are:

'(i) a reduction in the length of compulsory education and a redefinition of the aims of schools and universities;
'(ii) a more rapid recruitment into productive work;
'(iii) the development of a more general education, rooted in vocational experience and problems;
'(iv) the chance for adults to return to their studies at any time in their working lives.'

Multi-media systems which would allow adults to study whilst they are in employment and which could make available a much wider range of resources, materials, and personal contacts than many conventional schooling systems will have an obvious part to play in creating a climate in which such political changes might realized. They would also form an important component in any established comprehensive system of '*éducation permanente*'.

From one point of view, a concept of flexible lifelong educational facilities must appear to be a luxury which only the richer countries could afford to contemplate. But, in fact, the four conditions cited above are just as applicable within the context of developing countries. It is probable that many of the developing countries would now be better off had their educational systems been based on these assumptions; this is particularly the case in countries which have inherited the Western school systems which are inappropriate to their needs. Flexible, open-entry teaching systems geared to a developing country's economic needs, administered with a minimum number of trained teachers, and without the need for expensive school buildings, could have an enormous impact on educational and social economic development. The various projects already underway in 1966, described by Schramm [6], illustrate the interest already being taken in using the new media in such countries. More recently, the continuing interest is apparent from the number of visitors from such countries received at the Open University who are considering adopting the Open University system, and some of its actual teaching materials, for their own use.

We have placed considerable emphasis in this paper on the teaching of science at the Open University, because at the present time this system is the only fully comprehensive one in operation. This concluding section looks at the implications of such a teaching system in the light of the purposes and effects of science education in modern society.

One of the conclusions that can be drawn from an appraisal of the Open University is that integrated multimedia distant study systems must be large-scale projects if they are to be cost-effective. By 1973, more than a third of all students studying science in British universities are likely to be Open University students—a sobering thought for those who are responsible for the courses these students will study.

Another feature of the Open University teaching situation is its public and exposed position. Because of its scale and novelty, as well as for the more obvious reason that part of the teaching is done 'in public'—on open-circuit television and radio transmissions using normal networks and channels and with, in consequence, 'eavesdropping' audiences that normally exceed the 'target' audience by a factor of ten or a hundred—the Open University courses are attracting a great deal of attention in academic circles. There are already many indications that the course materials and the methods and systems being developed in the O.U. are having repercussions in many places, both in the United Kingdom and abroad, stimulating and 'triggering' processes that are likely to have far-reaching effects on science education.

An additional factor, which should not be underestimated, is the significant proportion of practising school teachers among Open University students. Some 1,500 teachers completed the science foundation course in 1971 and a similar number are registered in 1972. It became apparent before the end of the first year that many of these teacher students were already applying their new learning of science to their work in the classroom, and we had scores of enthusiastic letters describing how the new approach to science, the new scientific insights and the new methods of learning they had already 'picked up' as O.U. students were being used to good effect in their classrooms.

Thus, the scale of the operation, its public or 'open' character, and its penetration into sensitive parts of the educational world, place the educator in an institution like the Open University in a unique and peculiarly responsible position. All good educators feel concern about their own motives and about the wider purposes and implications of their work. The Open University teacher is obliged to think even more seriously about such questions as

Why teach science at all?

Why teach science on such a large scale?

In a situation of general unemployment, including graduate unemployment, and in which expenditure on (non-military) science is being cut back, what is the sense of flooding the market with still more science graduates?

What are the implications of the present and foreseeable future social and political circumstances for the content and approach of science courses?

An adequate discussion of this question would require a paper on its own. Any set of answers to such questions amounts to a statement of one's educational philosophy, and must reflect one's outlook and perspectives in the given national and international context. In the case of the Open University, the content and approach adopted for our science courses may be said to reflect a basic viewpoint which might be summarized by a few statements:

1. Whatever the constraints of the present economic and political situation may be, we believe that the long-term future of the U.K. requires a massive increase and improvement of scientific education at all levels, and that there is an objective need both for more and better educated *scientists* and for more and better *scientific education* of the community. We cannot fail to note the impact that scientific education on a massive scale has had on the economic and social progress achieved in the U.S.S.R. Although

the simple schematic formula: more and better science → more and better technology → a higher standard of living and a higher quality of life, is likely to run into difficulties when economic development has reached a certain level, it is arguable that we have not yet reached that level in most parts of the world, if anywhere.

2. In a period of human history in which decisions about science, and especially about applications of science, have far-reaching effects on the quality—and possibly even the existence—of life on this planet, it is no longer rational or responsible to teach science as if it happened in a social vacuum. Whether the student is an intending professional scientist, a teacher, or simply a citizen interested in understanding science, he must be taught the external relevance as well as the internal content of science.

The education of a scientifically literate public, capable of influencing scientific and technological decision-making through democratic processes is probably one of the conditions for human survival.

3. The present trend towards generalism in science teaching reflects the present stage of rapid development of science itself, in which specialist knowledge and skills become very rapidly outdated, and ability to adapt to changed structures of knowledge and ability to acquire and manipulate new knowledge become far more important. It reflects, too, the increased permeation of society by science and technology. An understanding of modern science and an ability to think and act scientifically are attributes no longer required only of professional scientists. In management, in the trade union and labour movement, in national and local government and administrations, in social institutions of all kinds from political parties to conservationist societies, effective action on many issues requires a far higher level of scientific education than was the case only a few decades ago. The trend is undoubtedly becoming accentuated, as can be seen from the far greater public attention being given in recent years to problems of pollution, resource conservation, population control, nuclear weapon testing, strategic arms limitation, chemical and biological warfare, and general disarmament.

The systems of science teaching at a distance that are being developed at the Open University lend themselves better, as it happens, to the generalist than to the specialist approach.

BIBLIOGRAPHY

Introduction

1. See, e.g., GOODMAN, P. *Compulsory miseducation*, London, Penguin Books, 1970.
2. COUNCIL OF EUROPE, *Permanent education*. Strasbourg, 1970.
3. See, e.g., WEISBERGER, R.A. (ed.). *Perspectives in individualised learning*. Itasca, Illinois, F. E. Peacock Inc., 1971.
4. SGETTLER, P. *A history of instructional technology*. New York, McGraw Hill, 1968.
5. BAEZ, A. V. *Improving the teaching of science with particular reference to developing countries*. Unesco STD/8/1A, 1967.
6. EPSTEIN, H. T. *A strategy for education*. Oxford University Press, 1970.
7. PASK, G.; SCOTT, B. *Learning strategies and individual competence*. Richmond, U.K., System Research Ltd., 1971.
8. MACDONALD ROSS, M. *Behavioural objectives and the structure of knowledge*, paper presented at the APLET Conference, Bath, March 1972.
9. BRIGGS, L. J. et al. *Instructional media: a procedure for the design of multi-media instruction*. Pittsburgh, 1967, American Institutes for Research.
10. POSTLETHWAIT, S. N.; NOVAR, J.; MURRAY, H. T. *The audio-tutorial approach to learning*. Minneapolis, Burgess Pub. Co., 2nd edn., 1969.
11. ADAMSON, H.; MERCER F. V., A new approach to undergraduate biology. *J. Biol. Educ.*, Sept. 1970, vol. 4, no. 3, p. 155-176.

Trends in the development of distant study systems

1. ZAV'JALOV, A. S. *Erfahrungen der Fernhochschule für Polytechnik Nordwest in Leningrad*, in: Peters, O. (ed.), *Texte zum Hochschulfernstudium*, Weinheim, Verlag Julius Beltz, 1971.
2. WEDELL, E. G. *The place of education by correspondence in permanent education*. Strasbourg, Council of Europe, 1970.
3. WEDEMEYER, C. A.; BERN, H. A. *Das Fernstudium an der Universität von Wisconsin*, in: Peters, O. (*op. cit)).
See also: WEDEMEYER, C. A.; NAJEM, R. E. *AIM: from concept to reality*. Syracuse, Syracuse University Press, 1969.
4. SCHRAMM, W. et al. *New educational media in action: case studies for planners*, vol. 1. Paris, Unesco/IIEP, 1967.
5. GAUDRAY, F. (ed.). *Multi-media-systems: international compendium*, Munich, Internat. Zentralinst. für das Jugend- und Bildungsfernsehen, 1970.
6. NISHIMOTO, M. *The development of educational broadcasting in Japan*. Sophia University, Tokyo, 1969.
7. ZIGERELL, J. J. *Chicago's TV college*, AAUP Bulletin, Spring 1967.
8. Report of the Vice-Chancellor, January 1969-December 1970, *The early development of the Open University*, Bletchley, 1972.

9. LeFranc, R. *Distant study systems in France.* Council of Europe internal publication CCC/ESR/(71)90.
10. Neil, M. W. *A systems approach to course planning at the Open University,* in: Romiszowski, A. J. (ed.), *The systems approach to education and training,* London, Kogan Page, 1970.
11. Bates, A. W. *The role of the teacher in a video-teaching system,* paper given at Unesco conference, *Auto-video,* Vichy, May 1972.
12. *Distant study in higher education: A German model,* (obtainable from O. Peters, DIFF, University of Tübingen, Federal Republic of Germany).

Also consulted:

Erdos, R. F. *Teaching by correspondence,* London, Longmans/Unesco, 1967.

Draft proceedings of the Ontario Round Table Meeting on *Educational communications and the New Technologies,* October 1971.

The influence of educational technology on the design of a multi-media system—the Open University

1. Kaye, A. R. *A set of learning tasks in the natural sciences,* Strasbourg, Council of Europe. (In press.)
2. Briggs (cited in Section 1).
3. Rose, S.; Pearce, L. Sulphur dioxide—a U.K. snapshot view, *New Scientist,* 17 February 1972.
4. Lewis, B. N. Course production at the Open University II—Activities and activity networks, *Brit. J. Ed. Tech.,* vol. 2, no. 2, 1972.
5. MacDonald Ross, M. *A report on developmental testing in 1971.* O.U. Institute of Educational Technology internal paper.
6. Lewis, B. N. Course production at the Open University III—Planning and scheduling, *Brit. J. Ed. Tech.,* vol. 2, no. 3, 1972.
7. Moss, D.; Chapman, P. Evaluation of Open University courses, (O.U. internal paper).

The future potential of multi-media distant study systems

1. See, e.g., Jankowich, *Satellite communication services for education in Europe.* Strasbourg, Council of Europe, 1971.
2. Beckwith, H. Innovations in industry likely to affect instructional technology during the next ten years, in: Tickton (ed.), *To improve learning,* Vol. II. New York, Bowker and Co., 1971.
3. Martinson, J. L.; Miller, D. C. Educational technology and the future of the book, Tickton (*op. cit.*).
4. See, e.g., Hickey, A. E. (ed.), *Computer assisted instruction: a survey of the literature,* 3rd edition. Newburyport, Mass., Entelek Inc., 1968.
5. Drouard, A., in: Mayne, R. (ed.), *Europe tomorrow,* London, Fontana, 1972.
6. Schramm, W. *et al. The new media: memo to educational planners,* Paris, Unesco/IIEP 1967.

Appendixes: Open University materials

1. Titles of currently available and planned Open University courses in mathematics, science and technology

Notes

1. The letters (A, D, M, S, T) before each course code number indicate the faculties involved in preparing the courses (A = Arts; D = Social Science; M = Mathematics; S = Science; T = Technology).
2. Courses are shown as having 'credit ratings' of 1, 1/2, 1/3 or 1/6; students can make up Science 1/2 credit courses by combining certain 1/3 and 1/6 components. The number of Course Units associated with each credit rating is shown below:

Credit rating	1	1/2	1/3	1/6
Course Units	34	17	11	6

Line of study	Number of course	Title of course	Credit rating	Available from year
Arts/Science/Technology	AST282	Science and the rise of technology since 1800	1/2	1973
Arts/Mathematics/Science/Technology	AMST283	Science and belief (from Copernicus to Darwin)	1/2	1974
Mathematics	M100	Mathematics foundation course	1	1971
	M201	Linear mathematics	1	1972
	M202	Topics in mathematics	1	1973
	M283	An algorithmic approach to computing	1/2	1973
	M241	Statistics	1/2	1974
	M231	Calculus and complex variables	1/2	1974
	M321	Applied differential calculus	1/2	1974
Mathematics/Science/Technology	MST282	Elementary mathematics for science and technology	1/2	1972
	MST282	Mechanics and applied calculus	1/2	1972
Mathematics/Social Science/Technology	MDT241	Statistics	1/2	1974
Science	S100	Science foundation course	1	1971
	S22-	Comparative physiology	1/3	1972
	S23-	Geology	1/3	1972
	S24-	An introduction to the chemistry of carbon compounds	1/3	1972
	S25-	Structure, bonding and the periodic law	1/3	1972
	S26-	The earth's physical resources	1/3	1974
	S2-1	Biochemistry	1/6	1972
	S2-2	Geochemistry	1/6	1972
	S2-3	Environment	1/6	1972
	S2-4	Geophysics	1/6	1973
	S2-5	Development and genetics	1/6	1973
	S323	Ecology	1/2	1974
	S321	The physiology of cells and organisms	1/2	1974
Science/Mathematics	SM381	Quantum theory and atomic structure	1/2	1974
Science/Social Science/Technology	SDT286	Biological bases of behaviour	1/2	1972
Science/Technology	ST281	Principles of chemical processes	1/2	1974
	ST285	Solids, liquids and gases	1/2	1973

Line of study	Number of course	Title of course	Credit rating	Available from year
Technology	T100	Technology foundation course	1	1972
	T241	System behaviour	1/2	1973
	T242	Systems management	1/2	1974
	T231	Engineering mechanics	1/2	1974
	T261	Integrative studies 1: design methods	1/2	1974
	T291	Instrumentation	1/2	1974
Technology/ Science	TS282	Electromagnetics and electronics	1/2	1972
	TS281	Introduction to materials (the solid state)	1/2	1973

2. Examples of Course Unit materials

2.1. Study guide sequence (S 100, Course Unit 21)

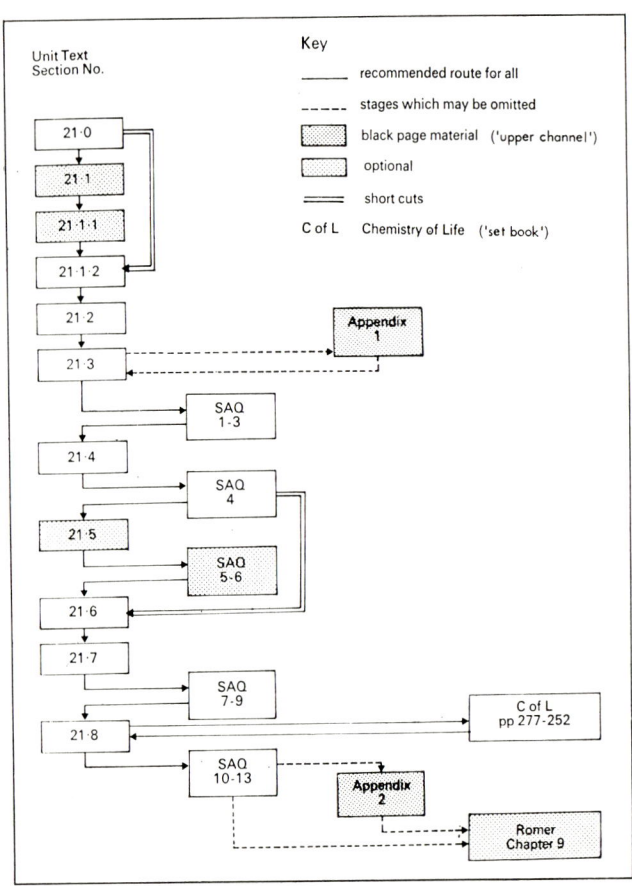

2.2. Objectives

The self-assessment questions (*SAQ*) in which these objectives are tested are given in brackets after each objective. An objective may also be tested by a question appearing in the text (*TQ*)

1. Define in your own words, recognize valid definitions of, or use in correct context the terms or expressions in column 3 of Table A.
   (*SAQs* 1, 2, 8, 15, 17, 22).
2. Write down all the two-dimensional structural formulae possible for a given molecular formula C_xH_y where x and y are integers up to and including five and twelve respectively.
   (*TQ*)
3. Write down and identify correct representations of methane, ethane, propane, ethylene and acetylene.
   (*SAQ* 3)
4. Recognize valid two-dimensional representations of structures of organic molecules which contain, in addition to carbon and hydrogen, one or more of the elements nitrogen, oxygen, fluorine, chlorine, bromine and iodine.
   (*SAQ* 4)
5. Given the molecular formula of a compound of the type described under 4 above, write down possible two-dimensional structural formulae.
   (*TQ*)
6. Describe reasons (in less than 300 words) why the compounds of carbon and silicon play different roles in nature.
   (*SAQs* 5, 7)
7. Identify structural isomeric relationships among a given series of structural formulae.
   (*TQ*)
8. Give reasons in order of importance why carbon forms more covalent compounds than all other elements combined.
   (*SAQ* 6)

9. Explain (in less than 50 words) the origin of the term 'organic chemistry'.
   (*SAQ 8*)
10. Use electron repulsion theory to deduce the approximate shape of a simple covalent molecule.
    (*SAQ 11*)
11. State the influence of bond angles in a molecular fragment if:
    1 bonding electron pairs are replaced by non-bonding electron pairs;
    2 one of the bonds is a double bond;
    3 there is a large difference in electronegativity between the bonded atoms.
    (*SAQs 9, 12*)
12. Recognize valid two-dimensional representations of the three-dimensional structures of organic compounds with less than ten carbon atoms.
    (*SAQ 13*)
13. Identify features in a molecule which reduce its structural flexibility.
    (*TQ, SAQ 17*)
14. Draw correct three-dimensional representations of ethane, ethylene, acetylene, glucose and a fragment of the primary structure of a protein.
    (*TQ, SAQ 16*)
15. Recognize geometrical and optical isomerism within a series of structural isomers.
    (*SAQ 14*)
16. Classify any object or simple molecule as being chiral or achiral.
    (*TQ, SAQ 15*)
17. Classify each of the organic compounds given in Objective 18 as a carbohydrate, amino acid, or neither of these.
    (*SAQ 15*)
18. Demonstrate knowledge of the structures of diethyl ether, glycine, alanine, tetrahydrofuran, glucose and mannose by identifying features which are present or absent in their structures.
    (*SAQ 15*)
19. Relate gross trends in melting points, boiling points, solubilities (in a given solvent) and colour within a given series of compounds to the concepts of polarization, polarizability, conjugation, hydrogen-bonding and molecular shape.
    (*TQ, SAQs 17, 18, 19, Home Experiment*)
20. Recognize simple structural features which may give rise to chemical reactivity.
    (*SAQ 21*)
21. Demonstrate an understanding of the reason why an enzyme normally accepts only one of a mirror-image pair of molecules for catalytic chemical change, by drawing relevant diagrammatic models and by recognizing valid analogies.
    (*SAQ 20*)
22. Recognize some typical molecular structures associated with each of the following fields of chemical technology:
    petroleum and petrochemicals
    dyes
    pharmaceuticals
    pesticides
    polymers
    detergents.
    (*SAQ 23*)
23. Give at least two examples of molecular design in chemical technology.
    (*SAQ 24*)
24. Identify concepts in the Unit that are relevant to:
    the role of carbon compounds in nature
    molecular shape
    chemical reactivity
    physical properties
    molecular design in chemical technology and its impact on society;
    and assess the extent of this relevance.

2.3. Table A

A List of Scientific Terms,* Concepts and Principles Used in Unit 10

Taken as prerequisites			Introduced in this Unit			
1	2		3		4	
Assumed from general knowledge	Introduced in a previous unit	Unit No.	Developed in this unit	Page No.	Developed in a later unit	Unit No.
natural gas	Periodic		polarization	9	carbohydrate	13
flint, quartz	Table,	7, 8	valence shell	12	protein	13
evolution	covalent and		structural isomers	14	polysaccharides	13
melting point	ionic com-		double, triple and multiple		enzymes	15
boiling point	pounds	8	bonds	15	polymers	13
dye	electronic		valency number	18		
anaesthetic	structure	7	non-bonding electronic pairs	18		
antibiotic	carbon		organic chemistry	21		
pesticides	tetrachloride	8	bond angle	24		
DDT	noble gas	8	stereochemistry	24		
nerve gas	halogen	8	conformation	26		
toxicity	Pauli exclusion		saturated and unsaturated			
detergent	principle	7	molecules	26, 27		
soap	quantum		stereoisomerism	28		
	numbers	6	geometric isomerism	28		
	electron spin	7	polarizability	29		
	solvent	9	monosaccharides	31		
	wavelength	2	amino acid	34		
	aqueous	9	chirality	35		
	ions	9	plane of symmetry	36		
			achiral	36		
			plane-polarized light	37		
* Excluding the names of chemical compounds.			optical isomers	37		
			hydrogen bonding	41		
			conjugated system	42		

2.4. Conceptual diagram

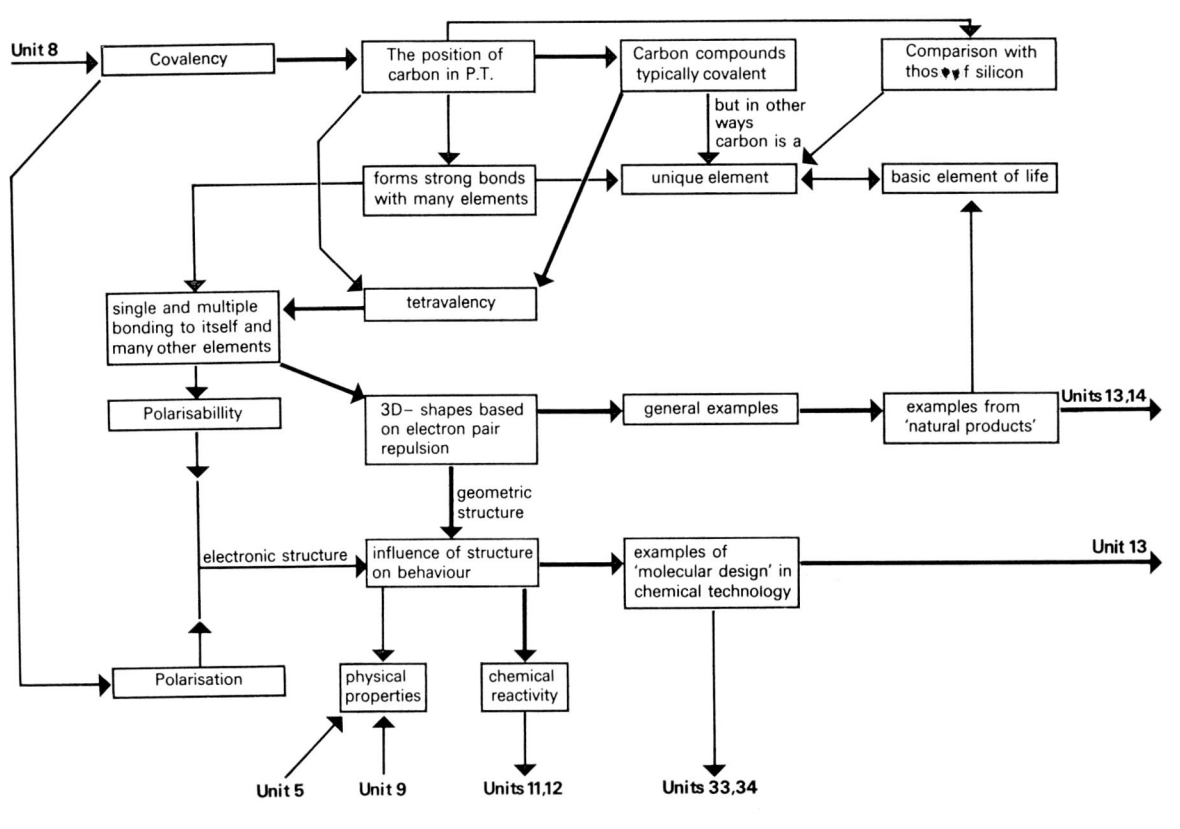

2.5. 'Activities-oriented' pages from a correspondence text of S 100 (Course Unit 6)

Experiment

View an ordinary light bulb with your spectroscope; comment on what you see. A spectroscope is a device for dispersing light. You will notice that it has a slit at one end that you point towards the light. The light coming through the slit is dispersed and focused with the lens at the other end. What you see is a separate image of the slit for each of the wavelengths arriving at the instrument. In the present instance what you see is a typical *continuous spectrum*, as the colours change continuously without a break. If you view the outside light from the sky you will see a similar continuous spectrum which is the *solar spectrum*.

continuous spectrum

solar spectrum

DO NOT POINT YOUR SPECTROSCOPE IN THE DIRECTION OF THE SUN FOR THE LAST PART OF THIS EXPERIMENT.

Light dispersed in this way is also separated in terms of energy. You probably know this already. For example, you will be aware that ultraviolet radiation (wavelengths shorter than visible light) causes sunburn whereas infra-red radiation (wavelengths longer than red light) only warms you.

What does this suggest to you about which colour, red or blue, is associated with the largest energy?

Blue.

As you saw in Unit 2 electromagnetic radiation from an atom is emitted in discrete amounts. These can be regarded as 'packets' of energy. What is true about electromagnetic radiation emitted from atoms is true of all electromagnetic radiation. These 'packets' of energy are called *photons*. The energy of a photon depends upon the frequency of the light involved. From your answer above, for example, the photons of blue light have larger energies than those of red light. The relationship between the frequency of the radiation and the energy of the photons is

photon

$$E = hf$$

where $E =$ energy of photons measured in joule (cf. Unit 4)

$f =$ frequency of radiation (Hz)

$h =$ a constant called *Planck's constant*, with value 6.626×10^{-34} joule seconds.

Planck's constant

The relation between wavelength and the energy of the photon is then given by combining two of the equations above as follows:

$$E = hf$$
$$c = f\lambda$$

So $\quad E = \dfrac{hc}{\lambda}$

The energy of the photons is less at longer wavelengths.

Experiment

Read the instructions in the Home Experiment Notes for this Unit on the use of the spirit burner and the spectroscope. Now light the spirit burner. Dip the wire into some common salt (sodium chloride). Examine the flame through the spectroscope. While you are watching the spectrum, place the wire with the sodium chloride on it in the flame. If you have access to sodium street lights (these are yellow ones), mercury street lights (these are bluish-white ones), a fluorescent light or neon advertising tube, examine the light from these sources. Draw a diagram of what you observe.

Exercise

On Figure 9 draw (horizontal) lines to represent the energy levels for the hydrogen atom that are involved in the spectral lines listed in Table 1. From the table you can see that the largest jump involved for these four lines is 4.843×10^{-19} joule. A suitable scale is 1 cm = 1×10^{-19} joule. Draw a line 18 cm above the bottom of your page (the reason for this will become clear later). This line will represent the energy level to which the electrons are jumping. This represents the lowest energy level involved in the Balmer Series. It is then a matter of placing another line at a position corresponding to 3.027×10^{-19} joule above this line. The third line will be at a position corresponding to 4.086×10^{-19} joule above the lowest energy level. When you have completed the diagram in this way with five lines (the lowest, plus the four 'jumps'), what you have is an *energy-level diagram* for the electron energy levels in a hydrogen atom.

energy-level diagram

Your diagram should look like this, without the arrows.

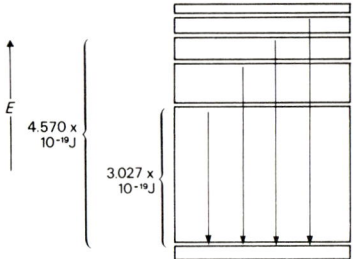

Figure 10

Energy levels for the Balmer series.

The Balmer lines are the result of the electron jumping from the upper energy levels on your diagram to the lowest level you have drawn. Jumps could be represented by arrows as shown in the answer diagram, where the arrows represent jumps from higher to lower levels and hence the emission of photons, leading to the emission spectrum. To represent the jumps that take place on absorption of photons (resulting in an absorption spectrum) the arrows would go in the opposite direction.

There is in fact an energy level below the lowest level that gives rise to the Balmer lines. This level is lower by 16.34×10^{-19} joule and is the lowest energy level in the hydrogen atom. It is called the *ground state* energy level, and when the electron is in it the atom is said to be in its ground state. The electron will then be closer to the nucleus than in any other energy level. If the ground state level is added to the energy-level diagram, it will now look like Figure 11.

ground state

All levels except the ground state energy level are called *excited states*.

excited state

6.5.3 The Lyman series

A series of jumps is clearly possible, each finishing in the ground state level. Each of these jumps would result in the emission of a photon, giving another series of spectral lines.

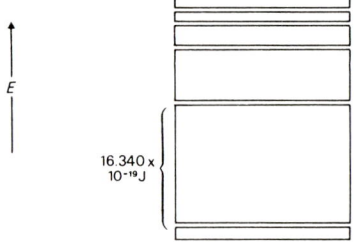

Figure 11

Energy levels for the hydrogen atom.

Exercise

Put the ground state energy level into your energy-level diagram and indicate the jumps down to this level by arrows.

Calculate the energy changes for the lowest four of these 'jumps' and from these energy jumps (ΔE) calculate the frequencies of the spectral lines that would result. Put the values you get in Table 2 and show what the spectrum would look like by drawing lines on the calibrated blank below the table.

The answer is in the next margin overleaf.

Table 2

Line	1st	2nd	3rd	4th
ΔE (J)				
f (Hz)				

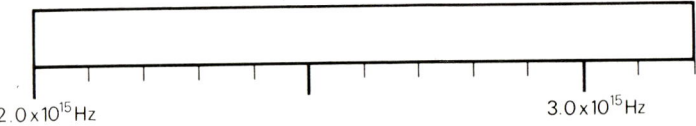

2.0×10^{15} Hz $\qquad 3.0 \times 10^{15}$ Hz

This series of lines is called the *Lyman series* and occurs in the ultra-violet region of the electromagnetic spectrum.

2.6. Self-assessment questions

Question 21 *(Objective 20)*

Group the following compounds according to those which might be expected to exhibit similar chemical behaviour.

(i) $CH_3.OH$

(ii) cyclobutane ($H_2C-CH_2-CH_2-CH_2-CH_2$ ring, i.e. cyclopentane: H_2C-CH_2 / H_2C CH_2 / CH_2)

(iii) 1,4-dioxane ($H_2C-O-CH_2-CH_2-O-CH_2$ ring)

(iv) benzyl alcohol (benzene ring with CH_2OH)

(v) $CH_3.O.CH_3$

(vi) C_5H_{12}

(vii) $CH_2=CH_2$

(viii) cyclopentene ($HC=CH-CH_2-CH_2-CH_2$ ring)

(ix) $CH_3-\underset{\underset{O}{\|}}{C}-CH=\underset{\underset{OH}{|}}{C}-CH_3$

(x) $CH_3-\underset{\underset{O}{\|}}{C}-CH_3$

(xi) benzene ring

Sections 1, 4, 5

Question 22 *(Objective 1)*

Indicate by a tick in the appropriate column whether each description on the left can be associated with the electronic property of polarization or polarizability.

	(a) polarization	(b) polarizability
(i) bonds whose electrons are responsible for the absorption of visible light	( )	( )
(ii) bonds which have a symmetrical distribution of electrons but are nevertheless relatively easily broken	( )	( )
(iii) *intra*molecular bonding which induces strong *inter*molecular cohesive forces.	( )	( )

3. Abbreviated lists of home kit items for some current Open University science and technology courses

The science foundation course (S100)

Instruments and apparatus:	Spectroscope, plotting compass, flowmeter, stopwatch, stereoviewer, slide viewer, McArthur microscope, chemical balance, colorimeter, pump for SO_2 in air determination, laboratory burner, thermometer, tuning forks.
Glassware:	Various (including test-tubes, syringes, microscope, slides, etc.); also retort stand equipment.
Chemicals:	69 different chemicals.
Miscellaneous:	Atom models, litmus paper, carbon electrodes, magnet, graph paper, logarithm tables, etc.

The technology foundation course (T100)

Instruments and apparatus:	Power supply/meter kit, logic tutor, analogue/control kit, sound level indicator ('Noisemeter'), mechanical ball-operated computer ('BOBCAT'), thermometer.
Glassware:	Beakers, test-tubes, etc.
Chemicals:	Five different chemicals (used in conjunction with samples of different metals in an experiment on corrosion).
Electronic and electrical:	Various resistors and diodes, a solar cell, a thermistor, potentiometer, etc.
Miscellaneous:	Crystal simulator, various connection leads, fuses, bulbs, protractor, graph paper, etc.

Comparative physiology (S22-)

Instruments and apparatus:	McArthur microscope, slide viewer, spatula, forceps, scalpel holder and blades.
Glassware:	Petri dishes, pipette, measuring cylinder, slides, coverslips, watch glasses.
Chemicals:	Various.
Specimens:	Macerated celery, cucumber seeds.
Miscellaneous:	Prepared slides, camel hair brush, filmstrips, graph paper, etc.

Geology (S23-)

Instruments and apparatus:	Petrological microscope with tripod and transformer, hand lens.
Chemicals:	Acetic acid, amyl acetate.
Glassware:	2 pipettes.
Specimens:	22 thin sections of rock specimens, 25 actual rock specimens, 7 minerals, 7 fossils.
Miscellaneous:	Solder wire, plasticene, tracing paper, graph paper, etc.

An introduction to the chemistry of carbon compounds (S24-)

Instruments and apparatus:	Retort stand, folding magnifier, spirit burner, spatula, test-tube rack, thin layer chromatography kit, forceps, thermometer, laboratory burner nozzle.
Glassware:	Test tubes, syringes, pipette, stirring rods, beaker, funnel, flasks, etc.
Chemicals:	33 different chemicals.
Miscellaneous:	Atom models, styrofoam balls, protective eye glasses, filter paper, wire gauze, logarithm tables, etc.

Structure, bonding and the Periodic Law (S25-)

Similar to that for organic chemistry, with the exception of the chemicals (55 different inorganic chemicals).

The biological bases of behaviour (SDT286)

Instruments and apparatus:	Tachistoscope, galvanic skin response (GSR) meter with leads, electrodes and gel stopwatch, goldfish discrimination-learning tank folding magnifier, thermometer, scalpel handle and blades, spatula, forceps, tray.
Glassware:	Petri dish, medicinal dropper.
Chemicals:	Acetic acid, quinine bisulphate.
Specimens:	Sheep's brain, worms.
Micellaneous:	Plastic gloves, playing cards.

Electromagnetics and electronics (TS282)

Instruments:	Current meter, multi-meter (assembled by student from components), generatorscope (combined signal generator, oscilloscope, and multiple power supply).
Miscellaneous:	Various diodes, transistors, resistors, potentiometer, solar cell, magnet, cable, crocodile clips, etc.
Tools:	Supplied by student (soldering iron, wire cutters, screwdrivers, etc.).

Integrated multi-media systems for science education (excluding television and radio broadcasts)

by S. N. Postlethwait
Purdue University, West Lafayette, Indiana 47907, U.S.A.

and Frank V. Mercer
Macquarie University, North Ryde, NSW 2113, Australia

Summary

All learning is begun at the interface between the learner and his environment. Media systems permit us to provide the student with a simulation of original learning environments and, in some cases, improve the learner's perception through the mechanisms of magnification, amplification and time lapse sequences. The present capability of technology to store and retrieve audio and visual stimuli provides almost unlimited opportunities to develop learning porgrammes involving all basic communication tools. These tools fall into four basic categories:
1. Tangible items
2. Printed materials
3. Projected images
4. Audio.

Various devices serve as the container for storage and retrieval of one or more of the communication tools either singly or in combination. An example of the latter is illustrated by a movie which is the storage of audio in combination with a series of projections. Many devices are available that synchronize audio with slides, film strips and/or movie clips through signals which activate the projection mechanism.

Most classroom use of multi-media systems centres upon some particular device or system of devices with the instructional programmes adapted to these specific vehicles. Those learning systems which include all four communication tools have the greatest flexibility and potential for creating a simulated environment for the greatest effectiveness in learning. Some examples of multi-media systems include the Audio-Tutorial system, Purdue University, U.S.A.; the Macquarie University system, North Ryde, Australia; the Southin-Chambers system, McGill and Sir George Williams Universities, Canada; the Personalized system of Instruction, Arizona State University, U.S.A.; the Guided Design system, West Virginia University, U.S.A.; and a Swedish Course in Applied Electronics, Royal Institute of Technology, Sweden.

The key component of any multi-media system is the software or intellectual content of the multi-media product. This component is directly related to the capacity and skill of the developer of the materials. At the present time hardware is more readily available than software. Production of high quality instructional materials must occur before the use of technology can become widespread. This will require strong financial support and extensive co-operation among the members of the academic community. If materials are to be widely used, their production must involve input from an equally wide source of expertise. Co-operation among schools on an international basis can substantially reduce the high cost of the development of the learning programmes. Some effort should be made to standardize the format for these programmes and thus reduce the difficulty of interchangeability. A practical solution which will facilitate exchange is to develop self-instructional materials covering relatively small units of subject matter. These units or minicourses can provide the flexibility necessary for arranging instructional programmes in a variety of combinations suitable to many different learning programmes. Models of international exhanges should be set up and the problems carefully investigated. Programmes currently in progress should be fully supported and encouraged by the institutions and governments concerned. The technology is abundantly available. The main difficulty relates to our ability to produce appropriate materials.

Media—what are they?

The expression, 'a multi-media approach to education', has come into prominence during the past decade, and it generally implies the use of technology in some non-traditional approach to learning. One expects the approach to involve a learning centre or study room equipped with electronic devices which make adjustments for individual differences and cause students to achieve at a high level and at a very rapid rate. The teacher is replaced and students proceed independently with little or no personal contact with other human beings. This image is unfortunate for, while some of these features are possible and important, the use of media does not necessarily replace the teacher, make education less expensive, or result in greater learning. Many of the educational procedures which have been utilized in the past were based on sound learning principles and the use of multi-media does not replace them. Many time-tested practices in traditional instructional approaches are very necessary ingredients in a successful multi-media programme. The confusion arises primarily from a rather superficial understanding of what media can and will do and how the role of the instructor and student may change when these materials are being used.

The definition of a medium as 'an agency, such as a person, object, or quality, by means of which something is accomplished, conveyed, or transferred' [47] [1] is helpful for placing the relationship of teacher-medium-student in proper perspective. In other words, the medium is a means to the end and not an end in itself. Education is primarily a human-to-human process where the student benefits from the wisdom and activities of scholars, teachers and peers, whether the learning activity is in the instructor's physical presence or not. If the learning activity is not in the presence of the teacher, then some medium must be employed as a 'bridge' between the teacher and student to 'convey, transfer or involve' the student in an informative experience. Until recent years the 'agency' of transfer has primarily been books, journals and other printed materials. More recently, many other devices have been made available to the teacher and the student so that the 'transfer' can involve more sense and provide the student with a greater range of experiences. The teacher can employ many avenues (multi-media) to convey his message to the learner, and the programme he designs permits the student access to the instructor in a great many more dimensions than through the printed word only. The nature of the message is limited mainly by the imagination of the originator, the capacity of the technical devices to faithfully record and transfer the

1. For bibliography, see page 214.

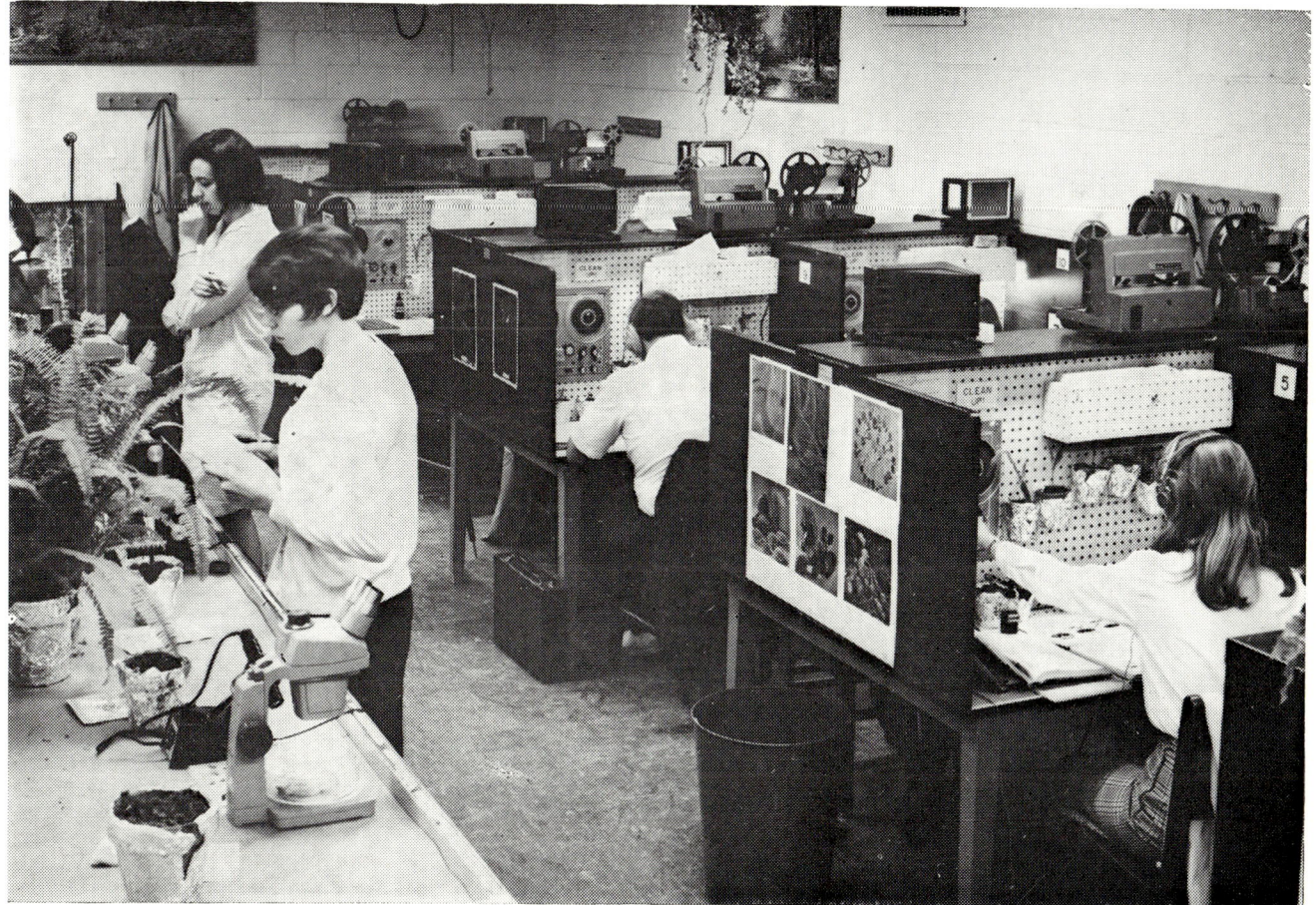

FIG. 1. A 'multi-media system' is one which employs a variety of media involving primarily three senses—seeing (visual), hearing (audio), and feeling (tangible), and combinations of these.

information, and the ability of the recipient to interpret the stimuli he receives. This view of a multi-media programme places the responsibility of the outcome of its use squarely on the shoulders of the two intelligent components of the system, the teacher and the student. One expects the results of media use to be dependent on the cleverness of the originator of the message and the capacity of the recipient to interpret the message.

Media, as defined in this paper, primarily utilize three senses—hearing, seeing, and feeling or combinations of these (Fig. 1). These basic tools of communication between individuals are not new, even though the devices for storage and retrieval of the stimuli are new and in some cases sophisticated. When one uses the basic avenues of hearing, seeing and feeling to communicate a message, the nature of the storage and retrieval vehicle is of consequence only in that it must convey accurately the thoughts of the originator and the recipient must be conveniently and effectively exposed to those thoughts. Thus a movie and a video tape are essentially equivalent since each is a combination of audio and visual stimuli and both are therefore basically the same medium from the standpoint of effectiveness in the transfer of the message. However, each has some special characteristics which may affect the feasibility of its use and the conditions under which it can be used most effectively.

Media—a classification scheme

Media are classified in a variety of ways to suit the needs of the classifier and his audience [10].

One simple scheme which the authors have found useful is to classify media on the basis of the senses stimulated in the learner. This results in four major categories of classifications:
1. Audio or sound media
2. Visual media
3. Tangible items
4. Audio, visual and tangible combinations.

A fifth category is sometimes added, i.e., the interaction of student with student and student with teacher; however, in the strict definition of the term *medium* as *an agent of transfer,* the fifth category, while very important in a good learning system, is not a true medium. Below are listed some of the items which would be included under these categories:

1. Audio or sound media
    A. Remote access
        (1) Radio
        (2) Telephone
        (3) Dial-access tapes
    B. Local access
        (1) Tape players
        (2) Disk players-phonographs.
2. Visual media
    A. Printed
        (1) Books
        (2) Manuals, study guides
        (3) Journal articles, pamphlets
        (4) Magazines, newspapers
    B. Flat-prints—graphics
        (1) Photographs
        (2) Charts, diagrams
        (3) Murals
        (4) Maps, posters
    C. Projected images
        (1) Slides, filmstrips
        (2) Silent movies
        (3) Overhead projections.
3. Tangible items
    A. 'Real things'—specimens (living or preserved), skeletons, objects, equipment, collections and materials for experimentation
    B. Models, mock-ups, reconstructions, miniatures and cutaways
    C. Museums, field laboratories, and simulation devices.
4. Audio/visual/tangible in various combinations
    A. Intrinsically associated audio/visual
        (1) Movies
        (2) Television
        (3) Computer
    B. Individually operative
        (1) Synchronized tapes/disks and projection devices (2 x 2 in.-slides, filmstrips)
        (2) Tapes/disks co-ordinated with printed materials, graphics, etc.
        (3) Tapes/disks co-ordinated with tangible items (models, specimens, etc.)
        (4) Tapes/disks co-ordinated with projections, printed materials and tangible items (audio-tutorial programmes)

(5) Telephone/radio co-ordinated with projection, printed materials and tangible items (tele-lecture).

1. *Audio or sound media.* The technology for capturing and retrieving sound in simple and inexpensive devices has made giant strides in the last few years. Since the days of Edison, there has been continuous improvement in sound technology. The development of transistors in recent years has made sound available so inexpensively that potentially, a very large proportion of the world can be exposed to any information we have the interest and capacity to package and the wisdom to distribute. The devices for transfer of sound have been classified as either remote access or local access. This access difference in the sound delivery system may have considerable impact on the nature of the instructional programme.

A. *Remote access.* Remote-access sound-delivery systems are primarily controlled by the source of origin rather than the recipient. This is especially true of radio; however, some control can be exercised by the student over telephone and dial access materials. Remote-access systems generally are less expensive and instructional materials are maintained in a central storage centre.

(1) *Radio.* Some use of radio for instructional purposes was begun as early as the 1920's but little use has been made of this medium until recently. Radio has the obvious advantage of delivering education to remote areas and has been used extensively in Sweden, Republic of Korea, Japan, United Kingdom and Australia. The Radio School of the Air in New South Wales, Australia, maintains contact among students who are too widely scattered to physically meet together for any instructional programme. Each student is equipped with a transceiver so that he may communicate directly with the teacher and, at specified times each day, all co-operating students communicate with the teacher and each other. Apparently the primary contribution to each student is the identification with one another and with the teacher. In the United States, several schools and colleges use FM stations and AM stations to broadcast lectures, with full credit being awarded to students who pass appropriate tests.

(2) *Telephone.* Perhaps one of the most neglected tools for remote-access instruction is the use of the tele-lecture. The instructor lectures into an ordinary telephone, but where several students are involved, amplifying systems are available to permit large audiences to hear the presentation; often these presentations are accompanied by the use of slides, movies or other appropriate materials. This technique is frequently used as a supplement to the regular instruction; however, entire courses may be taught in this manner.

(3) *Dial-access tapes.* The primary advantage of the dial-access system is that instructional materials are stored in a central facility. Even though remote, some dial-access systems permit control by the student for repetition, rapid advance or pacing. In principle the dial-access system is much like the local-access tape player arrangement.

B. *Local access.* The main advantage of most local-access systems over the remote-access systems is the student control and initiation of the programme. Self-pacing and independent study are the features most commonly emphasized by proponents of these media.

(1) *Tape players.* The audio tape recorder/player, first developed around the turn of the century, has been used for a great many educational activities in recent years. This has been due to the simplification of the recorder-player device and improved tape and tape mountings. At present, tape players can be purchased so inexpensively they are commonly used as gift items for children and included as components of many toys. Important features are:
(a) control of the pace of instruction by the student;
(b) convenient and almost foolproof operation;
(c) re-useability of audio tape for new programmes;
(d) simple recording procedures which do not require special technicians;
(e) a general trend to common usage of standard materials.

The magnetic nature of the system permits the utilization of inaudible cues so that other media can be programmed to synchronous presentation without disturbing the student. Tape recorders have great potential for providing equal opportunity education and continuing education programmes whenever and wherever they are needed.

(2) *Disk players-phonographs.* Perhaps the greatest instrument for recording sound has been the phonograph, invented by Edison in 1877. The basic principles involved in the original invention are still utilized in modern phonographs; however, the quality has increased to almost flawless reproduction of sound. This device has received only limited utilization in classroom instruction, perhaps due to its lack of playback or repeat

features and the requirement for professional assistance in the production of quality disks. Another limitation is the lack of potential for erasure and re-utilization of the original disks. It is likely that the phonograph will be completely replaced by the audio tape player for most instructional programmes requiring a sound component.

2. *Visual media.* A great many items are included in this category and many are so common that little or no comment is necessary concerning their use. Visual media have played an important role in the instructional process for a great many years and will continue to do so in the future.

A. *Printed materials.* Printed materials such as textbooks, manuals, study guides, journal articles, pamphlets, magazines and newspapers make an important contribution to the learning process for all students who can read. The proliferation of these materials is at a very rapid rate and many view with alarm the difficulty of storage and retrieval of such a large volume of information. Computers, microfiche, microfilm and several electronic devices are designed basically to store and retrieve printed information. Even though printed materials are so familiar as to be taken somewhat for granted, they will continue to play an important role in instructional systems for many years to come.

B. *Flat-prints, graphics.* Photographs, charts, diagrams, murals, maps and posters included in this category, obviously, are printed matter and might have been included in the preceding paragraph. They are separated here because they are commonly involved in a different instructional pattern from the printed materials listed in paragraph A. These media are usually larger and designed for use in conjunction with another medium such as a printed pamphlet, a taped presentation or a lecture. Again, these materials are so common as to require no comment except to emphasize that here is an old technology that is still a very functional means of communication.

C. *Projected images.* Photography has changed the world. It has made us familiar with places, things and information that no other medium could accomplish. It began about 1840 through the efforts of Daguerre and Talbot. Photographic procedures were greatly improved by Eastman in 1889 with development of flexible film. Edison is commonly given credit for developing the first projection device, which gave rise to movies and slide projection. Excellent equipment for production and projection of slides, filmstrips and movies has made photography a practical educational tool for nearly all teachers. Cameras for both movie and still frames are so completely automated and versatile that even amateurs can produce relatively high quality materials. Projection devices, too, have been improved by the addition of automatic loading features, so that even the most unskilled teacher can approach these devices with some confidence. Perhaps most important, it is becoming increasingly more practical to provide students with projection equipment so that they can use these materials independently and integrated with other learning activities. Both cameras and projection devices use cartridge-loaded film which reduce the wear and tear on the film as well as providing a handy container for storage and projection.

Overhead projectors are included in this category, although the materials are usually produced differently and are used in a different context from movies and slides. A great many lecturers use the overhead projector as a substitute for the blackboard. An important advantage is the pre-preparation of diagrams, charts and other images to be used in the presentation, thus improving the quality over the hastily-produced blackboard drawings. The popularity of the overhead projector has resulted in a proliferation of commercially-prepared visuals and a variety of overhead projection instruments to suit the tastes and needs of most teachers.

3. *Tangible items.* The extensive use of the lecture-laboratory approach to education over a great many years has conditioned most instructors to divide the learning process into two parts: theory and practice. This conditioning has caused both teachers and students to expect that the subject matter will divide and in fact to divide it into lecture content and laboratory content and to consider that the two are not necessarily related. In many cases, the lecture-laboratory arrangement has resulted in two separate courses (a lecture course and a laboratory course) on the same subject with no correlation between the two. Apparently this schism has developed in part because of the logistics of supplying real and tangible items to many students in a lecture hall arrangement. The format of lecture/laboratory is so strongly accepted as tantamount to education that alternate teaching strategies which reduce logistic problems for integration of tangible items into the learning programme are not readily ac-

cepted by either students or instructors. A great deal of redundancy and inefficiency in our learning systems has resulted from this perception of the educational system.

A. *'Real things'*. Specimens (living or preserved), skeletons, objects, equipment, collections and materials for experimentation are commonly accepted as useful in the learning process but normally relegated to a laboratory session. It is logical that, if these materials contribute to the student's concept of the subject, their use integrated with the other subject matter content would be more effective than if used for study separately. These materials are rather specific and there are little or no guidelines for their use except that the 'real thing' should be used unless practical considerations override the pedagogic impact.

B. *Models, mock-ups, reconstructions, miniatures, and cutaways*. The use of these materials in laboratory situations has been a common practice for many years and frequently can be used to emphasize specific features of real things and to communicate concepts that are otherwise somewhat obscure. These media are most frequently used in association with some other medium such as the real thing or printed materials.

C. *Museums, field laboratories and simulation devices*. In many specialized courses, these media are very necessary components to the development of a sound instructional programme. These media, too, may be used in conjunction with other media and their use is dependent on the judgement of the instructor and the availability of these materials to the teacher.

4. *Audio/visual/tangible combinations*. One medium is seldom used alone as a total instructional programme. Numerous devices have been invented which involve a combination of audio/visual/tangible media. Many of these devices have limited utility because of the unique structural features which permit utilization of only one instructional strategy. Most teachers are perplexed by the great array of these devices and fear their use because of the limited amount of software available and the high cost of production of new materials. The degree of flexibility is related to how intimately the audio and visual components are associated.

A. *Intrinsically associated audio/visual*. Movies and television are basically alike in that the audio and visual are totally synchronized so that both vehicles are used essentially in the same instructional format. The major differences are in cost of production of software, cost of hardware and to some degree the flexibility of projection equipment versus television monitors. Recent improvements in the delivery systems of both movie projectors and TV sets have increased their flexibility and mobility. We are on the threshold of even further improvements in cameras, playback mechanisms and cassettes to be used with both TV and movies so that even greater flexibility and wider use of these instruments can be anticipated in the future.

Within the coming decade a new audiovisual medium —cassette TV—will become readily available in many countries. Although the title of our paper excludes television and radio it seems to us that it does not exclude cassette TV. Our view is that cassette TV is in the same category as audio and video tapes. The videocassette is essentially similar to the audiocassette in being portable and carrying its own materials (i.e., independent of a central transmitting station). It would play a somewhat similar role to the audiocassette, but with an added dimension.

At the present time there appear to be four main systems being developed for cassette TV—electronic video recording, magnetic tape systems, plastic disk and holographic tape. Each system has its particular advantages and disadvantages but at this point in time it is not possible to say which is likely to be most useful for educational purposes. However, that is probably not of great concern to us. What is important for educators is to recognize that portable audiovisual cassettes will be yet another medium to be integrated into the multi-media systems for science education in the future.

It is not the purpose of this paper to discuss computer potential for instruction and it is sufficient to say that while it primarily uses audio and visual modes of communication, its flexibility and capacity for programming and involving students with media is far beyond its utilization at this time.

B. *Individually operative media*. Several devices have been developed for delivery of audio co-ordinated with other media but with the delivery systems not interdependent unless specifically desired. This feature is particularly important for the self-instructional programmes and independent study approach. A basic component of most of these systems is a tape recorder or a phonographic disk which mechanically or electronically programmes the operation of a slide projector, filmstrip projector or movie device but which can also be manually operated by the student with various overrides, and backup and repeat procedures. With this one basic medium, the audiotape, all other media can be selected and programmed as appropriate to provide the maximum range of flexibility for students. The primary limitation of this system is the imagination and creativity of the instructor who puts together the programme. Programmes produced in this way are multi-media in the truest sense since no medium is excluded as a component of the total programme.

The telephone, used in co-ordination with projectors, printed materials and tangible items to give what is commonly called a 'tele-lecture', has had limited use, but provides considerable potential for inexpensive access to top scholars who are off campus. Basically, this procedure is not different from the audiotape programme co-ordinated with projectors, printed materials and tangible items except that the telephone presentation is live, in some cases less polished, and gives the feeling of being somewhat more personal. The tele-lecture approach could probably be used more widely if better known. Perhaps it is used less frequently because of the impression that it is expensive. Compared to travelling costs or the cost of a continuous salary, the tele-lecture system can provide a very direct exposure to high quality instructors at a low cost.

Some examples of multi-media systems

The hardware or delivery systems are much advanced over the production of software. However, numerous efforts are underway in many institutions and in several countries to develop instructional programmes and to determine ways of best utilizing the new technology. Most of these experiments are independent operations and closely associated with the activities of one or a few individuals. In this paper it would be impossible to give a comprehensive review of all the worthy and exciting innovations using multi-media. A few examples are included here to give some idea of the nature and range of multi-media systems.

A. The Audio-Tutorial System (Purdue University, U.S.A. [55])

'The technological revolution is met with mixed emotions by both educators and students. The reason is relatively simple—people do not like to be replaced by machines and people do not like machines as a replacement for people. Education is more than an information dispensing and absorbing process. It requires a comradeship of sharing and exchanging of experiences and an excitement that grows from common interests and hopes between teacher and student. Most of us can trace our interest in a specific topic to the inspiration derived either directly or vicariously from some other human being.

'Does this mean that one must reject the utilization of all technical devices in the development of an educational system? Far from it! Properly used equipment may very well enhance the personal relationship between student and teacher. Some of our most effective and powerful lessons are learned from teachers whom we have never seen but with whom we have a fellowship derived through some spanning medium such as the written word. Clearly this written 'bridge' between teacher and student has been a great boom to all of mankind and, until recent years, has been our only link to many of the great teachers of the past. Fortunately, technology has expanded our potential for even more intimate access to great teachers and with greater facility. The dimensions of audio and visual now can be preserved and retrieved in fantastically convenient vehicles. The imagination of man for utilization of these devices individually or in combination with the written word and/or tangible items is one of the major limiting factors in education today.

'If one uses the education model proposed by Hopkins of a "student on one end of a log and a teacher on the other", the role of technological devices becomes more clear and less foreboding to many of us. The purpose of technology in this context is to "capture" to the greatest degree possible the events or activity between the "good teacher" and student in a one-to-one relationship, so that the product can be duplicated to accommodate many students in a close approximation of the original situation. With today's audio and visual devices it is possible to involve the student with nearly every exposure to the subject conceivable. The programme can contain tangible, printed, audio and visual materials in any combination in which the "good" teacher wishes to use them. The only limitation is the capacity of the teacher and student to relate to the simulated situation.

'In 1961, at Purdue University, a programme was begun which has been called the Audio-Tutorial system. The basic philosophy of this system is very simple. A "good" teacher is asked to assemble the items he would use to teach one student and, while sitting among these items, to record on audiotape the conversation he would have with one student as he tutored that student through a sequence of learning activities. The product —the tape, tangible items, visuals and printed materials can be duplicated as many times as necessary to accommodate any number of students. Obviously the programme produced in this way will be limited by the cleverness of the teacher but the corollary is also true— a clever instructor can intimately involve the student in important and useful learning activities. The student now has access to the clever instructor in more ways than through the written word. Subtle communication through connotations by inflections in the voice are provided by the audiotape and the tangible, visual and printed materials assembled can exhibit the full skill of a great teacher to involve a student in a sequence of learning activities or a "symphony of learning".

'The Purdue programme evolved slowly. The first programmes were mere lectures on tape and relied on audio as the sole medium of communication. Later, other media were added including tangible items (live plants, models and equipment for experiments), printed materials (textbooks, study guides and journal articles) and visuals (2 x 2 in. slides, 8 mm film and photographs). Study programmes were set up in booths in a learning centre which was open from 7.30 a.m. until 10.30 p.m. Monday through Friday. Students came in at their convenience and spent as much time as necessary for them to master the lesson. An instructor was available at all times to help students on an individual basis if necessary. This study session was referred to as an independent study session (ISS).

'Two other study sessions were included in the system: (1) a general assembly session (GAS), and (2) a small assembly session (SAS).

'The GAS was scheduled on a weekly basis for one hour. It included several hundred students in a large lecture hall and involved them in the kinds of activities best done in a large group. Specifically, this assembly was used for major tests, long films, help sessions and an occasional lecture.

'The SAS included eight students and an instructor and was scheduled on a weekly basis and used primarily for short written and oral quizzes. It served in an administrative role and for the identification of students with a specific instructor in the large unstructured course.

'The course planned in the above pattern provided a full range of learning activities and situations. The ISS permitted students to enjoy some important features of a learning system that are not commonly available in a conventionally taught course. Some of these included:

'1. *Repetition.* There is little question but that the nature of many objectives requires repetition for their achievement. However, repetition ought to be engaged in intelligently and adapted to the individual needs of a particular student. In a course with 500 students the teacher cannot possibly make the adjustments in repetition for individual student need. Only the student can determine in an intelligent fashion how much repetition is necessary.

'2. *Concentration.* Most classrooms are not organized to permit students to concentrate during their study. Students are distracted by one another, and other dissociated events which may be occurring tend to distract the student's attention from the subject at hand. The audio-tutorial system permits the student to isolate himself from the surrounding environment by covering his ears with the earphones and by the use of other media to reduce his awareness of his surroundings.

'3. *Association.* In a study of plant science the major objective is to learn about plants. It makes sense therefore that a study of plants should be conducted where plants are available for observation. Diagrams, charts, models, photographs, and other such devices should be a "means to the end" so that students' attion is directed to the actual plant. The audio-tutorial system provides an opportunity for the student to have an object available at the time he reads about it, does experiments, etc.

'4. *Appropriate sized units of subject matter.* People vary considerably in the amount of subject matter that can be grasped in a given amount of time. Programmers have demonstrated that most people can learn almost anything if it is broken into small enough units and the student can take time to become informed about each unit before proceeding to the next. Any programme of study therefore should provide each student with an opportunity to adjust the size of the unit to his own ability to assimilate the information, so that those who can absorb large quantities of information may do so in an unrestricted fashion, whereas others who must proceed more slowly are permitted to do so. The audio-tutorial system allows the student to proceed at his own pace and to break the subject matter into units commensurate with his ability.

'5. *Adaptation of the nature of the communication vehicle to the nature of the objective.* It is logical that no simple vehicle such as lecturing or a textbook can achieve the full spectrum of objectives for a complex subject. The student's experiences should not be confined to any particular vehicle as film, audiotape, textbook, or a lecture. In cases where the development of a procedural skill is necessary, there is no substitute for the student doing this procedure himself. A properly structured course, therefore, would carefully define objectives and not try to mould objectives to fit a favourite medium (lecture, for example) but instead would use the medium best adapted to the nature of the objective.

'6. *The use of multi-media.* Individuals differ in their responsiveness to different kinds of communication devices. Some people learn well through reading, some can learn best by auditory communication, and some can learn best by literally handling specimens and performing experiments. The audio-tutorial system thus provides an opportunity for subject matter to be covered in a great variety of ways with the student exploiting the medium which communicates most directly and effectively for him.

'7. *Finally, and most important, the integration of learning activities and situations.* It stands to reason that if learning events are to be complementary and to have some relationship, they should be brought into close proximity and properly sequenced. The conventional structuring of a lecture, recitation, and laboratory does not take this into consideration but rather may expose a student on Monday to a lecture concerning a given subject; perhaps on Wednesday the student does experiments related to that subject; on Friday a recitation will involve the student in some exposure to the subject; and then on Sunday night, late, the student may read on this subject from text. The audio-tutorial system permits the student to bring all of these learning experiences into an integrated sequence so that each learning event may enhance or complement the adjacent ones and thus result in a "synergistic" effect.'

The A-T System has been elaborated to provide further individualization through the use of minicourses and mastery. This effort began in 1969 when Dr. Robert N. Hurst joined the Purdue University staff to teach a zoology course using the A-T System. At this time the content of both a zoology and a botany course was divided into topical units which could be treated as reasonably complete areas of study. These units were called *minicourses* because of the similarity in concept, except for size, to the conventional course structures used at most college and universities. Each minicourse had its own set of objectives, instructional programme, tests and credit. Students were required to master minicourse content to a certain minimal level, and a study could be repeated until acceptable achievement had been accomplished. Basic core minicourses were required for students enrolled in both courses, credit for minicourses was transferable between the two courses, and a pool of optional minicourses was provided for individual interests. The use of the minicourse and mastery concepts has provided flexibility for students in both pacing of study and course content. The interchangeability of several minicourses between the botany and zoology courses serves as a model system for examining the problems for interchangeability of instructional materials on a larger scale. From the results of this experience it appears that self-instructional programmes made in the Audio-Tutorial format of minicourses make feasible and practical the possibilities of free exchange of materials between courses within a school, between schools, and even on an international basis.

The Minicourse Development Project at Purdue University, funded by the National Science Foundation, is now producing self-instructional materials in the Audio-Tutorial format for minicourses covering a core of biology. It is anticipated that materials developed in this project will be interchangeable for some subjects and will provide a further opportunity to investigate the interchangeability of instructional materials on an international level.

'The individual nature of human beings cannot be over-emphasized. Any good educational system must be based on the fact that "learning must be done by the learner". It must involve the student in the process and must always provide a high degree of flexibility and adaptation to individual needs. However, superimposed over this quest for individuality is the dependence of each student on teachers to guide, facilitate and stimulate him to engage in appropriate learning activities. Today's technology provides new dimensions to accomplish this and, when these new tools are used properly, they provide more intimate access to the teacher —they do not dehumanize!' [55].

B. The Keller Plan (Arizona State University)

As pointed out earlier (page 186) there is a tendency to interpret media as meaning television, cassette TV, radio, audio and video tapes and teaching machines—technological hardware—and to overlook human interaction, textbooks and the written word. This is unfortunate because some of the most exciting and potentially significant teaching innovations for science education of recent years make no use or very little use of technology as a means of communication between teacher and learner.

The source of many of these systems is the Keller Plan developed by psychologists Professor Fred S. Keller and Professor J. G. Sherman for a freshman general psychology course at Arizona State University between 1965-67 (preliminary development began in 1963 at the University of Brasilia).

Since then it has spread to over 200 colleges in the United States and Canada and is being used for teaching a variety of subjects including biology, electrical engineering, geology, mathematics, physics, philosophy, and psychology. In addition, other universities are developing instructional systems that are based upon the main features of the system—self-pacing and mastery.

Perhaps the best way to describe the method is to quote from Keller. The following is from a handout given to all students enrolled in General Psychology at Arizona State University (1967):

'This is a course through which you may move, from start to finish, at your own pace. You will not be held back by other students or forced to go ahead until you are ready. At best, you may meet all the course requirements in less than one semester; at worst, you may not complete the job within that time. How fast you go is up to you.

'The work of this course will be divided into 30 units of content, which correspond roughly to a series of homework assignments and laboratory exercises. These units will come in a definite numerical order, and you must show your mastery of each unit (by passing a 'readiness" test or carrying out an experiment) before moving on to the next.

'A good share of your reading for this course may be done in the classroom, at those times when no lectures, demonstrations, or other activities are taking place. Your classroom, that is, will sometimes be a study hall.

'The lectures and demonstrations in this course will have a different relation to the rest of your work than is usually the rule. They will be provided only when you have demonstrated your readiness to appreciate them; no examination will be based upon them; and you need not attend them if you do not wish. When a certain percentage of the class has reached a certain point in the course, a lecture or demonstration will be available at a stated time, but it will not be compulsory.

'The teaching staff of your course will include proctors, assistants, and an instructor. A proctor is an undergraduate who has been chosen for his mastery of the course content and orientation, for his maturity of judgement, for his understanding of the special problems that confront you as a beginner, and for his willingness to assist. He will provide you with all your study materials except your textbooks. He will pass upon your readiness tests as satisfactory or unsatisfactory. His judgement will ordinarily be law, but if he is ever in serious doubt, he can appeal to the classroom assistant, or even the instructor, for a ruling. Failure to pass a test on the first try, the second, the third, or even later, will not be held against you. It is better that you get too much testing than not enough, if your final success in the course is to be assured.

'Your work in the laboratory will be carried out under the direct supervision of a graduate laboratory assistant, whose detailed duties cannot be listed here. There will also be a graduate classroom assistant upon whom your proctor will depend for various course materials (assignments, study questions, special readings, and so on), and who will keep up to date all progress records for course members. The classroom assistant will confer with the instructor daily, aid the proctors on occasion, and act in a variety of ways to further the smooth operation of the course machinery.

'The instructor will have as his principal responsibilities: (a) selection of all study material used in the course; (b) the organization and the mode of presenting this material; (c) the construction of tests and examinations; and (d) the final evaluation of each student's progress. It will be his duty, also, to provide lectures, demonstrations, and discussion opportunities for all students who have earned the privilege; to act as a clearinghouse for requests and complaints; and to arbitrate in any case of disagreement between students and proctors or assistants....

'All students in the course are expected to take a final examination, in which the entire term's work will be represented. With certain exceptions, this examination will come at the same time for all students, at the end of the term. ... The examination will consist of questions which, in large part, you have already answered on your readiness tests. Twenty-five per cent of your course grade will be based on this examination; the remaining 75 % will be based on the number of units of reading and laboratory work that you have successfully completed during the term' [37].

Clearly, in such a system the role of the instructor is vastly different from his role in the conventional lecture laboratory format. He is no longer the dispenser of information through triweekly lectures, but the manager of a system. His role is to prepare the study guides, identify the specific objectives of each unit and plan the ways and means for achieving the objectives, which means he must also design the mastery test. Further, he is responsible for supervising the undergraduate tutors (proctors) and must deliver the occasional motivating lectures. Because he has more time to move about the class he is expected to come to know the students as individuals. His success as a teacher now depends to a considerable degree on how effectively he chooses and integrates the media, readings, research papers, and experiments, so as to allow the student to achieve mastery of the objectives as efficiently as possible. If learning is to take place, the learner must be responding; hence, a prime function of the teacher is to use the media to obtain the optimal response from the student. His principal job is the facilitation of learning in others.

The system, as Sherman points out [64], is based upon student response rather than teacher performance and encourages students to demonstrate their skills more freely. This comes about because the purpose of the unit mastery tests is not primarily to evaluate students but to identify and remedy errors. Since errors do not lead to bad grades but only to another chance at a perfect score, the instructor can demand mastery and students are willing to expose themselves for tutor comments and corrections.

The situation of the student in the Keller system is also different from his situation in the conventional system. He has freedom to work at his own pace on his own time and this means he has the responsibility of self-discipline. For this reason not all students in a class are taking the same units at the same time. Some may have completed half a semester's work before others have finished the first five units. Green points out: 'Some students make rapid progress in spite of failing many tests; others are more cautious and never fail although their pace is slow. (The usual equation "fast" equals "good" breaks down.) What counts is progress, no matter how fast or slow. The important thing is mastery of each unit. ...' [27].

Self-pacing is not without its troubles for those students who work best by meeting weekly deadlines. There are some students who prefer the discipline of lectures and laboratories. When other courses have time requirements on their assignments there is a tendency for students in a Keller course to postpone study.

Another innovative feature about the system is the use of undergraduates as tutors. These are majors in the subject having generally passed the same course earlier with a grade of A. Some instructors even use the faster members of a class to tutor the slower members. It might be argued that undergraduates could not have the expertise to be a tutor. Experience at MIT indicates that undergraduates are better tutors than graduate students, perhaps because they are closer to the tribulations of the freshman, and they will work for academic credit not money [28].

The system accepts the Skinnerian view that learning is enhanced by immediate reinforcement. As Green points out, 'most powerful is the mere functioning of a unit, accompanied by smiles and compliments from the tutor, by having one's progress entered on the wall chart, and by personal satisfaction—born of the consequences of early success—at a job completed, reinforcement is immediate, since tests are graded on the spot. We usually make the first unit pretty easy so the student can get a little reinforcement early. Threats and punishments are avoided except late in the term when final grades become a concern. There are no deadlines within the term, no pleading or urging; the contract is clear. We try to pour on the reinforcement when a student does something right. But the student must do his part before we do ours. We do not give away rewards' [28].

It is not our intention to discuss some of the problems and limitations of the method. Nor is it our intention to give a very detailed description of the method.

In summary, the Keller Plan, or the Personalized System of Instruction (PSI) as it is sometimes called, is a self-paced student-tutored, mastery-oriented instructional system. Its basic elements as outlined by Keller are:

1. The student moves through the course of study at his

own pace, commensurate with his ability and available time.
2. The student must perfect a unit of materials before he is permitted to advance to new material.
3. Lectures and demonstrations are used as motivators, rather than sources of critical information.
4. There is stress upon the written work in teacher-student communication.
5. Teaching assistants, or proctors, are used to permit repeated testing, immediate scoring, tutoring, and a strong personal-social element [37, 64].

C. The Southin-Chambers System (McGill and Sir George Williams Universities, Montreal, Canada) [65]

'*General.* Collaborative Studies in Science and Human Affairs is a course given co-operatively by McGill and Sir George Williams Universities. All students, whether enrolled formally at McGill or SGWU, are considered participants in one course. It is open to all university and college-equivalent students without pre-requisites.

'The course examines the historical, contemporary and future interactions between science and society. It is structured so that students can build on whatever specialized information they already possess and at the same time sample viewpoints and methods of disciplines initially alien to them. Students choose individually the learning procedures they think most appropriate: lectures, seminars, research projects, independent reading, community service, tutorials, educational media, etc. The course offers the following specific areas of study: What Is Science? Science and Ideology; The Scientific Community; Is Objectivity a Myth of Our Times? Measuring the Growth of Science; The Two Cultures; Scientific Responsibility; New Definitions; Technology and the City; Science, Technology and War; Alienation in a Technological Age; The Technological Order; Computers and Society; Biological Engineering; Drugs and Society; Medical Priorities and Policy; Sex and Society; Race and Society; Social Aspects of the Environmental Crisis; Population and Resources; Man's Environment: An Appraisal of the Damage; Survival of the Biosphere; Energy and Environment.

'*Staff.* In addition to course co-ordinators there are many graduate and professional course assistants, media assistants, student teachers and volunteer resource people from both university and community.

'*The module system.* Because of the inter-disciplinary, educationally innovative nature of this course, you are entitled to think of yourself as a special student. There are no required lectures, classes or standard term projects, and it is not unusual to find several events scheduled at the same time. The staff is prepared to give you special attention, but you must ask for it. Staff members are available during posted hours at the Resource Centre. In addition, each student is assigned an advisor who will help her/him with academic and administrative matters.

'The twenty-two basic modules listed above form the core of the course. A module is an autonomous self-study unit which a student does at his own pace. You choose the modules, decide when to do them, and present yourself for examination when you think you are prepared. Modular study is not so much "unorganized" or "free" as it is "choiceful". The key to avoiding confusion is to make choices in an orderly way.

'*Procedures.* The following are steps one takes to begin activities in this course, and illustrate what is meant by "doing a module". Other sections referred to (e.g., "see *Resource Centre* section") appear after the *Procedures* section.

'1. Go to the Resource Centre when convenient and open working file (see *Resource Centre* section). For a course fee of $ 12.00 you will receive a blue looseleaf binder (Coursebook) containing guides for the twenty-two basic modules, and a Calendar. Other course publications will be designed to be placed in the Coursebook. The course fee covers all expenses incurred in the course such as guest lecturers, films, media, printing and computer exams. No textbooks are required.

'2. Read this Course Description carefully and note any questions you may have. Attend one of the orientation meetings during September 18-22.

'3. Select and read one module guide in your Coursebook. Obtain and read the readings book for that module. In order to provide a smooth introduction to modular study, each staff member will conduct one Module Seminar during the months of October and November. At each session, the content of one basic module will be discussed.

'4. Present yourself when ready to be examined on the module you have studied (see Examination/Grades section).

'5. Steps 3 and 4 are what is considered "doing" a module. Repeat these steps for each module you wish to do for a credit. At least two modules must be "done" by December 9th.

'6. If you wish to earn a grade of "B" or "A", you must choose one or several of the projects described in the *Advanced Projects* section in addition to six modules. Note that a Project Report Form must be completed for each such undertaking, and that the mere completion of a project does not assure a "B" or an "A". The grade of "A" will be reserved for students whose work shows special merit.

'Examination/grades

'*Computer exams*. In order to complete a module, you must pass a computer exam based upon the required readings and module guides. You must register for these exams at the Resource Centre *one week* in advance. A schedule for exams will be posted.

'When you arrive, you will receive an examination with your name and I.D. number on it. No two examinations are the same and there is no standard answer code. Your answer card and question sheet must be turned in for credit to be awarded. Results will be posted in the Resource Centres after one week. Exams may be repeated three times. Pass in 75%, first try; 85%, second try; and 90%, third try. Failure on the third attempt will require study of a different module.

'*Final Grades*. Four modules must be completed for a "D" grade and six modules for a "C" grade. Advanced projects may be presented to the Co-ordinating Committee for grading. The file must be complete to receive a final grade.

'Six modules should be done when the advanced projects are presented for grading. Once submitted, the file will not be returned to the student for revision. A file will not be accepted later than two weeks before the end of classes.

'*Advanced projects*. Advanced projects, undertaken by those students who intend to earn a "B" or an "A", are limited only by the imagination and resourcefulness of the students, the resources available and the staff's demands for quality. The best procedure is to work on advanced projects while working on the modules. Projects are usually long-term and should not wait until six modules are completed—this may not leave enough time for a well developed project. PROJECTS MAY BE UNDERTAKEN INDIVIDUALLY OR IN GROUPS.

'*Project Report Form*. All projects, however varied, must be accompanied by a Project Report Form in three sections:

'1. Contract. Before beginning any advanced work, the student must contact the appropriate staff member and decide which basic modules (if any) are prerequisite and what will be necessary to satisfactorily complete the proposed project. This should be detailed in writing and signed by both student and staff member. There will be no penalty if the project is not completed.

'2. Report. The work will be reported in the form of a journal. Any relevant meeting should be listed with complete bibliographic data (author, title, publisher or journal, volume or year, page number), summarized in one short paragraph, and then thoughtfully related to the other work for the project. Any course-related actions engaged in by the student (such as survey administration, volunteer work, marches, sit-ins, letter-writing, campaigning) should be described and criticized for effectiveness and relevance. Any products of the above activities (survey results, clippings from magazines or newspapers, quotes from books or articles, films, artistic works, videotapes, audiotapes, slides, photographs, interview) should be attached to the report. The entire report should be edited to make an effective presentation.

'The Report should be descriptive in that it states what the student DID, and it should be critical in that it states what the student THOUGHT. It is up to the individuals who sign the contract to determine the explicit details and requirements.

'3. Evaluation. Upon completion of the project the contract, the Report, and any "products", should be returned for evaluation to the staff member who signed the contract. Frequent consultation with the evaluating staff member is recommended. At least two weeks should be allowed for the staff to comment on and summarize his reactions to the project as Honours, Acceptable, or Unacceptable. If a student is not satisfied with the evaluation, he may request withdrawal of the project for revision. If the staff agrees, the student may resubmit the project report.

'*Examples of advanced projects*. Most advanced projects require additional information not covered in depth by the twenty-two basic modules. Files of readings and study guides have been collected during previous years of the course, and this material is available when staff are present at the Resource Centre. The following is a list of activities that could qualify for advanced work in the course.

'For each type of activity there are usually one or more Staff Co-ordinators whose responsibility it is to help students plan their work.

'*Autonomous study groups.* THE AUTONOMOUS STUDY GROUPS are set up only at the instigation of interested students. Direction and advice are available from the instructors and resource people on request, but organization and supervision are the responsibility of the students. A minimum of five or six hours of actual discussion time is required for each topic chosen, and a critical and detailed report of the group's activities must be signed by all participants.

'A handout, "Organizing Autonomous Study Groups", available at the Resource Centre, describes a variety of techniques for effective group discussion.

'*The learning diary.* Students working alone are encouraged to keep a regular LEARNING DIARY. The diary is handed in only at the end of each term, but it is best to have the Co-ordinator evaluate it periodically. He will enter his comments, criticism and advice when requested.

'The events entered in a diary need not be sponsored by the course itself. That is, a student should attempt to relate information and attitudes derived from the course to events and situations occurring in his usual week's activities. Among the activities that might be recorded and discussed are:

Films, TV and radio programmes
Lectures in this and other courses
Cine-Club and course TV showings
Public lectures and speeches
Governmental policy
Relevant ideas, new commitments, etc. generated in response to specific course materials.

'It is important that the diary also include a complete record of the serious reading done for the course. Books, parts of books, articles in scholarly and scientific journals and, in some cases, items from the popular media should be entered in the following manner:

Date (of entry)
Complete bibliographic reference
Brief summary of content
Critical comments.

'Participation in projects, autonomous study groups, etc., may be entered in the diary, or credit can be awarded separately.

'*Seminars and short courses.* Course assistants, community resource people, and interested faculty have been invited to initiate seminars, short courses and research teams in subjects in which they have special competence. These may be organized in any way the initiator so desires. The responsibility for recruiting students from the course is also entirely in the hands of the initiator. These activities will be advertised at the Resource Centre with a complete description of subject, format, evaluation procedures, credit points offered, and schedule of meetings.

'*Directed research and projects.* A variety of research topics and projects based on the modules is possible. The reprint "Community Outreach in Biology Teaching", available at the Resource Centre, describes some of the projects undertaken by students in the course last year. Research and project teams are organized on student initiative but must submit to the supervision of the course staff or resource personnel. All projects must be approved in advance by one of the Research Co-ordinators.

'*Videotape production.* A production-research team consists of five students. All potentially interested students should plan to attend the first TV meeting held during the second week of term.

'*Newsletter production.* A course NEWSLETTER is prepared periodically and short reviews of relevant books and films, progress reports of major course activities, etc. can be submitted for publication.

'*Direct action.* Events such as teach-ins, eco-theatre, town meetings, anti-litter campaigns, etc. can be organized and staged for credit. A written contract must be approved in advance.

'*Skill exchange.* A class of several hundred students contains many who already possess useful skills which other students, if they were aware of them, would like to acquire. A bulletin board for advertising of SKILL EXCHANGE proposals is available. When these skills relate to the course (as would, for example, skills in taxonomy, microscopy, instrumentation, natural history, nature photography, laboratory experiments, and survey questionnaire design), credit can be awarded to both the learner and the teacher. Students should define the skill to be taught, write a mutually agreeable contract, and discuss it with the Skill Exchange Co-ordinator.

'*Action letters.* An ACTION LETTER is a detailed, carefully documented, and cogently reasoned letter to some public official whose responsibility it is to deal with the problem presented. The letter should not only document in precise detail the current situation but should also present a practical alternative policy, equally well reasoned and documented. Most of the modules in this course can readily be seen to include issues requiring some change in current administrative and governmental policy, and equally obvious are the "public servants", who should be advised of your findings and views. The procedure is as follows:

'1. Discuss your proposal (including the name you suggest as addressee) with the Action Letter Co-ordinator.

'2. Give your letter, *typed* and signed, together with an addressed and stamped envelope, to the Co-ordinator. He will make a xerox copy of the letter for your course file and mail the original.

'3. Discuss any reply you receive with the same Co-ordinator.

'*Term papers.* Under no circumstances will term papers be approved, credited or even considered.

'Sources of information

'*Resource Centre.* A unique and important feature of this course is the RESOURCE CENTRE, where students from both universities can meet easily and informally. This feature is especially important because the success of the student organized study groups and projects depends on drawing interested participants from as large and diverse a group as possible. The Resource Centre will serve as general discussion room office for all course staff, repository for module files and student records, advertising and organizing centre for study groups and projects, and administrative centre for the course.

'In addition to the regular staff hours at the Resource Centre, the professors, course assistants, and student teachers meet there collectively twice a month to discuss administrative and policy matters. These Co-ordinating Committee meetings are fully open, and students are especially welcome to attend and comment.

'Obviously, the Resource Centre is of central importance to the course. Students should make every effort to stop by as often as possible. Only in this way can students learn of the projects, books, films, lectures, etc. that constitute the course. The module files are available only at the Resource Centre and cannot be removed.

'Every student will have two files located in the Resource Centre. One, the working file, will be open and available at all times so that staff members and other students may leave messages or course announcements. A student may leave whatever material he wishes in the working file.

'The other is the records file which will contain examination scores, grades and projects presented for evaluation. This file will be locked, but may be examined on request.

'Staff members also have working files so that messages can be left for them.

'*Readings Books.* Each of the twenty-two module guides has an associated Readings Book containing about 200 pages of basic readings for that topic. The computer screening exams are based on these readings and none other. Several complete readings book sets are available on reserve at both the McGill and SGWU undergraduate libraries.

'*Module Files.* The Module Files contain collected copies and reprints of books, articles, newspaper clippings relevant to each of the modules. Many of these readings are at an advanced level and consequently very useful to those participating in advanced projects.

'*Lectures.* There is a course lecture at class time every Monday during the first term. Most of the lectures will be given by guest lecturers. Students may attend none, some, or all of these lectures. For those who want it, an exam will be given.

'*Cine-Club.* Cine-Club presents weekly screenings of a large variety of films based on course modules. These screenings are at class hours on Wednesdays.

'*Television series.* All student-produced videotapes will be screened Friday afternoon at SGWU. Arrangements have been made for the best thirteen shows to be presented on Cable-TV, Channel 9. The primary aim of this series is to stimulate discussion and further reading rather than to convey information. All students are encouraged to attend.

'*Resource People and Organizations.* A large number of volunteer Resource People and Resource Organizations have been recruited from among the faculties of the two universities and from the community. These people and groups have agreed to assist students in their study of various modules by consulting about the material itself, organizing seminars, sitting in on autonomous study groups, suggesting and over-seeing projects, lecturing, providing facilities and so on.

'Many resource people and groups have agreed to be listed on our open REFERRAL LIST at the Resource Centre. Others prefer to be contacted only through one of our Research Co-ordinators.

'*Newsletter.* A course Newsletter contains reviews of relevant books, movies and public lectures; information on major course projects and seminars; the critical evaluations of staff, facilities, and the course itself.

'*Student-teachers.* A number of the best students from the course last year are serving as unpaid student-teachers. They are formally enrolled in a separate project course, but receive their credit from assisting in "Collaborative Studies". At the beginning of the course the student-teachers may catalyze autonomous study groups, incorporate new students into on-going or pre-tested projects, assist with television programming, organize seminars, and the like. The student-teachers meet bi-monthly with the course staff to report and discuss formally the course activities with which they are involved. These meetings are also open to students and alternate with those of the Co-ordinating Committee.

'*Periodical literature.* Many scientific journals contain relevant commentary, editorials, and even major articles, which avoid technical jargon and are easily understood by non-scientists. Opinions expressed therein reflect the range of political, social and moral values of the scientific community at large. Technical and scientific statements in these sources are highly authoritative; however, their political and social analysis, while generally thoughtful and carefully considered, is no more reliable than that of any other publication. Therefore any reading done, in whatever source, should be critically evaluated in light of what you have learned in other course activities. Scientific journals likely to be of particular value include *Science, Bio-science, Environment, Science Forum, New Scientist* and *Science Journal, American Scientists,* and *Nature.*

'Since scientific journalism in the popular press has attained very high standards in recent years, you may also find that newpapers, such as the *New York Times,* contain useful articles. Nevertheless such essays are seldom as reliable in technical matters as those which appear in the scholarly literature and therefore should be read with a measure of caution. Popular magazine articles are sometimes extremely penetrating and thoughtful, but again they must be read with a critical eye to documentation and sources.

'*Evening events.* Occasionally during the year the general public will be invited to attend course-sponsored presentations, such as speakers, forums and films. As a rule, these events will be held at SGWU on Tuesday evenings. Notices will be posted in the Resource Centre and advertisements placed in the daily newspaper.

'Students enrolled in "Collaborative Studies" are encouraged to use this opportunity to challenge the speakers and to open a dialogue with members of the Montreal community.

'*Planning your course.* Although you may complete the modules in any order whatever, you will probably find that certain sequences and concentrations suggest themselves. Students primarily interested in pollution will want to start with modules in the "Environment" section. Some will wish to confine their concerns entirely to this section. However, since some of these issues are closely related to the population question, a student might then turn his attention to modules on birth control and abortion.

'Once you have selected a particular module and have read the material recommended in the Study Guide, then decide how to make use of this information. Would you care to participate in an autonomous study group? Then advertise at the Resource Centre to form such a group. Do you want to do some further research on your own? Consult the list of research advisors at the Resource Centre and work out a proposal. Would you like to join others on a videotape project? See the handout on the Television Series and then discuss your ideas with the television staff. Would you like...

'*Alternatives in education.* In the past some students were immediately prepared to exercise the freedom this course allows and were eager to work independently. Others, long accustomed to authoritarian teaching practices and passive acquisition of information, were confused when suddenly required to manage their own learning. For these students, many years of schooling have succeeded in teaching only the need to be taught.

'To smooth the transition into this course and at the same time to acquaint students with the sources of some of our notions on education, the first two weeks will be devoted to seminars and readings on Alternatives in Education. A separate handout, available at the Resource Centre, describes these activities in detail.'

D. The Guided Design System (West Virginia University, United States of America, and University of Windsor, Canada) [71]

Education, like many other professions, has its researchers and its practitioners. But what education does not have is enough teachers who translate the results of research into improved operations. As a result, much of the available research on learning has not been effectively incorporated into the present educational system. Of course the present system is functional, but design work could make it much more effective.

Dr. Charles E. Wales, West Virginia University, and Dr. Robert A. Stager, University of Windsor, performed this design work and developed a book on *Educational Systems Design*. Their work is based on a defined set of goals, intellectual operations related to Bloom's *Taxonomy of Educational Objectives* and appropriate psychological principles. It includes a concern for testing based on content-performance objectives and the use of mastery examinations. A major result of this design process is a new teaching-learning technique called Guided Design.

In Guided Design, students work in four- to seven-man teams which solve relevant, interdisciplinary open-ended problems that involve social-humanistic considerations. The students are guided through the decision-making process by a series of printed 'Instructions', which ask them to discuss each step and make a decision. Then they are given printed 'Feedback' which allows them to evaluate their judgement. Class time is primarily devoted to this student discussion and decision-making. The teacher's role is that of manager, stimulator, guide and a model of thinking behaviour. Each open-ended problem establishes the need for certain facts, concepts and principles which are treated as components used in the thinking process. This subject matter is studied outside of class using programmed or Audio-Tutorial instruction. Tests on this material are given in class on a mastery basis. Eight articles on the Guided Design process were published in *Engineering Education* (February-May, 1972).

There are three operating Guided Design systems at West Virginia University. The original model is a two-semester course in Freshman Engineering Design. The second application was in the Chemical Engineering Department, where the sophomore and junior course work has been converted to this model. The third system is a master's programme which prepares rehabilitation counsellors. In this programme the faculty combined all four courses in their first semester into an integrated Guided Design package. This concept has also been applied to a high school course concerned with environmental problems. Although multimedia support systems are an integral part of each of these models, the Guided Design concept goes *beyond* the acquisition of subject matter skills by focusing on the use of the decision-making process required to solve an open-ended problem.

E. The Macquarie University System for External Students (Macquarie University, Australia [2])

The Macquarie example has been included because it highlights one of the major challenges facing science educators in the seventies—the off-campus or external student and part-time student in the natural sciences.

In 1965 Macquarie University was established in Sydney, Australia, committed to accepting both on-campus and off-campus students. It began by offering off-campus courses in biology, chemistry, physics and earth sciences as well as mathematics, psychology and education.

Science education has at least two aims. One is to convey an understanding of some of the facts, principles and generalizations which are basic to the particular discipline being taught. The other is to convey an understanding of science as an investigation activity.

In order to offset the 'second-class status' that is often associated with part-time courses, the university decided to offer the same courses to both on-campus and off-campus students and to require both groups to sit for the same examinations and award both groups the same degree.

The problem of achieving these three objectives faced the university when it started teaching for the first time in 1967 and the different science schools tackled the problem in different ways. The following discussion refers primarily to biological science.

In the absence of experience of off-campus teaching of science—at that time Macquarie was a unique institution in offering a range of natural sciences off-campus—the approach to teaching in 1967 followed conventional paths. A weekly lecture-laboratory format was adopted for the on-campus students, while the externals received weekly correspondence materials covering the lecture information and the year's laboratory work was concentrated into two four- to five-day residential schools. This approach enabled some of the problems of off-campus science teaching to be identified. These are summarized in Table 1.

As may be seen, two major problems characterize the off-campus teaching of biology at Macquarie (and this might be true for external science students in general)—the twin problem of perspective and communication and the problem of laboratory work.

TABLE 1 [2]

Problems and approaches in off-campus teaching

Problems arising from off-campus study situation	Approaches
Perspective	Detailed syllabus supplied to students
Communications	
one way — transmitting information	Taped lectures, lecture outlines, study guides
	Taped / Printed practical instructions
two way — dialogue between student/staff	Assignments — written correction — feedback
	Taped comments to individual students
	Opportunity for student response on tape
instant aid — solving immediate difficulties	No general solution at present, telephone useful in special cases (internal students have access to teaching staff to clear up specific difficulties)
Timetable requirements — Residential school Allows intensive teaching over short periods — but little time for absorption (learning) or independent thought or action	Confine residential school to those aspects of the course which CANNOT be dealt with away from the university.
	Reduce amount of experimental work done at residential school through home experiments
Difficult to integrate practical work with lectures	Spread experimental emphasis throughout course — use of experimental kits (home experiments)
Problems arising from particular objectives	
To teach biology involves:	
(a) TRANSMITTING INFORMATION (Theory)	See communication, one way (above)
(b) TEACHING SKILLS (Practical work)	
of: observation — direct	Issue external students with microscopes, practical instructions, specimens. Audio/tutorial approach — off-campus
— indirect — through designing and carrying out experiments — collecting data	Experimental kits, practical instructions
	Assignments
	Residential school — laboratory work
of: interpretation — analysing data	Residential school tutorial
evaluating data	Feedback from corrected reports
interpreting data	Self instructing practical manual

Communication and perspective problem. It is not necessary in this paper to discuss Table 1 in detail. This has been done elsewhere [2]. The communication problem is threefold: (1) the communication of lecture-type material; (2) the communication or exchange of ideas between staff and student—that which occurs for on-campus students during discussions with staff and fellow students; (3) the communication of immediate information: the 'instant help' that the on-campus student can get by direct contact with staff or other students. The perspective problem arises from the isolation of the off-campus student.

The laboratory problem. Experience in 1967 showed that weekly laboratory periods are more effective than intensive laboratory schools. It is not possible to take a series of three-hour practical periods designed for internal students, string them end to end and so cram into a week the laboratory work for half a year.

One of the purposes of laboratory work is to illustrate the lecture materials. This is impossible to achieve when lectures and laboratory are up to nine weeks apart in time. The other major purpose of practical work—teaching biology as inquiry—also proved difficult to achieve in an intensive residential school.

The 1967 experience led to two main conclusions: (1) the communication of information and perspective to the off-campus student was not as effective as with the internal student; and (2) the amount of laboratory work would have to be reduced or else not confined to residential schools on campus.

It follows that the teaching of biology (science) to off-campus students requires a new teaching strategy.

The role of multi-media: communication. Table 1 also lists some of the means that have been adopted in an attempt to solve the communication and laboratory problems.

Some aspects of the communication problem have been vastly improved through the use of a variety of media—including textual materials, study guides, audiotapes of lecture material, plus visuals and tangible items used in lectures, some specimens and materials used for descriptive laboratory work illustrating lectures, two track audiotapes for feedback and discussion between staff and students. Each week the student receives through the post what is essentially a multi-media learning package. It is worth pointing out that the media includes very little 'hardware'. Television and radio are not used (not available); the main 'hardware' item is a tape recorder.

Practical work: experimental kits—study guide. The problem of practical work has been solved to a considerable degree by issuing each off-campus student with an experimental kit and a study guide containing the ways and means for carrying out both descriptive and experimental practical work. The kit includes a microscope, glassware, slides, appropriate stains, reagents and other expendable materials. Audio-tutorial units covering the descriptive work are included when appropriate. A more detailed account of the problems of practical work and experimental kits is available elsewhere [2].

With these aids, a large part of the practical work can be carried out at home. They have transformed the residential school by making more time available for tutorials, seminars and informal discussions between staff and student. They have increased the external student's sense of involvement and hence his motivation and they have extended the scope of the work available to the student. Although the equipment in the kits is simple, students can do more than in the conventional laboratory period, because they are not obligated to clean up at the end of two of three hours and do no more until the next laboratory period.

The problem of equivalent courses. These advantages have led to the use of kits in some on-campus courses at Macquarie University. This means that there is a convergence of the teaching approaches adopted for the on-campus and off-campus students. The further extension of this might be part answer to the problem of equivalent courses.

Indeed, the School of Chemistry at Macquarie has developed a detailed study guide-assignment approach, which is used by both the on-campus and off-campus students. There are no lectures for on-campus students. Assignments are completed at regular intervals by the student. Appropriate comments are made by the teacher and the assignment returned to the student. In this way the student, on-campus or off-campus, is able to gauge his progress and the teacher can assess the student's performance. However, experimental work in chemistry for off-campus students is confined to the residential schools.

Other problems. There are many problems in off-campus science teaching. In addition to the ones already mentioned, there are educational problems also. If it is assumed that the mastery approach is essential to the learning process, then how can mastery be demonstrated by off-campus students, particularly for certain aspects of practical work. Objectives can be specified, materials can be supplied in an experimental kit, but how can mastery be assessed when the student and teacher meet together only for a few days each semester at a residential school?

Study centres equipped with audio-visual aids and a tutor might be one answer, but this may be practical only in situations where student numbers are sufficient to justify the setting up of a centre and where sufficient skilled tutors are available in the community.

Television demonstrations might be helpful. This again assumes a large student body. But what of more specialized or advanced courses? It would be difficult to justify the use of television for a few students.

Another educational problem is the problem of the 'second class status' that is generally labelled against any form of extension courses, part-time programmes and external studies. The answer to this may lie in part in curriculum reform and in specifying objectives of a science education. The learning of techniques, for example, might be a matter for in-service or on-the-job training.

Answers to these problems are not obvious.

Experience at Macquarie demonstrates that a multi-media approach is helpful in off-campus science teaching. The authors believe that laboratory work, particularly in senior years, is the major problem to be solved for off-campus teaching. Whether or not improvements in existing media, the appearance of new media, or the development of better learning systems through the more imaginative use of media will further improve the off-campus teaching of science is a matter the future will determine. The challenge is there; it is up to us to meet it.

F. The Markesjo System (Royal Institute of Technology, Stockholm, Sweden [45])

'A Swedish course in applied electronics. In order to meet the big demand for new material for instruction in electronics at gymnasium level, a complete media course was prepared in Spring 1969 by the PE Group at the Royal Institute of Technology in Stockholm and in collaboration with a number of organizations.

'1. Extent and level. The electronic course in question comprises, according to the plan of studies, approximately 240 scheduled hours. The course is intended for the fourth form of gymnasium, i.e., the same level as for the first year 1969/70 by approximately 50 classes (i.e., about 1,000 students) all over the country—though not all of them were synchronized with the recommended time schedule.

'2. Packaging of the course. The course was divided into three parts, having 7, 7 and 5 weekly packages respectively. Every weekly package contains 12 hours according to the plan of studies (Fig. 2-3).

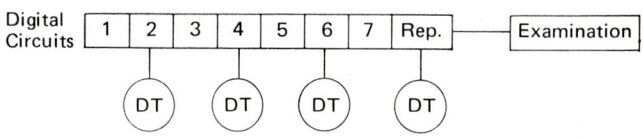

FIG. 2. The section of the course 'Digital Circuits' is divided into seven weekly packages plus one week of refresher work. DT = diagnostic test.

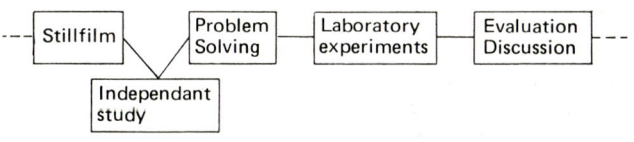

FIG. 3. Breakdown of the weekly package into various activities. (Stillfilm = taped radiovision)

'3. Choice of media for the motivation parts. Because of the costs involved, neither films nor TV could be used for the motivation-creating introductory programmes for every weekly package. We, therefore, chose colour slides with taped comments. In some cases movement could have offered more to the introduction programmes but, on the other hand, colour slides offer vast possibilities both for documentary reporting as well as for pictures illustrating principles. Another advantage with a series of slides is that individual slides can be replaced when revising the course. The course comprises 19 radio programmes, with an average of 15 slides per programme. The commentary is distributed by the Swedish Broadcasting Corporation to the audio-visual centres of the schools through its school broadcasting. The pictures are mailed to the participating schools (in three instalments).

'4. *Diagnostic tests*. For the checking and the statistical analysis of the diagnostic tests optically read and prepunched reply cards were used.

'As a result of the manual checking by the teacher of the reply cards with the aid of correct answer patterns (templates) before these cards are sent for computer checking, both the teacher and the students get an immediate feedback. The result of the diagnostic test can thus have a direct influence on the instruction in the classroom.

'The tests are programmed in such a way that the questions are shown on slides and at the same time are read out in a more detailed form through the programmed tape. The students answer the questions by marking the correct alternative on the reply cards.

'The teacher's manual checking takes place during a ten-minute break after the test and then a new tape, with comments on the answers, is played while the slides with the questions are shown once again.

'During this stage the teacher can stop at the questions that require more information and the students have the opportunity to ask questions related to the posing of the problems. Going through a diagnostic test of this type usually results in very lively discussions between teachers and students.

'After a week the class receives a summary of the results from the computer. This summary lists individual results, class results and average results for the entire country on the questions asked.

'5. *Text material*. For financial and practical reasons it was necessary to divide the text material in such a way that at least a part of it can be revised within a short time.

'The text material is therefore divided into three different categories: textbook, workbook and teachers guide.

'6. *Theory—solving of problems—application*. The constant aim has been to maintain as close a connexion as possible between theory, the solving of problems and application during the preparation of all the weekly packages. Every weekly package is, as a rule, built-up around a laboratory experiment demanding a certain amount of problem solving as preparatory work. The weekly theory section provides the background to the problem solving and thus the theory is verified directly during the practical experiments.

'7. *Results and follow-up*. Local timetable planning differences have made it impossible for a number of classes to follow the centrally recommended timetable.

'Approximately 20 classes have followed the recommended schedule and attitude tests, one for the students and one for the teachers, were held in these classes. The results of these tests show that the attitude of both students and teachers towards the media course is very positive.

'The students have found the diagnostic tests to be both instructive and guiding, and the teachers have often used these tests as topics for discussions.

'The course has proved to be a strongly motivating factor for the subject—this applies both to the teachers as well as to the students.

'The students have found the teachers' attitude towards the course to be very positive.

'One would assume that a media course containing a large amount of material intended for self-tuition would permit the students to absorb most of the essence of the material unassisted. The test questions show, however, that this is not the case. The teacher's role in the media course is essential but his duties are different. He can devote more time to the students for adapting and administering the given material and less time for dealing with the material itself.

'Course material in the form of ready-made packages and ready for immediate use saves so much time and work for the individual teacher that he can accept it without hesitation, though it may restrict his own influence as regards the choice of material on the subject. Through diagnostic and attitude tests, both teachers and students have the opportunity to influence the course in the long run. The teacher does not find that the comprehensive teaching material available to him restricts his role as teacher. On the contrary, he finds it a valuable form of assistance which leaves him free to give each pupil more individual attention.

'The course has now been running for three years and has had an important effect upon in-service training of the teachers (in both subject matter and learning methods). The course has also been widely used for retraining engineers in a number of Swedish electronic industries.

'In a coming project a sample of the engineers who have taken the course as well as their employers will be asked which of the given course objectives have been useful, which should be deleted and which new objectives should be included. In such a way the electronics course will really be adapted to the needs of the Swedish industry.'

The educational potential and implications of multi-media

The Carnegie Report [10] discusses in considerable detail the educational potential and implications of the new informational technology for students, faculty, institutions and society. The report views the implications of the new technology as follows:

'Off-campus instruction of adults may become both the most rapidly expanding and the most rapidly changing segment of postsecondary education.

'Fewer students may study on campus; and more may elect to pursue their studies off campus and get credit by examination. This will reduce enrolments below the levels they otherwise would reach.

'Students in small colleges will have more access to a greater variety of courses and greater library resources. The big campus will have fewer advantages on these scores.

'The library, if it becomes the centre for the storage and retrieval of knowledge in whatever form, will become a more dominant feature of the campus. New libraries should be planned with the potential impact of technology in mind.'

This report is an extraordinarily significant publication. In the present authors' view it is the definitive 'text' on instructional technology in higher education.

'New buildings should be built with adequate electronic components. They should also be planned for 24-hour use.

'Some new colleges and universities may be constructed with a central core area and with satellite campuses scattered around within commuting distance. The core area will provide access to knowledge; the satellite campuses will provide a greater sense of community because of their smaller size.

'New configurations will take place to the extent that students are dispersed as consumers and as some faculty members and many technicians are concentrated as producers.

'New professions of multimedia technologists are being born.

'Students will need to be more familiar with the use of certain technologies—particularly the computer—as they begin their college training.

'Prospective high school teachers and prospective college and university teachers will need to be trained in the use of the new technologies for instruction. Many of these prospective teachers who are in college now will still be teaching in the year 2000 when the new technology will be in general use in educational institutions.

'Universities and colleges will be able to trade-off in their overall budget making between funds for construction of buildings for on-campus instruction and operating costs of off-campus instruction.

'Less remedial instruction will be necessary on campus. Students will come to college better prepared or will receive their remedial instruction in off-campus courses or through independent learning assisted by the new technology.

'Good systems for informing and advising students will become more essential and more complex as additional options are made available and as more instructional opportunities are located off campus.

'Many more and better tests will be required to evaluate the progress of students who learn through the new forms of instructional technology.

'Some of the informational technology, thus far, seems better at training skills than at general education. The better it is at training skills, the more general education may suffer as a result—particularly if students move off campus and content themselves with skill training. But instructional technology represented by such media as television and film, can also contribute to general education and to the teaching of concepts.

'Some equivalent of the university press, or an expanded university press, may eventually be necessary to produce videocassettes and other instructional software that can be used with the new technology.

'Copyright laws will need to be reviewed by Congress to adapt them to the new carriers of information.'

This is an impressive list which bears out Eric Ashby's view that the fourth revolution in education 'is patented by developments in electronics, notably those involving radio, television, tape recorder, and computer' [10].

Achieving the Potential of Multi-Media. At the present time much of the hardware of the new technology is available and within a short time videocassettes will be available. Yet the new technology is having only a limited impact. It is falling short of its potential. The Carnegie Report attributes this to a shortage of software.

'The fourth revolution will not mature in a fortnight. In fact, it now seems to be faltering.

'The principal deficiency is in the availability of computer programmes, video and audiotapes, printed learning modules, films and other "software" of instructional technology. This deficiency exists for six important reasons.

'1. Instructional technology is not uniformly welcomed by the academic community.

'2. Faculty members who are interested in designing learning materials for the new instructional technology usually are not properly rewarded for their efforts.

'3. Although the physical equipment of instructional technology exists, there is little compatibility of components in models of some mediaware made by different manufacturers. This incompatibility forces non-technologists to guess which models are most likely to dominate the educational market before proceeding with the design of learning materials.

'4. There is ongoing debate over the relative virtues of learning materials produced for local campus use and those produced for national distribution. We believe that a combination effort will be most fruitful. On the campuses, faculty members should be encouraged to develop new instructional programmes with the assistance of educational technicians and media specialists. Beyond the campuses, teams in consortia of institutions, in learned and professional societies, and in the communications and information industries should be similarly engaged. The quality of the programmes thus developed will vary widely for the next decade or so, but as more programmes become available it will be increasingly possible for faculty members to be selective on a qualitative basis. If a programme fully satisfies the instructional objectives of a course or segment, it will be a valid choice for classroom use regardless of its point of origin.

'5. Few faculty members have the combined interests and expertise in subject matter, media development, and learning theory that the design of high-qualtity instructional materials requires. Some campuses do not have this combination of expertise available even in different individuals.

'6. Faculty members have been disenchanted by persistent findings in many studies that the learning effectiveness of instruction provided by technology is not significantly different from that of "good professors and teachers using conventional modes of instruction". As Anthony Oettinger points out, these findings tend to "fly in the face of common sense", and " ... confirm limitations of formal research on schooling rather than deny the impact of technology on learning". Such findings overlook advantages of technological applications that may not be measured in current research. In interpretations of the findings, sufficient care is not always taken to make clear that while "no difference" does not necessarily mean "better", it also does not necessarily mean "worse". The studies that produced these findings have been of great value and have properly restrained unbridled enthusiasm, but they should not prevent educators and manufacturers from efforts to design and use effective learning materials.

'The greatest deficiency in instructional technology at the present time is caused by the inadequate supply of teaching and learning materials of good quality suitable for use with the new technologies' [10].

Current trends in the use of multi-media

It is clear that we have just begun to exploit the potential of some of the new technological devices for storage and retrieval of information. These communication tools are slowly being incorporated into our educational system but effective use may require some major changes in procedures, attitudes, and personnel related to the educational process. With both fear and hope, a few innovators have rather timidly accepted the challenge to change educational practices and to keep pace with the changes in other areas such as communication, transportation and medicine. The effect these isolated efforts will inevitably have on both students and faculty is indicated in the Carnegie Report:

'In assessing impacts of technology on the campuses, we expect students to benefit from it because it offers them greater flexibility and more alternatives in their learning experience. Faculty members, we find, now tend to be resistant or apathetic in their attitudes toward instructional technology. But there are signs that they, too, have much to gain if the new media are introduced to colleges and universities in appropriate ways' [10].

Flexibility and alternatives are the key words in this quotation and the impact, which at first was primarily on learning rates and self pacing, is now being extended to flexibility in content and alternatives to the classroom. The modular movement is one concept which makes this 'new look' in education possible.

1. *The modular (minicourse) movement.* The 1950's saw the introduction of programmed instruction based upon the concepts of learning through reinforcement and behaviour modifications. Several teaching principles characterized programmed instruction—small steps, active student involvement, guided learning experiences, immediate confirmation. These same principles also formed the basis of the more sophisticated systems such as the various audio-tutorial systems of the sixties. The latest development in this trend of innovative instructional methods is the modular (minicourse) movement of the seventies [53].

The systems of the sixties placed greater stress upon learning through the mastery of specific objectives adopted of combination of large and small steps, increased student involvement by a wider use of media and, in fact, could rightly be called multi-media systems. In general, their use was tied to a rigid faculty teaching-timetable.

It is difficult to define a module; possibly we should not try since a variety of modules is likely to evolve within the next few years. Perhaps the simplest definition is an operational one—a module is a self-contained, independent, self-paced unit of work programmed to a set of objectives.

The current state of modular teaching has been discussed in some detail in a recent publication [31] of the Commission of Undergraduate Education in the Biological Sciences, 1971. And an article by Creager and Murray [11] sets out the components of a module as:
1. Statement of purpose
2. Desirable prerequisite skills
3. Instructional objectives
4. Diagnostic pre-test
5. Implementers for the module
6. The modular programme
7. Related experiences
8. Evaluation post-test
9. Assessment of the module.

It is not difficult to see how the module is related to the programmes of the sixties. There is a finer degree of resolution of the components, but the educational principles are essentially the same.

What is new about the modular movement is the separation of the course content into self-contained units or modules and the individual module as an independent self-pacing instructional unit. The significance of these features are described more fully in the section discussing minicourses.

Murray described the individual components of a module in the following way (although his account refers to biology it applies equally to all the sciences):

'*Statement of purpose.* The statement of purpose should be an elaboration of the title of the module which relates the content of a specific module to the interests and needs of the student. One way of developing this statement is to describe the significance of the module as seen through the eyes of the author. Views on its significance from some student users might also be helpful.

'*Desirable prerequisite skills.* If particular skills are needed before beginning a module, they should be stated explicitly. Generalized skills, however, need not be stated. Prerequisites should be worded in such a way as to be of maximum help to the student in deciding whether he is prepared to undertake the module. It is best not to go for "over-kill" in stating prerequisite skills.

'*Instructional objectives.* The clear specification of instructional objectives is a vital element in the development and use of teaching modules. Many instructional objectives can be stated in behavioural terms. In defining such behavioural objectives, it is helpful to consider the following questions:
1. Who will exhibit the behaviour?
2. What observable performance is the learner expected to exhibit?
3. What conditions, objects, or informational sources are given?
4. Who or what sets the terms of the learner's performance?
5. What responses are acceptable? Are there any special restrictions on these responses?

'*Diagnostic pre-test.* A diagnostic pre-test may be appropriate for some modules. This instrument should provide information useful in determining whether or not the student is prepared to undertake a particular module. It is not to be presented as a "hurdle" to be jumped before beginning a module. Outstanding performance on the pre-tests may indicate that the student need not take the module.

'*Implementers.* Lists of equipment and supplies should accompany each module. They should be supplemented with notes on acquisition, where this information is not readily available. Helpful procedures for the preparation of materials and solutions and for the maintenance of living organisms or special equipment should be included' [11].

In some modular programmes 'implementers' are included in a separate instructor's manual. This is part of the module, but is not made available to the student. Such an arrangement is desirable if modules are to be exchanged between different institutions or colleges.

'*The modular programme.* Teaching modules may be presented in a variety of forms. Some may be short "how-to-do-it" booklets. Others may be prepared as integrated, multi-media instructional units using audiotape, film loops, Kodachrome slides, etc. In any case, the student should be learning by the actual manipulation of biological objects, rather than merely reading, watching, and listening.

'If an *audiotape* is to be used, it is important to note that the central function of the tape is to integrate learning activities, not to deliver a lecture. Care should be taken to avoid tape presentations that involve the student in listening only. Authors of tapes are encouraged to explore alternate sources of audio material. Some possibilities that have been considered are taped comments and presentations by resource persons, the recording of biological sounds (human and non-human), taped conversations, and dialogues. Past experience has indicated that a *tape-guide* should accompany an audio presentation; this may take the form of an outline, a listing of key words, or diagrams related to the tape.

'Where *visuals* (35 mm slides, diagrams, charts, films, TV tapes) are used, variety is important. They should not be locked into a fixed format. Keep in mind that students have come to expect quality production of visual materials. One should avoid complex diagrams, lengthy films or slide sequences, undigested tabular material, elaborate models, and irrelevant photographs. McLuhan may be correct when he says, "the medium is the message"—a distracting profusion of visual materials may convey chaos, while interesting variety may encourage creativity.

'The modular approach allows for the inclusion of diverse *tangibles* (biological materials, equipment, etc.). Available funds can be applied to the purchase of a variety of materials, equipment, and supplies instead of being used to purchase many replicates of the same tangibles.

'It is desirable to include a *glossary of terms,* preferably listed in the order in which they occur in the module' [11].

This description of the modular programme highlights the potential for the use of media in the structure of a module. To stress this point a lits of media that can be used by module developers is shown in Table 2.

TABLE 2. Media and tangibles appropriate for modules

Audio tapes	Textbooks
Films loops	Research papers
35 mm slides	Radio
Diagrams	Television
Charts	Experiments
Video tapes	Specimens
Data	Models
Field trips	Computer

The media are chosen that best enable the student to achieve a particular objective. This means that a developer can draw upon almost every aid that is available.

Murray's account of the modular programme does not indicate the importance of the study guide. The statement of purpose, prerequisite skills and most of the components of the module are set out in the study guide. It is a workbook that includes what the student should learn and suggests ways and means of going about it.

'*Related experiences.* Some modules are likely to include other learning experiences, outside the direct context of the modular program. The module may direct the student to lectures, reading assignments, small group discussions, and visits to museums, zoos, and natural areas. In some cases, lists of references for further reading may be supplied or an indication of related modules may be provided. In other cases, opportunities for individual independent study (i.e., depth studies, projects, etc.) may be suggested. In any case, such experiences should not be viewed as supplemental to the module, but as related learning activities.

'*Evaluative post-test.* In addition to specified instructional objectives, an evaluative post-test is a key element of each module. This instrument must reflect the instructional objectives of the module. A variety of post-test forms should emerge; these may include pencil and paper tests, oral quizzes, or, for modules focusing on a particular skill, performance tests. Such tests should provide an index of the learner's accomplishments.

Assessment of the module. The developers of a module are encouraged to include an assessment of each module. Assessment procedures should be designed to test the modules, not the individuals using them. An abstract of the field data collected in developing, using, and testing a module should be made available to those considering adopting it. This abstract might include:

1. Number of students involved in the field tests;
2. Average completion time for the modules;
3. Degree of competency obtained by students on instructional objectives;
4. Any other data that summarized the field experience in the development and utilization of a particular module' [11].

Obviously there is plenty of scope for variation in the way modules are developed and in the format of the modules themselves. The extent to which media are used and the choice of media can be varied at will in an effort to produce a module that enables the student to master the objectives as efficiently as possible.

Modular instruction is rapidly gaining acceptance, as is witnessed by the various approaches that are being developed in the United States under such names as Minicourses, ATP's (Audio-Tutorial Packages) Conceptopaks, BioTechs; in Canada, at McGill, modular courses are offered in Biology and Social Change, Linguistics and Chemistry; in New South Wales, Australia, at Macquarie University; the Nuffield Inter-University Project in England. In all these experiments teaching and learning are being based on self-instructional units—call them what you will.

2. *Portability and interchangeability.* Multi-media instructional programmes in science education are certain to include a large self-instructional component consisting of separate units (modules-minicourses). An important characteristic of these multi-media instructional packages is portability. This is likely to have far-reaching, possibly revolutionary, consequences to education. The advantages of portability as far as on-campus instruction is concerned have been predicted for teachers, students and institutions using a modular approach by Creager and Murray [11]:

'1. The use of modules provides the opportunity for organizing numerous sequences of experiences to reflect special interests of the teacher or the student.

'2. Self-instructional units allow the teacher to focus on student deficiencies in subject matter that must be corrected and also serve to eliminate the necessity of covering subject matter already known by the student.

'3. The modular approach provides a way of assessing the student's progress in learning.

'4. Modules reduce the routine aspects of instruction, leaving the teacher free to engage in personal contact with the student.

'5. The independent nature of self-instructional units facilitates the updating of study materials without major revisions.

'6. Modules can serve as models for teachers who wish to develop their own materials and insert their own individuality.

'7. Self-instructional units are easily exchanged between institutions.

'In the same context, the modular approach offers avenues for individualized study on the part of the student. The following list is offered of some, but again not all, of the possibilities inherent in such a scheme of study for the student.

'1. The student must be involved in the learning process so his commitment to the task is likely to be enhanced.

'2. A large pool of modules will permit students to explore portions of subjects of particular interest without having to enrol in a full course containing topics not relevant to their needs.

'3. The student has full control of the rate of study; thus, he can progress at his own pace.

'4. The student is not forced to cover materials which are already familiar to him.

'5. The consequences of failure are reduced. Each student can master each module completely before proceeding to the next.

'6. Each student can participate in the decision as to whether he has learned the subject matter adequately.

'7. It may be practical for some modules to be checked out and studied at home, resulting in a saving for both the student and the university.

'8. Each student has the opportunity to develop a sense of responsibility for his own learning.

'And finally, there are advantages to the institutions which must not be overlooked. Some of these follow.

'1. Modules make it possible to define the content of a course so that members of the biology department can avoid inadvertent duplications or omissions and members of other departments can determine what portions of courses might be useful to their students.

'2. Modules can provide for dividing the responsibilities of course preparation among members of a team, thus achieving the main advantage of team-teaching without many of the disadvantages of team-teaching.

'3. Modules are exportable from one campus to another so that the expense of preparation can be shared among institutions. Many can benefit from the modules prepared at any one institution.

'4. In the evaluation of the productivity of teachers, an institution could treat the preparation of modules in the same light as publications are now treated.'

The new technology also means savings on capital investment in buildings. Fewer classrooms and dormitories are needed. Campuses and campus services need not be so extensive.

Portability is the key to off-campus instruction and makes possible the prediction of the Carnegie Report that by the turn of the century some 80 per cent of off-campus teaching will be carried out through the new technology.

Portability of high-quality self-instructional units may be the answer to the stigma of 'second-class status' that is commonly associated with extension and part-time courses. It is going to be much more difficult to sustain such an attitude if all students use self-instructional units from the same resource centres. This does not mean a stereotyped training. On the contrary, it means greater diversity—students are free to choose programmes to meet their special requirements. The only stereotyped thing about the multi-media systems will be the high quality of the individual instructional packages.

Portability means flexibility, and as Chancellor Boyer of the State University of New York [7] puts it, this means 'flexibility as to where a student studies, when he does it, what he studies, and how much he needs to complete his education.'

Software for use in multi-media systems is currently being produced in many isolated locations and with little or no communication among the developers. This results in much duplication of effort and costs. It is desirable and important to avoid the stereotyping of instructional materials yet a great need exists for the development of interchangeable materials and capitalizing on the expertise and ideas of others.

The problems of transfer of instructional programmes are many. Printed texts and manuals have been exchanged for many years, yet not without some difficulty and with much modification and the use of supplementary materials. Programmes involving the integration of additional media in the structuring of materials for learning increase the hazards. One solution which appears to provide the necessary flexibility is the reduction in size of the transfer unit from whole courses to minicourses or modules. This approach permits adjustment in several important ways, as listed previously. The following points should however be emphasized:

1. When necessary, these small units can easily be modified to accommodate cultural differences without a major reorganization of the whole course.

2. Large 'pools' of minicourse (module) programmes can be established from which appropriate minicourses can be selected by the local instructor for use in his specific situation and to suit the special needs of his students. This 'pool' can provide for exchange of programmes on a state, national or international basis and can be contributed to by experts from a great variety of sources. Schools which cannot afford the faculty or for some other reason cannot cover a full range of subject matter can use materials from the 'pool' to expand their course offerings at relatively less expense.

3. Evaluation of each programme can establish its effectiveness and determine its suitability for a given student population.

Portability and interchangeability of instructional materials can provide new dimensions for science education. It is a goal worth striving for and current technology has advanced to a level that makes it possible and practical.

Achieving the potential

'Since a grossly inadequate supply of good quality instructional materials now exists, a major thrust of financial support and effort on behalf of instructional technology for the next decade should be toward the development and utilization of outstanding instructional programmes and materials' [10].

This recommendation of the Carnegie Report emphasizes the most urgent problem for achieving the potential of educational technology.

In science education the software problem is more acute than in other fields of education because of the practical work involved. It is not our intention to get involved in a discussion of the role of the laboratory in science education, nor to consider ways and means of adapting multi-media to the teaching of practical work. Our concern is to draw attention to the problem and to stress that a considerable amount of development, educational and evaluative research will need to be undertaken to determine how best the new technology can be used for the teaching of laboratory work. The problem is particularly acute when the objective is a self-instructional portable minicourse that is to be used off-campus.

Actually the problem of practical work is only another aspect of the software problem and this is critical. Enthusiasm for multi-media should not blind us to the fact that the key component of any multi-media system is the software component. At the present time hardware is more readily available than software and the situation is likely to deteriorate further because greater effort is being devoted to the development and production of hardware than software. Videocassettes are likely to become widely available within the next two to three years, but—'Are there enough freshman science courses ready to exploit this new medium?' The answer quite obviously is, No!

Although, at the present time, experience in the production of multi-media units for the sciences is not extensive, sufficient is known about the state of the art to indicate that it is a slow process and therefore costly. Estimates of the time required to produce a multi-media unit (about a week's student contact time) from inception to final product range from 100-500 man-hours. The production of the 36 units of the first-year science course of the Open University involved some 12-16 academic staff plus technical staff, typists, artists, laboratory technicians.

Doubtless, as know-how improves, the production time will decrease, but, even if halved, production would still remain a slow process. This being so, it is difficult to see how science schools can go it alone, and the smaller the school the truer this is. In fact, small schools, which would benefit most from flexible multi-media units, are not likely to be in a position to produce their own teaching programmes.

The production of high-quality instructional materials 'must' occur before the use of technology can become widespread. This will require strong financial support and extensive co-operation among members of the academic community. If materials are to be widely used, their production must involve input from an equally wide source of expertise. Many relatively trivial difficulties can be avoided when the developmental team is representative of the areas for which the materials are to be used. Co-operation on an international basis can be achieved best by the establishment of international materials production centres. The Minicourse Development Project at Purdue University, U.S.A., has involved developers from Macquarie University, Australia, as regular members of the development team. Materials produced under these conditions can be transferable between the two schools with a minimum of difficulty.

Teacher training programmes will be important to the success of any exchange of instructional materials. Model systems should be operated especially for this purpose where teachers can participate in an ongoing successful programme before attempting to set up their own. Each new programme should be carefully established with all factors rigorously controlled to insure its success. Innovative programmes which were based on sound principles have often failed because of tangentially related factors which override and frustrate the activity. The innovation then receives unwarranted criticism and further prejudice is built up against the new technology.

A clearing house should be established to collect mediated programmes where teachers can examine and evaluate these materials for possible use in their courses. Some effort should be made to standardize the format for these and thus reduce the difficulty of interchangeability.

Models of international exchange should be set up and the problems carefully investigated. Programmes currently in progress should be fully supported and encouraged by their institution and their governments.

It is estimated that by the year 1990 electronic technology will be generally introduced into higher education and by the year 2000 it will be in general use [10]. This estimate may be a bit conservative if research and development continues at its present pace and the new generation continues to press for reforms and changes in the establishment. The limitation at present lies in the development of high-quality software. It is up to us, and our creativity, to exploit this technological abundance to achieve our goal—*better education for all.*

BIBLIOGRAPHY

1. AAAS. Science education in 'new' colleges. *Science education news,* April 1971.
   Includes a description of Governors State University, where the College of Environmental and Applied Sciences will be one of the four colleges opening in 1971. The calendar comprises six sessions per year, each of eight weeks, for improved flexibility. Interdisciplinary work is to be encouraged. The traditional lockstep approach to earning a degree has been abandoned. The instructional elements will be learning modules. Each student will have an individualized instructional programme. The staff will be augmented by 'professors' from the community. The conceptual base in College of Environmental and Applied Sciences is built upon biological, physical and earth statistics, 'urbanology', economics, etc. Learning modules deal with such topics as 'Limits of tolerance' and 'Bioenergetics'.
2. ADAMSON, Heather; MERCER, F. V. A new approach to undergraduate biology. *J. biol. educ.,* September 1970, vol. 4, no., p. 155-176.
   Two separate articles deal respectively with (1) problems and techniques in teaching external students, and (2) kits and the open lab for internal students. The methods developed at Macquarie University cater for students living up to 700 miles away. For the introductory course, external students receive at weekly intervals the same lecture material as internals (printed or tapes), and they attend vacation courses, intensive in lab work and tutorials. Lab work is done in groups of about ten, under supervision. Kits are provided for home experiments. The audio-tutorial aspects follow those of Postlethwait. The authors believe that their methods may be relevant 'in countries where education is a burden on the national economy'. The techniques have been adapted to design an 'open-lab' approach for internal students.
3. BAEZ, Albert V. *Innovative experiences in physics teaching.* Harvard University, 1971. (Duplicated document.)
   Summary description of a concentrated course for non-specializing students in Physics S-1 during the summers of 1967-71 inclusive. Innovations included 'autolectures', musical intervals and micronotes.
4. BANKS, B. Programmed learning in a secondary modern school. *Visual education,* October 1968, p. 39-41.
   Account of experimental use of programmes on tape in the mathematics department of Ridgeway School, Kent (United Kingdom). Best mode is student-controlled in groups of two or three. The tapes are integrated into a multi-media system. The results point to a logical extension of self-instructional methods, with teacher as support resource.
5. BANKS, B. An experimental auto-instructional course in mathematics. In: A. C. Bajpai and J. F. Leedham, eds., *Aspects of educational technology IV.* London, Pitman, 1970.
   The experimental course started at Ridgeway School, in Kent (United Kingdom) over three years previously has spread to many other schools, involving nearly 2,000 pupils. 220 tasks are available and ICL are programming a computer to help the teacher select tasks suitable for each individual child. Both teachers and pupils are finding the scheme acceptable.
6. BERMAN, Arthur I. *Balanced learning, the personal approach.* Copenhagen. Draft chapters for book in preparation, 1971.
   A discussion of modern educational technology methods and devices, in which a distinction is drawn between 'high technology' (e.g., computers, TV) and 'low technology' (e.g., audio tape recorders), leads to a detailed description of the autolecture.
7. BOYER, E. L. A different type of college: It comes to the student. *U.S. news and world report,* 4 October 1971.
   A popular article describing the trend for use of technology to carry education to 'off campus' students.
8. BRETZ, Rudy. *The selection of appropriate communications media for instruction: a guide for designers of Air-Force technical training programmes.* Rand Report R-601-PR, February 1971.
   Distinguishes between the *appropriateness* and the *practicability* of a given medium for a given instructional task. This report deals only with the following eight headings: audio/motion-visual; audio/still-visual; audio/semi-motion (e.g., telewriting); motion-visual; still-visual; semi-motion; audio; print. The selection of the most *practicable* medium from a range of *appropriate* media is to be dealt with in a subsequent report.
9. BUTTLER, Michael. Books or what? *Education in chemistry,* September 1970, vol. 7, no. 5, p. 215.
   The large comprehensive textbook has been dying for some time, and is being replaced by a very wide variety of multi-media material usually designed for a specific course, such as one of the Nuffield science projects. It is too early to say whether these multi-media packages will be more successful than the old textbooks. One danger is that the pupil has no formally laid down map of the subject.
10. CARNEGIE COMMISSION ON HIGHER EDUCATION. *The fourth revolution: Instructional technology in higher education.* McGraw-Hill Book Co., 1972.
    This report discusses in great detail the educational potential and implications of the new informational technology for students, faculty, institutions and society. It predicts the rate and amount of technology likely to be used in educational systems by the year 2000 and makes recommendations on how the potential can be achieved.
11. CREAGER, Joan G. and MURRAY, Darrel L. *The use of modules in college biology teaching.* Washington D.C., Commission on Undergraduate Education in the Biological Sciences, 1971.
    This publication resulted from two conferences on modules sponsored by the Commission for Undergraduate Education in Biological Sciences. It discusses the rationale for use of modules, recommends a format for module materials, and includes some examples.
12. DARLING, Richard L. Media centers. In: WIMAN and MEIER-HENRY, *Educational media: theory into practice.* Colombus, Ohio, Merrill, 1969.
    The ideal school media centre combines into one integrated service the facilities and services of traditional libraries and A-V programmes; the hardware as well as the software is

kept in the media centre. District media centres have been developed to supplement and support centres in individual schools.

13. DEALL, Louis. A brief description of a new type of secondary school programme. *The physics teacher,* February 1971, vol. 9, no. 2, p. 90-91.
The Modular Physics programme introduced at Clayton High School, Missouri, in 1967, covers four years. The laboratory is open and staffed at all times, and at least 40 per cent of the student's time is unscheduled, during which he works at his own pace, independently, on study and experimental work, divided into topics each of which is checked on completion. The percentage of students taking physics has increased substantially since the modular course was introduced.

14. DIAMOND, Robert M. A flexible approach to an independent study facility. *Educational technology,* December 1970, vol. X, no. 12, p. 29-31.
The system installed at the State University College, Fredonia, New York, combines dial access and audio-cassette check-out with a flexible approach to the provision of specific capabilities at the 56 stations. A large collection of programmed texts and review sequences is available on topics including sciences and mathematics.

15. DOWDESWELL, W. H. *The inter-university biology teaching project.* Bath (United Kingdom), The University, 1969 (Internal Document).
This project is being conducted jointly by the Universities of Bath, Birmingham, Glasgow, London and Sussex, with support from the Nuffield Foundation. Work is planned in modules. Three types of courses are involved, namely: *Bridge courses* (between school and university); *Technique courses* (e.g., on aseptic techniques); and *Main courses* (e.g., on enzymes). Self-instructional methods are used, employing AV aids and programming techniques. Costly installations such as study carrels and CCTV are avoided.

16. DRIVER, S. C. Programmed individual study methods in the learning of sciences. *Australian science teachers journal,* May 1970, vol. 16, no. 1, p. 27-33.
Deals largely with activities at the University of Melbourne, in biology, physiology, biological chemistry (3rd year) and biochemistry. Though based on Postlethwait's methods, the term 'audio-tutorial' is not used, as it is considered unduly restrictive; the 'programmed' aspect is highlighted instead. The contrast with programmed ERV is emphasized and the lock-step nature of ETV is noted. The term 'structured' learning is introduced as synonymous with 'programmed' learning. The taped commentary must be prepared as a talk to an individual rather than as a lecture to a class.

17. DRUGER, Marvin. Using media to individualize biology teaching. *Australian science teachers journal,* May 1970, vol. 16, no. 1, p. 17-21.
Describes the use of audio-tapes for teaching 800 first-year biology students at Syracuse University. The technique is essentially the same as that of Postlethwait. Advantages: instruction is individualized, learning is self-paced, the learning sequence is more coherent, more effective in consequence of immediate feedback, space can be used more efficiently. Examinations were time spaced rather than concentrated.

18. EASTMAN, Stewart W. Biology in an individualized school. *The American biology teacher,* December 1970, vol. 32, no. 9, p. 533-536.
Account of system operated at Oak Grove High School, San Jose, Calif., where biology, chemistry and physics are offered as part of a total programme. The problems to be solved are, how to make the most effective use of staff, how to organize the use of rooms to accommodate specific activities rather than teachers, and how to develop learning packets with the necessary self-directing characteristics. Three teachers together with two assistants handle up to 75 students per period (studying biology, chemistry and physics) in a six-period day. Students ask for help only when a problem arises.

19. EDLING, Jack V. Individualized instruction the way it is: 1970, *Audio-visual media,* Spring 1970, vol. 4, no. 1, p. 12-16.
Summary of results so far obtained from a survey sponsored by the United States Office of Education. Individualized learning appears to be spreading rapidly. An interesting example is the DeKalb County Schools, Decatur, Georgia, where there is an individualized science instruction programme with instructional resources centralized, thus enabling provision of excellent equipment and PhD-level staff. 'New knowledge, rather than extra resources, is vital to programme initiation.' Aduio-tape is the medium most used.

20. ELTON, L. R. B. *et al.* Teaching and learning systems in a university physics course. *Physics education,* March 1971, vol. 6, no. 2, p. 95-100.
The course makes use of duplicated lecture books and taped lectures, with or without accompanying slides, to supplement or replace live lectures. Investigations were conducted on the use of PI with and without videotapes; results are encouraging for PI. Lab experiments have been devised for the purpose of separating out their effects on specific educational objectives. The lab is organized on 'self-service' lines, with tapes, films, etc. available.

21. FENNER, Peter; ANDREWS, Ted F. *Audio-tutorial instruction: a strategy for teaching introductory college geology.* Washington, D.C., CEGS Programmes Publication, no. 4, 1970.
A large number of scientists and science educators have collaborated with CEGS in preparing this monograph, which discusses the rationale and philosophy of 'individually guided' learning (repetition, concentration, association, appropriately sized units of subject matter, use of media, integration of activities); the history and evolution of the A-T system, illustrated by Postlethwait's work at Purdue University; notes on A-T geology programmes in several United States institutions; and problems and pitfalls of the A-T mode—e.g., when taping lectures it is even more important to avoid being boring than when the lecture is presented live. Appropriate sequence of activities is extremely important. Study carrels should be placed in an area permitting experimentation.

22. FULTON, Harry F. Individualized *vs* group teaching of BSCS biology. *The American biology teacher,* May 1971, vol. 33, no. 5, p. 277-279, 291.
The samples consisted of 20 students selected at random from the 8th grade biology class at University High School, Iowa. Those selected in 1967-68 learned as individuals,

working with an adaptation of the BSCS 1964 (Blue version), in which reading material was interspersed with questions, activities, laboratory investigations. Each chapter was followed by a summary and a series of questions. Students proceeded by verbal 'contracts' with the teacher. Consistently greater gains were made by the individualized students.

23. GARNER F. F. Audio-Visual materials in the classroom. *Improving college and university teaching*, Spring 1971, vol. XIX, no. 2, p. 114-6.
Emphasizes that teachers are unlikely to use A-V aids unless they are readily accessible. In addition to this simple inertia problem, if it is difficult to obtain the equipment, the device will attract more interest than it warrants, possibly more than the lesson it is intended to present. Though the most convenient place for equipment is in each classroom, there is a risk that it will disappear; hence a central store is needed. Maintenance and repair difficulties are also pointed out. It is unfortunate that universities lag behind elementary and secondary schools in their use of modern teaching techniques.

24. GAUDRAY, Francine, ed. *Multimedia systems: international compendium*. München, Internat. Zentralinst. für das Jugend- und Bildungsfernsehen, 1970.
Compilation prepared as working paper for a meeting held in conjunction with the Council of Europe in April/May 1970. Eleven projects, in Brazil, Canada, France, Federal Republic of Germany, Japan, Poland, United Kingdom and United States, are described, and characteristic details are tabulated.

25. GOMES, L. C., et al., *New teaching methods at Brasilia*. Brasilia, Unesco Project Brazis-9, 5 October 1970. (Direct communication.)
Description of an adaptation of the Keller method used for teaching physics to undergraduates.

26. GREEN, Ben A., Jr. *A self-paced course in freshman physics*. Cambridge, Mass., MIT/ERC, 24 April 1969.
A 'personalized' course based on the ideas of F. S. Keller. Five essential attributes of the system are: self-pacing mastery as a criterion for advancement; use of lectures for motivation rather than transmission; emphasis on written performance; use of undergraduate tutors within a test-grading contest. A main task of the instructor himself is to define course content. Keller is quoted to the effect that 'teaching machines, tape recorders, computers, ... moving pictures and TV could be used. ... But these are luxuries, based on only partial recognition of our problem, and they could divert us from more important considerations.'

27. GREEN, Ben A., Jr. Physics teaching by the Keller Plan at MIT. *American journal of physics*, July 1971, vol. 39, no. 7, p. 764-75.
Extension of the Keller self-pacing system to a number of courses additional to the freshman course of the first experiment. Includes brief descriptions of physics courses at other universities also sharing two features of the Keller plan, namely, self-pacing and mastery as a criterion for advancement. This system of teaching is considered feasible only for relatively small numbers of students; it is 'destined not to spread quickly'.

28. GREEN, Ben A., Jr. Is the Keller plan catching on too fast? *Journal of college science teaching*, October 1971. Expresses the fear that the Keller system of self-paced learning may be spreading faster than its principles are being understood. One reason for the rapid spread is the low cost: the system uses sophomores as tutors, working under an experienced teacher. It is recommended that student numbers should be kept small when the scheme is being tried out initially.

29. HARBECK, Richard M. The science teacher and educational technology. *Educational technology*, January 1970, vol. X, no. 1, p. 33-35.
The non-hardware accomplishments of educational technology and the systematic approaches which lead to these accomplishments can have little meaning to teachers accustomed to working in a structure within which it is presumed that all must learn by the same mode at the same rate and that some proportion must necessarily fail. A major goal of educational technology is to assign students realistic goals and to provide teachers with tested alternatives, or modes of instruction, which can be prescribed or advised on the basis of the needs, style of learning and environmental conditions of the individual learner.

30. HARRISBERGER, Lee. Self-paced individually prescribed instruction. *Engg. educ.*, March 1971, vol. 61, no. 6, p. 508-10.
Asserts that adoption of the IPI system (alternatively termed 'the process approach to teaching') would result in 90 per cent of engineering students obtaining A-level achievement. The whole course is divided into 'mini-courses', or units, each having a clearly stated objective which must be stated in performance terms. (Resemblance to the Boy Scout merit badge system is pointed out.) A set of 'assessment items' and a learning activity package (various media) are provided for each minicourse. Most IPI courses are developed by repacking available materials.

31. HILLS, P. J. Science teaching and educational technology. *School science review*, March 1971, vol. 52, no. 180, p. 493-99 and June 1971, no. 181.
The swing from 'dogmatic science teaching' to 'teaching for understanding' is discussed in relation to development in A-V aids and PL and recognition of the need for individualized instruction. It is only when PL techniques are applied to the content of instruction that individual needs can be considered.

32. HOWES, Virgil M., ed. *Individualizing instruction in science and mathematics*. London, Collier-Macmillan, 1970. Selected readings (from *Science, Science education*, etc.) on programmes, practices and uses of technology.

33. HOWSON, A. G.; ERAUT, M. R. *Continuing mathematics*. London, NCET, 1969 (Working Paper No. 2).
A proposal for a systems approach to the mathematical education of sixth formers specializing in the arts and in social or life sciences. Materials are to be supplied in units of 6-10 hours work, each unit including: a 'discovery' part, involving individual and group learning, both teacher-assisted and independent, and using A-V resources; a multi-media programmed learning part designed to achieve mastery without unnecessary repetition of studies; and a part devoted to applications and problem solving. Students' kits are to include texts, workbooks, and A-V resources such as film loops. Great attention is to be paid to the developmental testing of the material.

34. INTERNATIONAL BUREAU OF EDUCATION. Educational technology. *Bull. IBE,* 4th Quarter 1970, Year 44, no. 177.
    This issue of the Bulletin is devoted largely to a bibliography of educational technology.

35. JAFFE, S.; BLUMSTEIN, Z. *Programmed A-V instructional system for Class-II personnel (in meteorology).* In: WMO/IAMAP, Symposium on Higher Education and Training, Rome, 1970.
    A-V aids prepared by professionals stimulate and retain the attention of the students and present the information in a clearer sequence than a course prepared by the instructor. The combination of programming and A-V aids described provides a method suitable for teaching groups of students. A recorded exposition is used in conjunction with 35 mm transparencies presenting information, questions and replies. The tape carries the signals for changing the transparencies. The A-V presentation is interspersed with discussion, questions, etc.

36. KAYE, Brian H. *Investigation of the utility of a self-paced first-year physics-course and the allied problem of off-campus teaching in Northern Ontario.* Laurentian University, Grant Request dated 24/2/71 (Direct communication).
    Detailed statement of a proposed investigation involving the use of videotapes; the services of five research associates were to be used. (However, the request was rejected.) Advantages stressed include bilingual, repeat, retraining and off-campus possibilities. Industrial scientists and technologists have participated in preliminary trials (at post-graduate level). (Ref. to work at Bucknell University, Pa., at MIT, and co-operation with University of St. Etienne, in France; with Professor Joubert.)

37. KELLER, Fred S. Goodbye teacher. *J. applied behaviour analysis,* Spring 1968, no. 1, p. 79-89.
    Description of the 'individualized', 'personalized', 'programmed', 'self-paced' system adopted at MIT for first-year physics etc.

38. KIEFFER, Jarold A. Toward a system of individually taught courses. *Liberal education,* October 1970, vol. LVI, no. 3, p. 443-53.
    Describes the objectives and methodology of teaching college courses on an individual rather than a mass basis, and suggests means for moving towards such a system. Instead of attending lectures, the student would come to a 'learning station' where a range of A-V and other facilities would be available. Students would contact the instructor for discussion and testing when they felt ready to move on to the next phase. 'Canned' lectures are not in themselves always satisfactory. With this individualized system, 'bright' students could reduce course time (and hence cost).

39. KLOPFER, Leo E. *Individualized science in focus.* Paper presented at the 64th Annual Meeting of the Illinois State Acad. Sci., 24 April 1971.
    Discussion of the design of a school science programme up to 8th grade (U.S.A.) and intended to help the individual child to understand himself and his immediate environment. Five goals are listed, namely: self-direction, self (comparative) evaluation, positive attitude towards science, investigative skill, scientific 'literacy'. (Individualized instruction kits have been developed, accompanied by instructional audio-tapes; the two lowest levels of the material are expected to be available commercially in January 1973, from Imperial International Learning Corp., Box 548, Kankakee, Illinois 60901.)

40. KLOPFER, Leopold E., and WEBER, Victor L., Jr. *IPI Science: A teaching revolution in the making, Reprint 51.* Pittsburgh University, Learning Research and Development Centre, June 1969.
    Report of IPI science programme in elementary science and mathematics. The six goals are: mastery of instructional content; individual optimum rate of working; active involvement; partial self-selection and self-direction of work; self-evaluation; individually adapted learning materials and techniques.

41. KURTZ, Edwin B. *Individualized approach to teaching biology.* Chicago, AAAS Meeting, December 1970.
    Brief note on five courses in biology, developed and implemented at Kansas State Teachers College, starting 1968. In addition courses have been offered on how to construct individualized science instruction programmes based on behavioural objectives. There are no lectures in the biology courses; students attend laboratory (with staff in attendance) once a week on an arranged basis; they are required to show competence in an assigned unit before moving on to the next unit. Student response and success has been excellent from the lower 75 per cent of the class.

42. LAMP, B. News and views from further education and technical colleges. *Visual education,* January 1970, vol. 9, p. 11, 13.
    Deals with audio-visual aids and programmed instruction in technical and industrial training. (Continued in March 1970, p. 37-41.)

43. LEITH, G. O. M. Experiments in visual education and programmed learning. Part 5: Educational technology. *Visual education,* October 1968, p. 13-17. Programmes employ A-V techniques in conjunction with printed material.
    Best learning results are obtained when a few small steps of a linear programme are followed by a large step used as a review. Pupil differences as regards anxiety and other personality factors are recognized as affecting results.

44. MACQUARIE UNIVERSITY. *Audio tutorials in university teaching.* Sydney, Macquarie University Centre for Advancement of Teaching, May 1968. Bulletin No. 2 in the series *Teaching and learning in the university.*
    Describes the development of Postlethwait-type tutorials in chemistry at Oakland Community College, Michigan, and in other colleges in the United States and Canada. Suggests the production of film loops on physics subjects by the students themselves as an interesting educational possibility.

45. MARKESJÖ, Gunnar; GRAHAM, Peter. *An electronics course with integrated feedback.* Stockholm, Royal Institute of Technology, Department of Education, January 1971. (Report No. PE-9)
    Describes a 'process-controlled' electronics course covering 240 periods of 45 minutes, which was used for some 50 classes in the 4th year of grammar school in 1969-70. Part of the work is self-paced. Built-in test points permit different types of feedback. Answer cards are collected immediately after each test. The multi-media course material includes experimental kits, tape-slides, work-books. This type of

'process-controlled' education is to be extended to other subjects and to take the future use of TV and CAI.

46. MEYER, G. R. and POSTLETHWAIT, S. N. Australian high-schools use audio-tutorials in field biology. *The American biology teacher,* February 1970, vol. 32, no. 2, p. 96-101.
Adaptation of Postlethwait's method for field work with high-school pupils in NSW. Boxed kits were designed specially.

47. MORRIS, William, ed. *American heritage dictionary.* American Heritage Publishing Co., Inc., and Houghton-Mifflin Co., Boston, 1969.

48. NATIONAL COUNCIL FOR EDUCATIONAL TECHNOLOGY. Educational technology and the training of teachers. *Brit. J. educ. technol.,* May 1971, vol. 2, no. 2, p. 99-110.
Based on a memorandum sent to the United Kingdom Area Training Organizations. Describes the design of the 'systems approach' to learning, a rudimentary form of which is already employed by teachers. The term 'over-all programming of learning activities' is used for practical applications of the 'systems approach'. As the application of educational technology in schools becomes more widespread, teachers will be more concerned with planning, advising and evaluation. The large-scale development of self-instructional materials for non-specialist mathematics courses at post-O-level is envisaged.

49. NOVAK, Joseph D. Relevant research on audio-tutorial methods. *School science and mathematics,* December 1970, p. 777-784.
On the basis of Ausubel's theory of learning, it is recognized that careful attention must be paid to the sequence in which concepts are elaborated in the design of audio-tutorial instruction. An important aspect of audio-tutorial instruction is that it affords a systematic way of monitoring learning time.

50. OETTINGER, Anthony G. *Run, computer, run.* Cambridge, Mass., Harvard University Press, 1969.
An essay on 'the mythology of educational innovation' by a Professor of Applied Mathematics at Harvard who is 'convinced that educational technology holds great promise', provided it is 'not force-fed, oversold and prematurely applied. Deals mainly with the schools attended by the vast majority of American children. Concludes that both the intellectual and the hardware aspects of educational technology are at present inadequate. Considers the feasibility of individualization in schools to be doubtful, largely on account of the increased demands on instructor's time. Discusses the use of computer display terminals to illustrate processes in maths and science; points out that the display can be viewed by individuals or by large groups.

51. OETTINGER, Anthony G. The myths of educational technology. In: Virgil M. Howes, *Individualizing instruction in science and mathematics.* London, Collier-Macmillan, 1970.
Emphasizes that the systems approach to education, CAI, and individualization are still very far from providing ideal education at the present time, though they may deliver useful results in the long time. Our present and most pressing problem is the lack of an empirically validated theory of teaching, and in fact we even lack a useful set of empirically validated principles of a theory of teaching. The encouragement of pluralism and diversity in educational techniques, and even in goals, is advocated.

52. PAGE, Colin Flood. *Technical aids to teaching in higher education.* London, Society for Research into Higher Education, 1971.
Report of a Working Party on Teaching Methods in Higher Education, Brunel University, December 1970. Reviews evidence gained from research on the use of A-V media in teaching practice. Covers pictures, moving pictures, recorded sound, PL, mixed systems, 'new' developments, and the use of computers. Factors such as motivation, conditioning, physical environment may have great influence on the successful use of technical aids; pertinence and timing count more than technical perfection.

53. POSTLETHWAIT, S. N. Time for microcourses. *Library-college journal,* vol. 2, 1969.
This paper discusses the potential flexibility for individualization through the use of small units of subject matter. These units originally were called 'microcourses' and later were called 'minicourses'.

54. POSTLETHWAIT, S. N.; NOVAK, J.; MURRAY, H. T., Jr. *The audio-tutorial approach to learning.* Minneapolis, Burgess Pub. Co., 2nd ed., 1969.
Detailed description of an individualized learning system introduced at Purdue University making use of a complete range of instructional aids. The student is provided with a taped presentation of the programme; but those activities which by their nature cannot be programmed by the audio-tape are retained and presented in other ways. Examples of the latter: guest lecturers and long films are presented in general assembly sessions, and regular small discussion meetings are held. An instructor is on duty during the open hours of the 'learning centre'.

55. POSTLETHWAIT, S. N. Students are a lot like people. *University vision,* January 1972, no. 8.
The basic philosophy of the audio-tutorial system is described and some advantages and disadvantages are discussed.

56. RASMUSSEN, L. V. Individualizing science education. *Educ. technol.,* January 1970, vol. X, no. 1, pp. 53-56.
Records advantages of individualized instruction learning noted at four Duluth schools. These are said to include economical use of lab equipment, encouragement of a 'discovery' attitude and of work in small groups. Students work to 'contract', i.e., written performance objectives, and they have a list of available resources to choose from. The classroom is about three times as large as the traditional classroom so as to accommodate three teachers and about 90 students. One teacher has concluded that a predetermined sequence of learning experiences is not as critically important as is often supposed—freedom of choice may sometimes be more important. Instructional materials and their availability impose significant limitations on the individualized approach.

57. REID, R. L. Educational technology: the back-ground and the need. In: D. Unwin, ed., *Media and methods,* McGraw-Hill, London, 1969.
The student population is highly selected in terms of school success. With this group, the gains from PI, considered as a way of presenting material to be learned, are likely to be less than with other kinds of pupil. On the other hand, the

contents of university courses reach the extremes of complexity. This suggests that it may be from the control that is exerted over the writer rather than over the pupil, that the greatest benefit will be obtained. The 1965 Brynmor Jones Report found that, compared with A-V aids, PI was more truly an aid to the student in learning than to the teacher in teaching.

58. RICHMOND, W. Kenneth. *The concept of educational technology*. London, Weidenfeld and Nicolson, 1970.
A discussion based on numerous quotations. It is not clear that there is as yet any conceptual framework which entitles us to speak, as we do, of a developing educational technology. One way of regarding educational technology is as the fusion of programming with a range of aids; the systems approach is another way.

59. ROGERS, E. H. Programmed mathematics. *Visual education*, December 1970, p. 2-4 and January 1971, p. 9 and 11.
A remedial counting programme developed at Shelton Junior School, Derby, for children entering junior school from infant school makes use of a special abacus to provide a 'practical and visual link' between the concrete situation (e.g., handling beads, etc.) encountered in the infant school and the abstract symbols used later. Loop films are used to show the movement of the abacus beads from one place column to another. Self-teaching was organized with a set of expendable work-sheets. For junior school use, where the child is working at his own pace, topic introduction programmes have been made for new concepts: programmes dealing with graphical representation are described; loop films and consumable duplicated sheets are the media used.

60. SANDERCOCK, E. R. Audio-tutorials in action. *Australian science teachers' J.,* May 1970, vol. 16, no. 1, p. 23-26.
Adaptation of Postlethwait's method at colleges in Australia and the United States are described. At the University of Michigan, Flint, because of technical problems with tape recorders, lack of ancillary staff and a shortage of faculty time, the taped lectures and instructions were temporarily replaced by lectures and instructions on blackboards and cards located strategically. At Elizabethtown College corridor experiments have been installed in the Physics Department. The audio-tutorial system can be modified to make it suitable for use in secondary schools.

61. SCHMIDBAUER, Michael. *New educational technology*. Unesco/ICDE Series C: Innovations, no. 5, 1971.
The approach to educational technology in the present paper puts the prime emphasis on the *educational problem* to be solved, and on the *integrated use* of new media together with the traditional ones. Distinguishes between radio and TV instruction for use in conventional classrooms and radio and TV schools coupled, e.g., with correspondence courses, such as the Open University and the radio elementary schools in Australia and New Zealand. Audiences can be multiplied by telephone and telewriting networks—the latter is a good substitute for TV in, e.g., maths and engineering (used for physics in Wyoming and North Colorado schools). Two-way radio is also noted, as well as video-recording and still-picture TV (i.e., slow-scan, transmissible over telephone lines). DIAL access has been tried, e.g., at Beverly Hills Unified School District. A mechanized library service is needed to encourage both students and teachers to seek information; retrieval activities could become part of classroom instruction. At the heart of individualization of instruction is *programmed learning*. This can be combined with CAI Brentwood School; East Palo Alto has demonstrated drill and practice in elementary arithmetic by CAI. For problem-solving by CAI, UCLA has implemented a college-level statistics course. Simulation by CAI may be used, e.g., for laboratory experiments; the student can react and obtain immediate feedback. The free dialogue mode of interaction in CAI is only in the research stage (e.g., at Stanford University). The A-V tutorial is particularly relevant for science laboratory work (see, e.g., Purdue University). The film medium (8 mm) can be used not only to present teachers' ideas but also for the students to report their findings or opinions.

62. SCRIBNER, Eugene S. *A progress report on an individualized integrated-science course for senior high-school.*
Elk River Independent School District, March 1970.
Material is presented in mini-units including behavioural objectives, suggested experiments, AV aids, tests, etc. Reported subjective tests results are promising, objective tests not yet reported.

63. SERVAIS, W.; VARGA, T. *Teaching school mathematics*. Harmondsworth, Penguin/Unesco, 1971.
A survey of new approaches to mathematics teaching, in practice and theory, from examples of classroom applications to attempts to identify the processes by which mathematics learning takes place. Experiences from many countries are included. Teaching aids are dealt with under: drawings and diagrams; geo-boards; three-dimensional models; space frames, moving models, films; algebraic aids; logical materials; calculating machines; textbooks; the mathematics laboratory. (Live mathematic teaching flourishes best in a classroom specially designed to encourage the use of the material now available.) Opinions on PL vary widely from those who proclaim its advent as a new era in education, to those who regard it as a plague on mathematics education. None of these views is acceptable. PL may be regarded as the application of cybernetic ideas to education. Claims about the merits of PL of mathematics are too high in relation to both objective evaluations or investments.

64. SHERMAN, J. G. Georgetown University, Washington, D.C. 1971.
A presentation for the Conference on the Keller Plan given at the Massachusetts Institute of Technology. It includes a detailed description of the plan, its implications and deficiencies. For further information write directly to Dr. Sherman.

65. SOUTHIN, John, McGill University; CHAMBERS, D. W. Sir George Williams University, Montreal, Canada.
A description of their system is provided. The information describes the course content and the procedures and options available to the student to complete the course. For further information write to them directly.

66. SWEET, Walter C.; BATES, Robert L. *An audio-visual tutorial laboratory programme for introductory geology*. Ohio State University, Department of Geology, December 1969.

Final report submitted to NSF for Grant GY-1436; refers to the lab component of a 10-week broad survey course, comprising 35 one-hour lectures, 10 one-hour lab sessions and a 3-hour field excursion. Problems to solve included high numbers with overcrowding, below-optimum use of staff, limitation of student numbers. Individualized learning and self-pacing are the solutions. Individual carrels are equipped with projectors for slides and sound-movies—earphones, screen, desk-lamp, etc., and a 'distress flag'. A programmed lab manual is provided. Slides and films are loaded by lab personnel. An evaluation of the programme by Maccini showed that students considered it fulfilled its purpose quite well; achievement test results were encouraging. Appendices give detailed lists of equipment and supplies (with costs) provided for students, and reproduce the forms used in the evaluation.

67. TAYLOR, L. C. *Resources for learning.* Harmondsworth, Penguin Books, 1971.

    Mainly about methods of learning in secondary schools. Includes short accounts of the situation in America, U.S.S.R. and Sweden. Describes the Swedish IMU package for learning mathematics; and discusses the teacher's use and adaptation of packages in an illuminating manner. Carrel arrangements, while suitable for third-level students, are not likely to suit school children for long periods. The logistics of packages constitute a big practical problem for teachers. In present circumstances, A-V aids are no daily diet of teaching; they are reserved for infrequent banquets; cheaper aids for use by pupils rather than teachers should be encouraged. In science and mathematics an ample supply of drawup may be all that is needed to supplement the printed material. A small portable tape-recorder is the most imperatively required single piece of equipment.

68. THIAGARAJAN, Sivasailam, *et al.* Programming the human component in an instructional system. *Brit. J. educ. technol.,* May 1971, vol. 2, no. 2, p. 143-52.

    Describes and illustrates techniques developed in the Division of Instructional Systems Technology at Indiana University for using human beings for various functions within a totally programmed (validated) instructional system. These functions include information presentation (the traditional lecturer/teacher), providing feedback to the learner, and the 'Grouprogram'—a group discussion controlled by an operational programme usually presented by an audio-tape, in which each discussion topic corresponds to a frame of a conventional self-instructional programme.

69. TREZISE, Robert L. Report on a National Conference on Computer Applications to Learning. *Educ. technol.,* December 1970, vol. X, no. 12, p. 60-62.

    Cost questions are much to the fore, but precise answers are not available because of the complexity of the problems. 'Hardware' costs are 'high', but 'software' costs may be expected to be even higher. Even highly trained personnel must devote literally thousands of hours to the construction of relatively small programmes. The 'hardware people' feel they cannot get the 'software people' to tell them what they want. Systems people tend to form cults. The individualization of education emerges as a major goal. For 'drill-and-practice' purposes, computerized methods appear to be superior in many ways to workbook drill and practice exercise, or even to programmed drill and practice. The computer can actually assist the teacher to become more human by off-loading tasks such as finding the appropriate exercises for the various students at their particular levels and keeping track of their progress. With CAI, student evaluation can be used purely to determine the next educational step.

70. VICTORIA UNIVERSITY OF WELLINGTON. Teaching aids programme. *VUW Gazette,* 30 June 1971, no. 8, p. 8.

    Brief note on workshops for university staff members organized on 'film as a teaching aid' and on 'multi-concept teaching' (multi-media teaching?). The Department of Chemistry of VUW has had considerable experience in the use of film, and the Department of Zoology is demonstrating the individualized study method using audio-tapes and 35 mm slides in carrels available at any time between 9 a.m. and 5 p.m.

71. WALES, C. E.; STAGER, R. A. *Educational systems design,* available from C. E. Wales, West Virginia University, 1970.

    Paper based on year of course design, development and operational performance at West Virginia University. It describes a step by step approach to designing a better educational system through research work of educators and psychologists. Background principles needed for implementing the ideas are described in the paper.

72. WILLIAMSON, Stanley E. A critical appraisal: the potential for educational technology in science education. *Educ. technol.,* January 1970, vol. X, no. 1, p. 66-69.

    Among the arguments that can be adduced against educational technology are: that programmed learning is highly structured and takes away initiative from the student 'by placing him in a lock-step learning process'; that teachers are not prepared to make effective use of the products of educational technology; that educational technology is expensive, and that available programmes are insufficient in number and lacking in quality; that science classrooms are not designed to make maximum use of educational technology. Key questions to be answered before introducing educational technology are: what are its *unique* possibilities; how does it relate to what is known about good teaching; what is its use under control; how useful is it overall.

73. WIMAN, Raymond V.; MEIERHENRY, Wesley C. eds., *Educational media: theory into practice.* Colombus, Ohio, Merrill, 1969.

    A group of papers dealing with learning and communication media utilization and media centres, and related topics.

Educational technology in the professional training of science teachers

by A. Perlberg
Teacher Training Department and Laboratory for Research and Development in Teaching and Learning at the Technion, Israel Institute of Technology, Haifa, Israel

Summary

Educational technology is defined in this paper as a systematic way of designing, carrying out and evaluating the total process of learning and teaching in terms of specific objectives and employing a combination of human and technival resources to bring about more effective instruction.

The rationale for using educational technology in the training of science teachers is twofold: first, it will enhance the effectiveness of training processes and, second, it will provide teachers with models which show how technology is used in learning processes.

The following selected practices of educational technology in teacher education are discussed in some detail in this paper: application of systems approaches to the development of new teacher education programmes, use of behavioural objectives and performance criteria, learning modules utilizing multi-media systems for individualized self-instruction, laboratory simulations systems, microteaching, minicourses, observation systems of appraisal and instructional processes, systematic observation techniques, and the presentation of college-level courses related to teacher education through new media.

The paper also discusses the problems of research evidence on the effectiveness of educational technology in teacher education and concludes with a discussion on the growing recognition the world over of the need for education in general and teacher education in particular to be based on a more systematic approach such as is embodied in the concept of educational technology.

Introduction

The scope of educational technology

For the purposes of the discussion of the role of educational technology in the training of science teachers, we have adopted a broad definition of educational technology which transcends media and devices such as television, films, overhead projectors, programmed books, computers and other items of 'hardware' and 'software'. We have chosen the definition of educational technology as a 'systematic way of designing, carrying out, and evaluating the total process of learning and teaching in terms of specific objectives, based on research in human learning and communication, and employing a combination of human and non-human resources to bring about more effective instruction' (Tickton, 1970, 1971). [1]

This broad definition was chosen for two important reasons, one negative and one positive. It would seem that the narrower definition, which refers mainly to technical media is perceived by many teachers as relating to using hardware or 'playing with gadgets' and so justifiably, has little appeal for the educational community. The broader definition which refers to attempts to bring order into the instructional process, through the application of more systematic, theoretically based and to some extent scientific principles would seem to have a greater appeal, especially to science teachers.

The second reason for choosing the broader definition is that it includes the most recent, and in the opinion of many, one of the most promising innovations in teacher education—the adoption of a systems approach in the development of new models and programmes in teacher education.

The training of science teachers

This paper discusses the use of educational technology in the professional training of science teachers. However, the description of experiences, developments and research evidence, upon which the discussion draws, is not unique to science teachers, but refers to teacher education in general. In only a few instances cited in the literature was a specific programme using educational technology geared especially to science teachers, and even then, the basic principles involved could easily have been applied to teachers in other disciplines as well. The limited number of reported experiences in the training of science teachers and the common problems involved, require us to draw inferences from the general body of knowledge and to apply them to the particular problems dealt with in this paper.

---

1. For bibliography, see page 236.

The effectiveness of educational technology in teacher education—research evidence and testimonials

In a most comprehensive survey and analysis of research and development in the area of 'Teacher education and the new media', Schueler and Lesser (1967) concluded that: 'The literature on media use in teacher education is more often composed of testimonials for its use rather than empirical reports of its effectiveness. ... Possibilities for new media in teacher education have neither been verified through research and experimentation (at least by the kind of research that by its nature and process can be considered meaningful) nor yet had sufficient study for proper conclusions and application to be drawn. It is particularly discouraging to note that most functions of new media which are largely operational in character and therefore capable of the greatest immediate effect on teacher education are those which though most studied and promoted, remain largely of unproved effectiveness.'

Five years have passed since Schueler and Lesser made these critical statements and though some recent reports on research projects verifying the effectiveness of various aspects of educational technology have been published, the basic problem of a comprehensive system of educational research that will stand up to rigorous analysis of critics in research methodology is still unresolved. It must be admitted that the problem is not unique to research in educational technology and teacher education, but concerns the whole field of research in teacher education, instruction and teacher effectiveness. Wittrock and Wiley (1970), Rosenshine (1971), and Gage (1972) are but a few who have indicated that due to methodological weaknesses and difficulties encountered, much of the research on instruction was unsuccessful in proving the effectiveness of innovations and new instructional strategies. Their criticisms have implications which apply equally to the use of educational technology in the field of teacher training.

This paper will describe various practices and innovations in educational technology and teacher education. It should be stressed at the outset, that many of the conclusions about the promise and effectiveness of these innovations stem from testimonials and a logical analysis of their features and the relationship to various principles and theories in psychology and communication, rather than from solid research evidence. While many of the innovations are recommended for adoption, it is imperative that rigorous research should accompany the development and adoption stages in order to determine their optimal uses.

Scope of the discussion

The need to provide a panoramic view of the problems involved, within the scope of this paper, has permitted the citing of only the most pertinent issues and practices in the use of educational technology in the professional education of teachers. Moreover, all of the examples on the use of educational technology in science teaching discussed during the conference and presented in this volume could be applied also to the professional education of science teachers. Therefore, this paper will concentrate mainly on areas and practices which were not discussed by the other authors. However, the full picture of the possible uses of educational technology in the professional training of science teachers must be derived from all papers presented in this volume. The order in which the issues are presented does not reflect the literature's consensus but rather the author's preferences, even in cases where the topics are discussed at greater length. The discussion is bound to be brief and readers interested in particular topics are referred to the bibliographic surveys and literature cited.

Educational technology in teacher education— rationale and purposes

Teacher education is an area of professional specialization similar to the education of other professionals. It is therefore safe to postulate that attributes of educational technology which enhance the process of communication and instruction in any area of specialization will also do so in teacher education. While professionals in other specialities are confronted with the processes of educational technology mainly during their formal training, teachers are expected to be involved intensively with educational technology throughout their careers. It is therefore surprising that teacher education has been slower than other fields to incorporate educational technology into its instructional process. Oddly enough, teacher educators, and even experts in the use of educational technology and media, often preach to prospective teachers on the importance of educational technology and its uses, while they themselves are employing the most conventional instructional methods.

It has often been said that teachers teach in the same way as they themselves were taught. Prior to becoming teachers, they have had about sixteen years of formal education and it is unlikely that during this period they were subjected to instructional processes based on the principles of educational technology. Hence, assuming their previous learning experience to have been stronger than the new theories they have more recently acquired, it is unrealistic to expect them to incorporate educational technology into their own teaching philosophy.

Teacher education programmes which utilize educational technology should serve a dual purpose. First, they should increase the efficiency and effectiveness of the instructional and educational processes in the programme. Second, and more crucial, a programme for teacher training should itself be a model for teaching embodying the most effective and most innovative procedures and concepts of instructional technology. Because such programmes are inherently an example of what they purport to teach, they should embody the very qualities of teaching which it is sought to instil in their students. These programmes must constitute the penetrating edge of new developments in the art and science of teaching. Thus, for example, the concept of individualized instruction and the use of individualized learning packages and multi-media systems should first be applied in teacher education programmes.

Educational technology in teacher education—selected trends and practices

The application of a systems approach to teacher education

The Model Teacher Education Project initiated by the United States Office of Education in 1967 (Le Baron, 1969) is probably the most comprehensive example of educational technology in its broadest definition. This is basically the application of a systems approach to teacher education. The systems approach in this respect is 'a process for relating a program or its parts to the goals envisioned for that program, for using information derived from operation to adjust the program towards its goal orientation, and for designing and selecting alternative approaches based on the particular characteristics of the operating environment' (Le Baron, 1969).

These models represent the first systematic efforts to develop comprehensive programmes of elementary teacher preparation. The models not only provide for analysis and replanning of a total programme of teacher education, but also constitute a vehicle for exploring new concepts and approaches. In many traditionally-oriented teacher education programmes, innovations such as multi-media, simulated observation, improved subject matter preparation, and new courses in human relations, have been piecemeal and fragmentary. The following sections will describe some of the main features of the systems approach in teacher education.

1. *The use of behavioural objectives and performance criteria.* A behavioural objective or performance criterion states the specific actions which the student will be expected to perform, or the knowledge he will have to demonstrate as a result of his training experience.

Educational objectives have been studied throughout the history of education, while work in programmed learning and instructional technology in recent years has heightened interest in the definition of precise objectives. Contributions by Bloom et al. (1956), Krathwohl et al. (1964), Mager (1962, 1968), Gagné (1965a, 1965b), Glaser (1962, 1965) and Popham (1970) have been very influential in this trend. Even though the theory, research and development of educational objectives has been taught in education courses, they are rarely applied to teacher education curricula.

The Model Teacher Education Project derived its series of significant teaching behaviours from a thorough analysis of the teaching process itself. Based on this careful analysis, behavioural objectives and criteria by which to evaluate the effectiveness of their training model were described, tested and implemented. The criterion measure specification as a means of evaluating results of training is one of the prime achievements of the systems approach and the new models.

These models are referred to in the relevant literature as performance-based or competency-based teacher education programmes. The competency-based concept is by no means limited to the Model Teacher Education Project. A recent survey by the American Association of Colleges for Teacher Education (1971) shows that performance-based teacher education is a concept which is currently being applied in a growing number of institutions and teacher education programmes.

2. *Learning modules utilizing multi-media systems for individualized self-instruction.* The traditional basic curriculum unit is a course in which a semester-long segment is organized in some ascending order of difficulty within a complete programme. The organization of material during the course is predetermined by the teacher or the textbook. Courses seldom include pre-testing as an entrance requirement or the individual sequencing of experience. Rarely are there direct ties between academic study and the on-going work of the teacher.

Teacher training programmes seldom attempt to provide for individual differences among teacher trainees. General discussions of the need for individualization of teacher training have been published (Combs, 1965; Haines, 1963). Preliminary efforts to individualize teacher education programmes have been made (Binnion *et al.*, 1963; Gennaro and Boeck, 1968; Wood, 1969). However, few teacher training curricula have either the resources or the necessary flexibility to custom-make and modify more than minor aspects of their training procedures to coincide with abilities, needs, and the experience of the individual prospective teacher (Schueler and Lesser, 1967).

The instructional module, as the basic unit of a curriculum in recently developed teacher education models is the most important current attempt to relate to the problems of individualization of instruction in teacher education. The module is organized around a single objective, including an instructional objective and criteria measurements. Student readiness to attempt the module is determined by pre-testing. Remedial experience is provided either during the learning module or as a result of failure. The student paces himself according to his ability to handle and incorporate the material of the module. The module may involve individual instruction, lecture attendance, interaction with groups of students or any combination of these elements. The student is allowed to skip modules by passing a pre-test. Various teaching techniques such as multi-media packages utilizing programmed texts, audio- and video-tapes, films, slides, computer-assisted instruction, sensitivity training, microteaching, simulation, are employed in the module. If the student desires, provision is made for discussing the module with a 'clinical' teacher. Instructional modules have been developed for the Model Teacher Education Project as well as for other projects in the cognitive, affective and behavioural domains. For a detailed description of multi-media self-instructional systems in science education see McAda, 1966; Meierhenry, 1966; Postlethwait *et al.*, 1969; CEGS Programs Publication No. 4, 1970; Novak, 1970; Creager, 1971; Ehrle, 1971; Hoffman and Druger, 1971; Hurst and Postlethwait, 1971; McDonald and Dodge, 1971; Kaye and Pentz, 1972; and Postlethwait and Mercer, 1972.

Although educators have long talked about individualized curricula, they have not, until recently, lived in a technological world which could envision and provide the means for such innovations. Almost all successful forms of individualized instruction are dependent on a very low instructor-student ratio, with instructors who are not only competent but who are also committed to individualization and personalization. Quality control has been similarly limited. Curriculum theory has demanded direct linkages between behaviourally-stated objectives, instructional alternatives and evaluation processes. The actualization demand has not really been possible. For example, no classroom teacher, no matter how dedicated and committed to individualization, can himself manufacture enough tests to track progress adequately and adjust instruction to the varying rates of progress of his students.

Today, however, we have the technological means for large and complex information-storage and retrieval systems and the capacity to develop management systems. Student characteristics can now be co-ordinated with their achievement and instructional strategies, while at the same time maintaining reasonable levels of quality control.

Simulation in teacher education

Training in a laboratory, preceding actual field practice, is an accepted procedure in many branches of learning. For scientists the laboratory has always been the means for both learning and producing new information. New curricula in all branches of science education emphasize more then ever the importance of involving students in laboratory activities. The laboratory concept has also been accepted in teacher education; certain schools were designated as laboratories for experimenting with new curricula, instructional materials, and as practice grounds for new teachers. Regular schools were designated as laboratories for neophyte teachers to gain experience through the traditional method of practice teaching. However, while the laboratory school has been an important feature in teacher education, it has never provided a safe and flexible practice ground for experimentation and training. The conflict between the need for experimentation and training and, on the other hand, obligations to the students and their parents, has always been a source of difference between teacher trainers, curriculum innovators and among directors and/or teachers of laboratory schools.

The common assumption that placing a neophyte in the classroom as an observer or a participating student-teacher is the best preparation for teaching, has been undergoing rigorous examination from within (Smith et al., 1969) and outside of the profession (Conant, 1964). In a review of research on simulation in teacher education, Cruickshank (1970) is critical of the present student teaching systems. He states that pre-service personnel are assigned to supervisory teachers and laboratory experiences in an unsystematic manner with assignment based on locale rather than more important variables. Further, teachers in service are not prepared, and far more important, not trained to supervise the neophyte trainee adequately. Some attempts have been made to alleviate this latter problem, but these have usually been of dubious value.

The simulation laboratory system introduced in teacher education appears to have great potential in resolving the difficulties and shortcomings of the present student teaching systems. Cruickshank (1966), one of the leading researchers and developers of simulation in teacher education, has defined simulation as the 'creation of realistic games to be played by participants in order to provide them with lifelike problem-solving experiences related to their present or future work'.

In addition to the general disappointment with the shortcomings of student teaching and other forms of laboratory experiences, Cruickshank cites several reasons for a widespread utilization of simulation techniques in the future pre- and in-service preparation of teachers and other professional educational personnel. For example, there is a growing recognition that theory can be taught best in the context of reality to which the theory relates. Smith *et al.* (1969) argues for the use and study of actual behavioural situations using 'protocol' material and simulation. The National Council for Accreditation of Teacher Education (1967), realizing the importance of simulation, has proposed that each advanced programme in education include direct and/or simulated experiences.

Cruickshank (1970) summarizes some of the major advantages of simulation techniques: in simulation, a realistic setting can be established in which theory and practice can be united, so that the trainee is forced into some form of action. It is a 'safe' procedure protecting children from experimentation. Unlike student-teaching, simulation can also be a controlled situation, placing the teacher trainee in carefully selected situations. The simulation technique can be applied to training in many areas of the educational field, such as special education, inner-city school situations, open-classroom, etc. Another important aspect of simulation is its relevancy for the student teacher; it focuses on the real world and the role of the teacher in society, while at the same time allowing the student-teacher to act as he would ordinarily choose to do in real life. 'Simulation seems to work. Although simulations in teacher education are not abundant and are relatively untested, some evidence exists that they are worthy of further study and use' (Cruickshank and Broadbent, 1968).

The initial work utilizing simulation in teacher education was done at the Teaching Research Laboratory of the Oregon System of Higher Education (Kersh, 1962). In order to provide the opportunity for students to practice teaching under close supervision and without embarrassment or censure, the classroom simulation, a replica of a sixth-grade classroom, was developed, employing motion picture forms and printed materials. Student-teachers were oriented to a hypothetical school, community and classroom in which they would teach. Each student-teacher was provided with a set of cumulative record cards for twenty-two children in a sixth-grade class.

As each classroom problem was presented, subjects were requested to act out a response, whereupon an experimenter selected and projected appropriate filmed feedback sequences. The intention of the simulation was to shape the student teacher's behaviour and to get him to react to classroom situations in ways judged to be optimal. Each problem sequence was repeated up to ten times until the student teacher demonstrated an acceptable response.

Kersh (1965), Vlcek (1965, 1966) and Twelker (1966) have repeated with variations the original study conducted in Oregon. Cruickshank and Broadbent (1967, 1968, 1969) have adapted the basic simulation concept and developed their own Teaching Problems Laboratory (a package containing written and audio-visual material describing a classroom setting). Lehman (1966) and Urbach have designed simulation exercises in the training of science teachers.

In the most comprehensive survey of the literature on Simulation in Preparing School Personnel (Cruickshank and Broadbent, 1970), the authors describe the intensive research and development that has been carried out in the area of simulation in teacher education. They conclude, however, that while development in this area continues to take great strides, simulation has yet to prove itself both to professional colleagues and to the public at large.

Microteaching

Microteaching, a recently developed procedure in teacher education, offers a new model for improving teaching. It was designed to overcome many of the shortcomings of traditional teacher education programmes, and increase our understanding of the teaching-learning process (Allen and Ryan, 1969; Cooper and Allen, 1971). It is a laboratory technique in which the complexities of normal classroom teaching are simplified. The trainee teaches a class of three to five students. The lesson is reduced to several minutes and is used for the practice of one particular teaching skill —lecturing, questioning, leading a discussion, or using instructional aids, etc. The lesson is recorded on videotape and the trainee may hear and see himself immediately after the lesson. The analysis and suggestions of a supervisor who attends the lesson or observes the videotape, and other sources of feedback, assist the trainee in reconstructing the lesson. The sequence is practised usually in a microteaching laboratory in a teacher training institution or in an in-service training programme in regular schools.

1. *The elements of microteaching*
The micro element is the cardinal feature of microteaching, which systematically attempts to simplify the complexities of the teaching process. Before attempting to understand, learn and perform effectively the task of teaching, one should first master the component skills of that task. Short lessons concentrate on one skill. When the trainee is proficient in this skill, it may be combined with other mastered skills in longer lessons. The short micro-lesson, by reducing the subject matter to be covered, eases the training process.

Technical skills of teaching. A repertoire of teaching skills such as lecturing, questioning, or leading a discussion, is the second important feature of microteaching (Berliner, 1969). For example, Allen et al., (1969) classify teaching skills under five general heading of response, questioning, increasing student participation, creating student involvement, and presentation. Three or four skills are further delineated within each category.

The feedback element is the third important feature of microteaching. At present, 'feedback' in the supervision of student-teachers is ordinarily based on a supervisor's recall and selective note-taking. His overall impressions provide the basis for subsequent analysis of the student-teacher's performance. But subjective factors enter into his assessment as they do into the student's ability to recall details of the lesson, and in the absence of objective criteria the student may covertly or overtly oppose the supervisor's evaluations and suggestions.

A portable battery-powered videotape recorder, a camera, and a small television monitor comprise the basic videotape recording system. It permits the recording of classroom activities with minimum disturbance to teacher and students. The videotape recorder enables instant and accurate feedback of classroom interaction (verbal and non-verbal) in the teacher's natural environment, the classroom or laboratory, and provides a basis for reliable analysis.

Safe practice grounds. Providing a safe practice ground is another important feature of microteaching. As a rule, the first teaching experience student-teachers have is in the methods courses, but usually only a few students have the opportunity of playing the teacher. Moreover, teaching peers (fellow-students) provide an unreal classroom atmosphere. It is generally assumed that the main teaching experience of future teachers will be gained during the student-teaching period (practice teaching) in an experimental or regular school, where pupils are required to study an approved curriculum, hence supervisors are careful to ensure that experimentation does no 'harm'. The student-teacher has no opportunity for 'try-outs'.

The microteaching laboratory seems to provide a safe practice ground for student-teachers since they teach relevant subject matter to genuine pupils, and because such teaching is a laboratory exercise tension is reduced and training is focused on the acquisition of teaching skills and instructional techniques related to pupils with different abilities, aptitudes, characteristics and needs. An opportunity is also provided to master curriculum material.

The teaching 'models'. Acquiring new behaviour patterns by observation and imitation is recognized as one of the major learning processes for humans and animals (Young, 1969; Borg *et al.,* 1970). Tapes or films of teaching models are an important facet of the learning process in the microteaching laboratory, and provide the trainee with many opportunities to study desired patterns of teaching behaviour. There are, of course, many styles of good teaching, and, hopefully, trainees will develop their own individual style using the models as a guide.

The research laboratory. The microteaching laboratory simplifies the act of teaching and provides opportunities for controlled research to a degree never before possible. According to Allen and Ryan (1969), the following areas of research seem to make the most effective use of the microteaching settings:

1. In-house studies designed to optimize the procedures and sequences in the microteaching situation;
2. Research in modelling and supervising techniques;
3. Task analysis of the teaching act and the investigation of the relationships between teaching behaviour and student performances;
4. Aptitude treatment interaction studies to try to provide optimal training procedures for teachers with different abilities, interests and backgrounds.

The list is not exhaustive, and the innovative educational researcher can find many additional subjects to be tested in the microteaching laboratory.

2. *Microteaching in pre-service teacher education.* Several alternatives have been adopted by colleges and universities in introducing microteaching procedures into the teacher education curriculum. Variables such as the content of teaching skills, the length of micro-lessons, size of classes, the teach-critique-reteach cycle, patterns of supervision, and the use of models, are still the subject of extensive experimentation and research in the search for optimal procedures. A few examples of different approaches to the use of microteaching in a teacher education programme are being carried out at the Laboratory for Research and Development in Teaching and Learning, Technion, Israel (Perlberg *et al.,* 1972), University College of Rhodesia (Gregory, 1971), in the Federal Republic of Germany at the Center for New Learning Methods of the University of Tübingen (Zifreund, 1972; Brenner and Bühler, 1972; and Krumm, 1972), Osnabrück (Castrup, 1972), Reutlingen (Wagner, 1972) and Schwäbisch Gmünd (Heilig, 1972); in Sweden at Malmo (Bierschenk, 1972) and Gothenburg (Brusling, 1972); in the United Kingdom at the University of Stirling (Perrott and Duthie, 1969, 1970) and (Perrott, 1967, 1972); University of Liverpool (Beattie and Teather, 1972; Teather, 1972), Eastbourne College and Sussex University (Britton and Leith, 1971); in Northern Ireland at New University of Ulster (Mcaleese and Unwin, 1971); in New Zealand at University of Canterbury (Nuthall, 1972); in Belgium at the University of Liège (Landsheere, 1971) in the United States in Baltimore Multi-State Teacher Education Project (Webb, 1968), and at the University of Chicago (Guelcher *et al.,* 1970); and in Australia (Turney, 1970) and (Owens and Hatton, 1971).

The microteaching procedures have also been incorporated into methods courses. Teaching skills and classroom interaction are illustrated with tapes of 'model teachers'. Tapes of students teaching or participating in class are brought before the group for analysis. Microteaching procedures provide the methods courses with new dimensions of audio-visual reality and experimentation. Emphasis has been placed on the improvement of specific teaching skills.

3. *Microteaching in in-service teacher education.* Teacher education is a never-ending process and pre-service training is only its first phase. In spite of their potential importance, in-service education programmes have little effect on teacher-pupil interaction. Accumulating more university credits or attending lectures organized by the school does not ensure a change in classroom teaching behaviours. Microteaching techniques, however, seem to offer a promise of effecting a real change. The same basic procedures including taping, analysis and subsequent behaviour modification are used in in-service education. If experienced teachers are expected to engage in these activities, they need to be confident, for obvious reasons, that

this training technique is not devised to monitor their activities for administrative purposes (Allen, 1966; Aubertine, 1967; Meier, 1968; Perlberg et al., 1968; United States Department of Health, Education, and Welfare, Office of Education, 1969; Klabenes and Spencer, 1970).

The minicourse is a self-contained package of training materials that can be used in any school or locality which has a videotape recording system. The package is self-explanatory, and relies on a system of self-evaluation and feedback. No outside supervision is required.

The minicourses which are being developed by the Far West Laboratory for Educational Research and Development utilize a basic microteaching technique (Borg *et al.,* 1968a, 1968b, 1970). They have been developed by using a systems approach over a number of years and have been field tested several times.

The minicourse instructional model has three main steps: (1) an instructional film describes specific teaching skills or strategies and a second film depicts a model teacher using the skills; (2) the trainee plans and teaches a micro-lesson, i.e., a short lesson to a small group of students, which is recorded on videotape; the trainee replays and evaluates the lesson; (3) the self-feedback assists the trainee to replan and reteach the same lesson to another small group of students; again, he evaluates his performance.

Research evidence (Acheson, 1964) suggests that self-evaluation of the videotape is adequate and that the trainee gains very little more from feedback supplemented by a supervisor. Further research by the Far West Laboratory indicates that teaching skills learned through the minicourse have been incorporated into the regular classroom behaviour.

The Far West Laboratory has probably been the most active group in advancing this technique. They have already produced and tested many such self-contained packages which can be used for both in- and pre-service teacher education. Titles such as 'Effective Questioning in a Classroom Discussion'; 'Thought and Language: Skills for Teaching the Child with Minimal Language Development', 'Verbal Interaction'; 'Individualizing Instructions in Mathematics (Elementary)'; and 'Organizing the Primary Classroom for Independent Learning and Small Group Instruction' are but a few of the already completed self-contained instructional packages.

Recently, the Center for Educational Research and Development of the OECD embarked on a project of international transfer of microteaching materials. Its first activity is the adaptation of Minicourse No. 9 (on high cognitive questions) to various countries in Europe and it is directed by Professor Elizabeth Perrott of Lonsdale College, University of Lancaster, Bailrigg.

In view of the great need for continuing in-service teacher education programmes and the shortage of supervisory personnel to conduct such programmes, minicourses have great potential value for teacher education.

Remote microteaching. The shortage of qualified university personnel and the distance of schools from universities may limit a college supervisor to a few visits, a practice known to be inadequate for the supervisory process. In a study at the University of Illinois (Perlberg *et al.,* 1968), a portable video recorder and its accessories were left at a school for several days and used by the student-teacher and his co-operating teacher to record and analyse each other's performance. Arrangements were made for the student-teacher to mail his taped lessons to the university, where they were analysed by his professor and discussed with the student and his co-operating teacher in a long-distance telephone conference, during which the professor was observing the videotape. Similar studies have been conducted by Meier (1968), Meier and Brudenell (1968) and by Cottrel *et al.* (1971).

4. *Microteaching in developing countries.* There are not as yet many published reports about experimentation with microteaching techniques in developing countries. Several 'trials' have been made by Dwight Allen, G. Dalgalian,[1] Evance (1970) and their colleagues in a few countries in Africa and South America. From impressions communicated to the author verbally, it appears that these attempts were successful and that teacher educators in these countries believe that microteaching techniques could greatly improve pre-service and in-service teacher education.

Some observers involved in international education have been sceptical, especially in view of the high cost of technical equipment and the problems involved in its operation and maintenance. While the importance of instant and accurate feedback through videotape re-

---

1. Although we cannot list as yet any publications, extensive work is being done by G. Dalgalian at the University of Dakar, Senegal. Further information concerning his activities may be obtained from him at the Centre de Linguistique Appliquée, Dakar-Fann.

corders has been stressed throughout this section, it should be stated emphatically that microteaching can be practical even without television equipment. Several other techniques to obtain oral feedback have been used and, in the absence of sophisticated technical equipment, the emphasis should be placed on the micro- and laboratory elements. The importance of reconciling and balancing theory and practice in professional teacher education programmes has been stressed by this author. It applies to technologically advanced countries and, even more, to developing countries. Experience has shown that the value of training educators from developing countries in the traditional teacher education programmes of already developed countries is highly questionable. On the other hand, the reality and practicability inherent in microteaching techniques could have a great impact on the quality of education in developing countries. It should be borne in mind that the microteaching model described represents a general framework that can be adapted to the needs of each case and each country.

While the present shortage of technological devices should not deter the introduction of microteaching, it can be assumed that rapid technological developments will in the near future greatly reduce the cost and simplify the operation and maintenance of technical devices. Small, portable, battery-operated, 'cassette' audio- and video-tape recorders will probably come within the reach of many educators the world over. The basic problem at present is to train teams who will be able to introduce these innovations into their local school systems, particularly in developing countries.

5. *Microteaching as an innovation.* While microteaching has been described as an innovation in teacher education, in fact certain of its aspects have long been in use in various training situations. Moreover, it is based on long-standing educational theories. For example, the 'laboratory' concept as a training procedure preceding actual field practice is an accepted part of many branches of learning (e.g., science laboratories). It was accepted also in teacher training, where certain schools were designated as laboratories for testing new curricula and instructional materials as practice grounds for new teachers. The microteaching innovation lies in the establishment of specific educational laboratories, rather than in the designation of regular schools as laboratories, with the many limitations they impose on experiment.

The micro concept which underlies programmed learning and computer-assisted instruction is based on previously established learning theories; in both it is assumed that learning is more effective if a complex skill is divided into its components and learned step by step before it is undertaken as a whole. Other examples of established theories adopted by microteaching are feedback, reinforcement and observational learning through a 'model'. However, the developers of microteaching should be credited with incorporating all these well-known principles into a systematic training system.

As is the case with many innovations, microteaching, too, has both its strong and sometimes over-zealous supporters and its critics. In a presentation at a microteaching conference at the University of Massachusetts, Earl Seidman (1969) raised several pertinent questions about microteaching and its future. Like several other critics, he is greatly disturbed by the fact that microteaching is strongly influenced by behaviouristic psychology: 'microteaching trains teachers to perform in ways those who are running the program think is good. ... Are we involved in a program which trains rather than educates?' Seidman is alarmed that such great emphasis is being put on training in teaching skills which 'presuppose the idea that the teachers' role is to control the students and to direct the class. ... If we treat our interns mechanistically, how will they treat their students?'

It is not our intention to elaborate on the problem of the place of behaviouristic psychology in education and teacher education. It should therefore be stressed that microteaching is only intended to be one aspect of a teacher education programme. It does not intend to substitute the theoretical foundation in the teacher-training programme. It corrects unevenness between the over-emphasis on theory and the neglect of effective practice. Practice in a microteaching laboratory begins with the basic teaching skills, but should not stop there; it should provide the teacher with opportunities to learn and practice 'clusters' of basic skills, high-order skills and instructional decision-making. Only those who have mastered the basic skills can freely engage in exercises of high-order skills and instructional decision-making.

The present status of microteaching was succinctly summed up by Allen and Ryan in 1969 and is still relevant today: 'The questions that are raised by microteaching, at least at this point in its evolution, far exceed the answers it has been able to supply. Microteaching currently has the same promise and the same danger that newly-devised research and training techniques have always had: the promise of opening entirely new avenues, prospectives and alternatives to human exploration; the danger of locking too early on the first alternative which arose purely out of chance and convenience.'

Observation and appraisal systems of instructional processes

1. *Direct classroom observation.* Directly supervised experience in a real, non-simulated educational setting is the final, and in the opinion of many (Conant, 1964), the most important stage in teacher education. The student-teacher observes the classes and teacher to whom he has been assigned and gradually assumes (under the guidance of a supervisory teacher and a college supervisor) responsibility for teaching. Direct observation and practice in teaching should assist the student-teacher to develop sensitivity, powers of observation and the ability to analyse and appraise the instructional processes of other teachers as well as his own. The observation and appraisal of others is a relatively neutral activity. However, the ability to see one's own actions as others see them, evaluate and learn from them, and to modify if necessary one's own behaviour, is a highly personalized and complex process.

Group observation of students and teachers in school situations is highly regarded by teacher educators as another form of direct experience with the educational process. This type of observational experience is intended to provide prospective teachers with authentic examples of teaching-learning behaviours in basic curriculum and method courses. It could also serve to provide experienced teachers with examples and case studies for further analysis and discussion.

The effective use of direct observations presents a wide range of problems. The prerequisite is the selection of adequately co-operative teachers and schools within a limited geographic area, so that school visiting does not become a time-consuming and inefficient activity. Further, one must co-ordinate the plans of the college instructor regarding topics in his syllabus and the school's instructional programme. It is also necessary to co-ordinate the lesson plan and the predicted interaction which will provide the material for analysis.

What to observe, how to observe, and most important, how to record the observation of the classroom in a reliable way that will provide accurate feedback needed for analysis and appraisal must be delineated. Empirical evidence has shown the existence of great differences in the impressions and even facts recorded during classroom observations. This phenomenon causes many problems in situation analysis and the subsequent attempts to modify the teaching behaviours of those concerned. These problems are a few of the many arising from direct classroom observation.

2. *Observation and appraisal through media.* Educational technology can facilitate the finding of solutions to these problems through the use of media which enables both the recording and provision of an authentic record of what is happening in the classroom, and secondly by adapting systematic observation techniques to an analysis of records.

Various technological devices have been used to record classroom activities: still-pictures and slides, through regular or time-lapse photography; tape recordings; syncronized recording of audio-tape and slides; films; and videotape. The recent advent of portable and relatively inexpensive videotape recorders has greatly increased their use in teacher education and teaching research. There is a growing bibliography of research in teacher education which testifies to the effectiveness of the various media as compared with direct observation (Schueler and Lesser, 1967; Baker, 1970). The bulk of research evidence, however, cited by the reviewers on this subject comes from related fields and from the logical analysis of factors inherent in the use of these media.

It should be stressed that proponents of the use of media for classroom observation do not recommend the elimination of direct observation or claim that observation through media could be used as a sole substitute. Direct observation will always be an essential feature of teacher education. However, classroom observation through the use of media enables one to augment the limited number of direct observations that learner-teachers are able to experience. By providing additional cases for observation, this method also facilitates better utilization of direct observation.

Instructional and learning interactions between teacher and class, a teacher and an individual student, and students among themselves, when recorded on any media could be examined and analysed many times.

Recording on videotape can provide the most authentic audio-visual feedback at the same time reducing natural conflicts and stimulating discussions arising from the different perceptions of identical situations. The recording of educational interactions which can provide a basis for discussion and analysis in the most convenient time period is for both instructor and student the most effective method of teacher education.

Most of the evidence on the effectiveness of observing others comes from research on 'learning by observation' and 'imitative behaviour'. The several reviews on these subjects (Mowrer, 1960; Bandura and Walters, 1963; Flanders, 1968; Baker, 1970; and Bourdon, 1970) provide the generalization that observing other people may affect the observer's own interaction by way of adapting the observed behaviour. The developers of microteaching, minicourses and microcounselling have relied heavily on this research evidence and have provided the teacher with models or protocols (Gee and Berliner, 1972) of specific behaviours to be observed, adopted or imitated.

3. *Self-confrontation through film, audio- and videotape.* Various media, especially audio- and video-tape recordings and films, have been used for self-observation. A growing body of research testifies that the process of self-confrontation is instrumental in modifying behaviour (Nielson, 1962; Schueler and Lesser, 1967; Geerstma and Mackie, 1969; Kagan, 1967; Baker, 1970; Onder, 1970; and Fuller and Baker, 1970). Self-confrontation through media involves two elements. It provides the person with feedback relevant to his behaviour, and information for reinforcing a specific or changing it. According to Baker (1970), the notion of feedback, or reinforcement seems to be the most useful point of departure for a review of the literature on self-confrontation. It underlies the traditional learning theories of Thorndike, Gutherie, Hull and Skinner. Gagné and Bolles (1963) state that 'at present there seems to be no contrary evidence to the general conclusion that learning is facilitated by frequent, immediate and positive reinforcement'. Feedback of information relevant to one's change in behaviour is seen as the most important variable in producing that change. This concept is also consistent with the learning theories of Benne, Bradford and Lippitt (1964), information theory (Frick, 1959), and particularly the feedback model of learning developed by Miller, *et al.* (1960). As stated by Kolb *et al.* (1968), the relationship of feedback and behaviour change is: 'the more an individual can effectively utilize the feedback of information appropriate to his change project, the more successful he will be in attaining his change goal'.

According to Baker (1970) the concept of change of goal is important. The reason for self-confrontation involves the moving of an individual towards a goal, whether that goal is in the mind of the one confronted, or is as yet to be formulated in the mind of the individual. The former implies the notion of shaping and reinforcement, reward and punishment; the latter involves expectancy, ideal, discrepancy and dissonance. These two approaches, shaping reinforcement and dissonance reduction (Staines, 1969), seem to be the most basic approaches of the research in the area of film or videotape feedback. The former is best exemplified in the behaviour therapy model, and uses the more traditional concepts of reinforcement, reward, stimuli, etc. It tends to conceive of man as manipulated by his environment and reactive. The latter is best illustrated by the cognitive dissonance model, and uses such concepts as expectancy, discrepancy, dissonance, commitment, self-concept. It sees man as proactive, having within himself the capacity to recognize the discrepancy between his actual behaviour and what he desires and the inherent ability to commit himself to a valued goal (Zimbardo, 1969; Kolb, Winter and Berlew, 1968).

Viewing one's own behaviour does not in itself assure change. Baker (1970) cites numerous studies which indicate that a person's viewing of himself on videotape with no one else present might be very ineffective in terms of learning new behaviours. Baker and Fuller (1972) reviewed 147 studies which have implications for 'Self-confrontation Counselling of Prospective Teachers'. On the basis of that review and other analyses (Fuller and Baker, 1970) they propose that feedback should be unambiguous, trustworthy and accepted both overtly and covertly by the teacher. For example, interactions the teacher perceives as atypical are unlikely to be accepted covertly.

The teacher must have some concerns, goals or expectations for himself which are related to the content of the feedback, and some discrepancy must exist between these expectations and the performance which the teacher sees. The discrepancy should be moderate rather than so small that it is insignificant, or so large that the teacher gives up on the job of reducing it. However, the teacher must feel he has the capacity to change.

The feedback situation must be one in which the teacher feels at ease, must be a low threat situation and must be one in which the individual is free to react. The focus must be on aspects of the teacher's behaviour which are remedial, those which are susceptible to change rather than those which are immutable, such as tics, height, accent, firmly entrenched habits and so on, and should be provided by someone or something other than the teacher himself.

To summarize, gain is most likely for strong teachers given positive feedback by permissive helpers who focus on those aspects of the feedback which can be remedied. Obviously this does not happen when unselected teachers (including the least hopeful) see unselected videotapes either alone or with a supervisor (whose job is to evaluate) when the discussion centres, as it often does, on those behaviours which are less important in communicationg with pupils and most resistant to change.

Recommended are procedures for selecting teachers for the treatment, a good technical job of taping, and a low threat situation for both videotaping and feedback. Probably it is desirable to tape what the teacher considers his best performance, to allow him to react spontaneously to his videotape and to focus non-evaluatively on those behaviours which are important and remedial.

4. *Systematic observation techniques.* The importance of proper analysis of educational interactions, whether through direct observation or together with the utilization of media, cannot be sufficiently emphasized. In the previous section, the discussion focused on a clinical approach to the analysis of feedback portocol. The following will focus on the use of systematic observation techniques in analysing the educational interaction protocol. The use of these systems does not depend on media; in most cases they were designed for direct observation. The use of media, however, secures optimal results from these systems.

Observation systems were designed originally as a research device to systematically collect observable and objective data concerning human interaction in a variety of settings (Medley and Mitzel, 1963). About a hundred such systems, all of which have in one form or another been used for research purposes, were collected and edited in sixteen volumes by Simon and Boyer (1970). Although originated as research instruments, many have gradually been changed to training tools used to give information directly to the people observed. They are designed to help teachers, counsellors and group members gain insight into their own behavioural goals. Training in process observation is now a requisite part of many teacher education programmes, and training workshops in interaction analysis systems for in-service teachers are becoming increasingly common.

Another function of the classroom observation system is that of an evaluation instrument for personnel and curriculum materials. Here evaluation consists of using the observation instrument as a tool for describing the behaviour of personnel using new curriculum materials or for analysing the materials themselves. The descriptions generated by the analysis are compared to statements of goals for teachers' behaviour or to the materials. The use of an evaluative instrument changes the concepts of evaluation from rating on a good-bad continuum to comparisons of results and expectations (Simon and Boyer, 1970).

The best-known observation system is the Flanders Ten Category System (Flanders, 1970) for analysing teacher-student interaction. The effectiveness of the Flanders system as a training tool in modifying teacher behaviour has been investigated probably more than any other system and found to be effective. The Far West Laboratory (1971, 1972) has produced a mini-course consisting of a programmed text and a set of audiotapes to train teachers in the Flanders system.

Several of the systems have been designed specifically for science education or have been tried out in science education. Wright (1959, 1967) developed an instrument for studying verbal behaviours in a secondary school mathematics classroom. Parakh (1965, 1967) designed an instrument to study the reactions of teachers and students in biology classes. Gallagher (1967) constructed 'A Topic Classification System in Analysis of BSCS Concept Presentation'. There exists a growing recognition that systematic observation techniques are becoming a viable technique in teacher education and research in teaching.

Presenting college-level courses related to teacher education through new media

The use of television and other media to teach college courses as a substitute for conventional classroom presentation is spreading throughout the world. From the voluminous research on the use of television as an educational medium, it appears that teacher educators have not used it very intensively. Many students in teacher education programmes have taken a number of courses in the basic sciences, arts, social sciences and some of the foundation courses in education, such as introductory psychology, by means of television. Since the majority of education courses are more specialized, they are given in small groups and the idea of teaching them through television is not always economical. Another reason for the infrequent use of television in education courses is the failure of research to distinguish between effectiveness of media instruction and normal classroom techniques in teaching college courses (Schueler and Lesser, 1967) and the notion held by many teacher educators that teaching through media 'dehumanizes' the educational process.

The above observations on the use of television in teacher education are valid for other media as well. For example, programmed instruction has been tried out in many subjects and many levels of education. While the professional schools, e.g. medical schools, are experimenting with the adaptation of programmed instruction principles in the writing of textbooks, very little has been done in teacher education. The extent to which the prospective teachers have been exposed to programmed instruction during their college years has been confined to the basic sciences, arts and education foundation courses such as psychology. Because of the high cost, the more recent experimentation with computer-assisted instruction (CAI) has not as yet involved courses in teacher education and it is safe to assume that most experimentation will be done first in other fields. As a result of the conflicting research on the effectiveness of television and other media in presenting college and university courses, as compared with conventional methods, teacher education seems somewhat reluctant to use it intensively.

There are, however, several advantages in the use of television and other media in teacher education which have not been tested in the regular research experiments and which should be taken into consideration when a decision has to be made whether or not to use such media.

Through the proper use of the media, prospective teachers who study in teacher education colleges and universities could be exposed to some of the most influential thinking in the educational world. Proponents of new teaching methods advocate that teacher education should set an example in abolishing or at least balancing the proportion of teacher-centred methods (lectures) versus student-centred methods (discussions and independent learning). Nevertheless, the lecture as a teaching method in teacher education is still dominant. Exposing prospective teachers to stimulating, selected, renowned leaders in educational thought and practice would seem to be important and relevant.

The use of television and other media as a means of teaching college and university courses to students who cannot attend regular university classes has become an accepted practice. In various countries, universities have co-operated with local radio and television stations in transmitting series of courses such as College of the Air, Sunrise Semester, etc.

A more recent development in this direction is the Open University in the United Kingdom (McIntosh, 1972; Kaye and Pentz, 1972). The basic tenet underlying this development is the growing demand for higher education. Large populations in our society, hampered by social, economic and other factors, could in this way acquire regular, general and professional higher education. For many of those 20,000 students who enrolled in the Open University this was the one chance of obtaining higher education and if they complete their course of study and graduate, the system can be considered successful.

The Open University is based on the concept of a systems approach to teaching and learning, and operates through a variety of methods and media such as television, radio, written material, multi-media packages for individual learning, and regular instructional procedures, such as live lectures, discussions and tutorials in local centres.

One way to evaluate the importance of the Open University is from the state's point of view as noted by McIntosh (1972): 'The resources available for higher education in Great Britain are inevitably limited although pressure for expansion is great. The Open University provides the possibility of expanding higher education by an amount more than commensurate with its cost. Since the majority of students are in employment while studying it follows that the net cost to the country is proportionately less. If solutions are found to the problems posed by the undertaking, the benefits to this country, and to other less developed countries will be great.'

The authorities of the Open University claim that many of their students are prospective teachers and while at present most of the courses are in the basic sciences, arts and social sciences, there will in the future be courses in subjects pertaining to professional education.

The Open University and similar institutions in other countries are planning to utilize their system in in-service education courses for teachers. Various national and international organizations are exploring the possibilities of utilizing television through satellite systems in order to provide teachers the world over with in-service training programmes. It has often been stated that one of the main problems facing education is the upgrading of the quality of education through the in-service training of teachers who will continue to function for a decade or two. Television and other media could greatly contribute to the fulfilment of this task.

The use of supplemental media resources for pre- and in-service teacher education programmes

While the use of media as a substitute for conventional courses in teacher education programmes is relatively small, there is a growing use of supplemental media resources for pre- and in-service teacher education programmes. In a publication entitled *Mediated Teacher Education Resources* (1970), the American Association of Colleges for Teacher Education has compiled a rich bibliography of audiotapes, motion pictures, multi-media packages, slides, filmstrips, videotapes, gaming, and simulation, all of which were produced for use in teacher education programmes. This recent publication is one example of the many catalogues and bibliographies published by non-commercial and commercial organizations which cite a wealth of resources that could be used in pre- and in-service education programmes. The extent of their use depends, in general, on the individual teacher educator who has access to these resources and can integrate them into his regular courses. The evidence of their effectiveness in the learning process of teachers is a matter of continuing research which must be done on the use of media in general. However, the various compelling arguments which have been put forward in favour of the use of media in the general learning processes are valid here too.

Conclusions

The evidence adduced in this paper indicates a growing recognition among educators the world over of the need for education and teacher instruction to be based on a more systematic approach such as is embodied in the concept of educational technology. It is widely felt that only by these means can a solution be found to the many problems of education and its diverse goals be realized. As stated, this growing awareness of the value of educational technology is not always based on research evidence, but frequently depends on rational analysis of its underlying theory, on the recommendations of those who already employ such methods, on inference from related fields and finally on the desire of innovators to introduce new methods.

Relative to other groups in education, science educators have been more prone to accept concepts of educational technology and media. The intensive use of various technological media has played a dominant role in most innovative curricula in science education. This example has been followed by educators in other disciplines, though teacher educators have in the main been more reluctant in applying educational technology to teacher training. The reasons for this reserve are varied and often consist in psychological or emotional resistance to change. For some educators, technological media symbolize dehumanization of the educational process, which they regard as something to be resisted. Conservative educators naturally resist change and innovation, while many others are concerned at the heavy expenditure for equipment and the effort involved in converting to a new and hitherto unproven technique. For such educators, the dilemma lies in the wish to employ the new methods versus a lack of research evidence concerning their overall effectiveness and clear instructions as to how best to exploit them.

The problem of educational technology and whether or not it constitutes dehumanization of education, does not fall within the scope of this paper, nor do other psychological aspects such as resistance to change. It is essential, however, to touch upon the problem of educational technology and research evidence.

It would seem to us that teacher educators in general and educators of science teaching in particular cannot afford to sit passively and wait until conclusive research evidence to support the use of educational technology in teacher education is forthcoming. Most educational practices, including the new curricula in science education, have yet to be tested through rigorous research. They were introduced mainly because they were based on accepted psychological theories and pedagogical principles, and are in the process of being tested under practical conditions.

Evaluation of new curricula in the sciences is still going on and there is as yet no hard data to prove their effectiveness. It is suggested, therefore, that those concerned with the training of science teachers apply the same principles which guide the developers of new curricula in science teaching. The first principle is to undertake a rigorous analysis of present curricula and their relevance to the present and future needs of teacher education. Secondly, there is need to design a curriculum and instructional process based on a systems approach which might improve the educational process. New technological media would be used in the instructional process when necessary and possible. Special attention should be given to ongoing evaluation procedures to test the efficacy of such new systems. These principles are embodied in the broad definition of educational technology, and their bold and vigorous application will prove the best means, both of testing their fundamental validity, and at the same time of advancing education at a speed commensurate with the advance in technological means, and with the demands of a technological society.

BIBLIOGRAPHY

ACHESON, K. A. The effects of feedback from television recording and three types of supervisory treatment on selected teacher behaviour. (Unpublished doctoral dissertation.) Stanford University, 1964.

ALLEN, Dwight W.; Microteaching: a new framework for in-service education. *High school journal,* vol. 49, p. 355-63, May, 1966.

ALLEN, D. W.; RYAN, K. A. *Microteaching.* Reading, Mass., Addison-Wesley, 1969. French trans. by G. Dalgalian: *Le micro-enseignement,* Sciences de l'éducation, no. 6, Paris, Dunod, 1972.

ALLEN, D. W.; RYAN, K. A.; BUSH, R. N.; COOPER, J. M. *Teaching skills for elementary and secondary school teachers.* New York, General Learning, 1969.

AMERICAN ASSOCIATION OF COLLEGES FOR TEACHER EDUCATION. *Standards and evaluative criteria.* American Association of Colleges for Teacher Education, 1967. Washington, D. C.

AMERICAN ASSOCIATION OF COLLEGES FOR TEACHER EDUCATION. *Mediated teacher education resources.* Supplemental Media Resources for Preservice and In-Service Teacher Education Programs. W. C. Meierhenry (Editor). Washington, D. C., American Association of Colleges for Teacher Education, 1970.

AMERICAN ASSOCIATION OF COLLEGES FOR TEACHER EDUCATION. *Performance-based teacher education: an annotated bibliography.* Washington, D.C., American Association of Colleges for Teacher Education, 1971.

AUBERTINE, Horace E. The use of micro-teaching in training supervisory teachers. *High school journal,* vol. 51, p. 99-106, November 1967.

BAKER, Harry P. *Film and video tape feedback: a review of the literature, Series no. 53.* The Research and Development Center for Teacher Education, University of Texas at Austin, 1970.

BAKER, Harry P.; FULLER, Frances F. *Summary—self confrontation counseling of prospective teachers: a review of the literature and some prescriptions.* Research and Development Center for Teacher Education, University of Texas at Austin, 1972.

BANDURA, A.; WALTERS, A. H. *Social learning and personality development.* New York, Holt, Rinehart and Winston, 1963.

BEATTIE N. M.; TEATHER, D. C. B. *Towards a taxonomy of microteaching situations—implications and issues.* University of Liverpool, March 1972.

BENNE, K. D.; BRADFORD, L. P.; LIPPITT, R. The laboratory setting. L. P. Bradford, J. R. Gibb and K. D. Benne (eds.). *T-Group theory and laboratory method.* New York, Wiley, 1964.

BERLINER, David C. Microteaching and technical skills approach to teacher training. *Technical Training Report No. 8.* Stanford, California, Stanford Center for Research and Development in Teaching, October 1969.

BIERSCHENK, Bernhard. *Self confrontation via closed-circuit television in teacher education: results, implications and recommendations.* Paper presented at the International Microteaching Symposium, Tubingen, April 1972.

BINNION, Hada, et al. Experimenting with individualism. *Concern for the individual in student teaching.* A. C. Haines (Editor). Dubuque, Iowa, William C. Brown, 1963, p. 147-156.

BLOOM, B. S., Editor. *Taxonomy of educational objectives: the classification of educational goals, Handbook I: Cognitive domain.* New York, Longmans, Green & Co., 1956.

BORG, Walter R., et al. *The minicourse: rationale and uses in the in-service education of teachers.* Berkeley, California, Far West Laboratory for Educational Research and Development, February 1968.

BORG, W. R., et al. *Video-tape feedback and microteaching in a teacher training model.* Berkeley, California, Far West Laboratory for Educational Research and Development, December 1968.

BORG, Walter R.; KELLEY, Marjorie; LANGER, Philip; GALL, Meredith. *A microteaching approach to teacher education.* Macmillan Educational Services, Inc., 1970.

BOURDON, Roger D. Imitation: implications for counseling and therapy. *Review of educational research,* vol. 40, no. 3, 1970.

BRENNER, Inge; BÜHLER, Hans. *Aims and developmental procedures for designing teaching-skills.* Presented at the International Microteaching Symposium, Tübingen, April 1972.

BRITTON, R. J.; LEITH, G. O. M. *Systems development in teacher training evaluation studies.* An experimental evaluation of the effects of microteaching on teaching performance. November 1971.

BRUSLING, Christer. *An experiment on microteaching at the Gothenburg School of Education, Gothenburg, Sweden.* Presented at the International Microteaching Symposium, Tübingen, April 1972.

CASTRUP, K. H. *The use of microteaching and portable television recorders as sources of information in the organization of pre-school classes.* Presented at the International Microteaching Symposium, Tübingen, April 1972.

CEGS. *Auto-tutorial instruction: a strategy for teaching introductory college geology.* CEGS Programs Publication No. 4, Washington, D.C., 1970, 22 p.

COMBS, A. W. *The professional education of teachers.* Boston, Allyn & Bacon, 1965.

CONANT, J. B. *The education of American teachers.* New York, McGraw-Hill, 1964.

COOPER, James M.; ALLEN, Dwight W. Microteaching: history and present status. *Microteaching selected papers.* ATE Research Bulletin no. 9, September 1971.

COTTREL, Calvin J., et al. Assessment of microteaching and video recording in vocational and technical education, Phases I-X (a series of reports), Center for Vocational and Technical Education, Ohio State University, 1971.

CREAGER, J. G., Editor. Other programs involving modularized instruction. *The use of modules in college biology teaching,* Publication 31, Commission on Undergraduate Education in the Biological Sciences, March 1971, p. 60-74.

CRUICKSHANK, Donald R. Simulation: a new direction in teacher education. *Phi Delta Kappa,* vol. 48, p. 23-24, 1966.

CRUICKSHANK, D. R.; BROADBENT, R. W.; BUBB, Roy L. *Teaching problems laboratory.* Chicago, Science Research Associates, 1967.

CRUICKSHANK, Donald R.; BROADBENT, Frank W. *The simulation and analysis of problems of beginning teachers.* Final Report Project No. 5-0798, U.S. Department of Health, Education and Welfare, Washington, D.C., 1968.

CRUICKSHANK, D. R.; BROADBENT, F. W. *Simulation in preparing educational personnel.* ERIC Center for Teacher Education, 1969.

CRUICKSHANK, D. R.; BROADBENT, F. W. *An investigation to determine effects of simulation training on student teaching.* Presented at the A.E.R.A. Convention, Los Angeles, 1969.

CRUICKSHANK, D. R.; BROADBENT, F. W. *Simulation in preparing school personnel.* Washington, D.C., ERIC Clearinghouse of Teacher Education, 1970.

CRUICKSHANK, Donald R. Teacher education looks at simulation: a review of selected uses and research results. *Educational aspects of simulation.* P. J. Tansey (Editor). London, McGraw-Hill Publishing Co. (In press).

EHRLE, E. B. Project Biotech: a modularized answer to a critical manpower question. *The uses of modules in college biology teaching,* Publication no. 31, Commission on Undergraduate Education in the Biological Sciences, March 1971, p. 39-44.

EVANCE, David R. Microteaching: an innovation in teacher education. *Education in Eastern Africa* (Nairobi), vol. 1, no. 1, 1970.

FAR WEST LABORATORY FOR EDUCATIONAL RESEARCH AND DEVELOPMENT. *Interaction analysis: answer book.* Berkeley, Far West Laboratory for Educational Research and Development, 1971.

FAR WEST LABORATORY FOR EDUCATIONAL RESEARCH AND DEVELOPMENT. *Interaction analysis: coordinator handbook.* Berkeley, Far West Laboratory for Educational Research and Development, 1972.

FLANDERS, James P. A review of research on imitative behavior. *Psychological bulletin,* vol. 69, p. 316-337, 1968.

FLANDERS, Ned. *Analyzing teaching behavior.* Reading, Mass., Addison-Wesley, 1970.

FRICK, F. C. Information theory. *Psychology: a study of a science.* S. Koch (Editor). Vol. 2. New York, McGraw-Hill, 1959.

FULLER, Frances; BAKER, Harry. *Counseling teachers: using video feedback of their teacher behavior.* The Research and Development Center for Teacher Education, The University of Texas at Austin, July 1970.

GAGE, N. L. *Teacher effectiveness and teacher education: the search for a new scientific basis.* Palo Alto, Pacific Books, 1972.

GAGNÉ, R. M. *The conditions of learning.* New York, Holt, Rinehart and Winston, 1965.

GAGNÉ, R. M.; BOLLES, R. C. A review of factors in learning efficiency. *Human learning in the school.* John P. DeCecco (Editor). New York, Holt, Rinehart and Winston, 1963.

GAGNÉ, R. M.; GROPPER, G. L. et al. *Studies in filmed instruction. 1. Individual differences in learning from visual and verbal presentation. 2. The use of visual examples in review.* Pittsburgh, American Institutes for Research, 1965.

GALLAGHER, James J. A topic classification system in analysis of BSCS concept presentation. *Classroom interaction newsletter,* vol. 2, p. 2-16, May, 1967.

GEE, James; BERLINER, David C., *Protocols: a new dimension in teacher education.* 1972 (unpublished).

GEERSTMA, R. H.; MACKIE, J. B., *Studies in self cognition: techniques of videotape self-observation in the behavioral sciences.* Baltimore, The Williams & Wilkins Co., 1969.

GENNARO, E. D.; BOECK, C. H., A self-instructional laboratory for science teachers. *Science education,* vol. 52, no. 3, p. 274-277, April 1968.

GLASER, R., Editor. *Training research and education.* Pittsburgh, University of Pittsburgh press, 1962.

GLASER, R., Editor. *Teaching machines and programmed learning, II: Data and directions.* Washington, D.C., National Education Association, 1965.

GREGORY, I. D. Microteaching in a pre-service education course for graduates. *British journal of educational technology,* vol. 2, no. 1, January 1971, p. 24-32.

GUELCHER, William; JACKSON, Travis; NECHELES, Fabian. *Microteaching and teacher training—a refined version.* Teacher Education Center, University of Chicago, June 1970.

HAINES, A. C., Editor. *Concern for the individual in student teaching.* Dubuque, Iowa, William C. Brown, 1963.

HEILIG, Bruno. *A study in the analysis and training of teacher behaviour.* Paper presented at the International Microteaching Symposium, University of Tübingen, April 1972.

HOFFMAN, F. E.; DRUGER, M. Relative effectiveness of two methods of audio-tutorial instruction in biology. *Journal of research in science teaching,* vol. 8, no. 2, p. 149-156, 1971.

HURST, R. N.; POSTLETHWAIT, S. N. Minicourses at Purdue: an interim report. *The use of modules in college biology teaching,* Publication 31, Biological Sciences, March 1971, p. 29-38.

KAGAN, N. et al. *Studies in human interaction. Interpersonal process recall stimulated by videotape.* East Lansing, Michigan, Michigan State University College of Education, Educational Publication Services, 1967.

KAYE, A. R.; PENTZ, M. J. *Integrated multi-media systems for science education which achieve a wide territorial coverage.* Paper presented at the Unesco Conference on the Utilization of Educational Technology in the Improvement of Science Education, Paris, September 1972.

KERSH, Bert Y. The classroom simulator. *Journal of teacher education,* vol. 13, p. 109-10, March, 1962.

KERSH, B. Y. *Classroom simulation: further studies on dimensions of realism.* Final Report Title VII Project 5-0848, National Defense Education Act of 1958. Department of Health, Education and Welfare, 1965.

KLABENES, R.; SPENCER, C. The video in-service program helpful help for classroom teachers. *Educational technology teacher and technology supplement,* vol. 1, no. 1, p. 521-522, March 1970.

KOLB, D. A.; WINTER, S. K.; BERLEW, D. E. Self directed change: two studies. *Journal of applied behavioral science,* 1968.

KRATHWOHL, D. R.; BLOOM, B. S.; MASIA, B. B. *Taxonomy of educational objectives: the classification of educational goals,* Handbook II, Affective domain. New York, David McKay Co., 1964.

KRUMM, Hans-Jürgen. *Interaction analysis and microteaching for the training of modern language teachers.* Paper presented at the International Microteaching Symposium, Tübingen, April 1972.

LANDSHEERE, G. de. *Projects in course at the University of Liège Experimental Pedagogy Laboratory.* May 1971.

LE BARON, Walt, System Development Corporation. *Analytic summaries of specifications for model teacher education programs.* Washington, D.C., U.S. Office of Education, National Center for Educational Research & Development, 1969.

LEHMAN, D. L. *Simulation in science—a preliminary report on the use of evaluation of role playing in the preparation of secondary school student teachers of science.* Paper presented at the American Assoc. for the Advancement of Science meeting, Washington, D.C. December 1966.

MAGER, R. F. *Developing attitudes toward learning.* Palo Alto, Fearon Publishers, 1968.

MAGER, R. F.; BEACH, K. M. *Developing vocational instruction.* Palo Alto, Fearon Publishers, 1968.

MAGER, R. F., *Preparing objectives for programmed instruction.* San Francisco, California, Fearon, 1962.

MCADA, H. W. *A multi-media approach to chemistry laboratory instruction.* Unpublished doctoral dissertation. The University of Texas at Austin, 1966.

MCALEESE, W. R.; UNWIN, D., A selected survey of microteaching; *Programmed learning and educational technology;* January 1971, vol. 8, no. 1, p. 10-21.

MCDONALD, R. L.; DODGE, R. A. Audio-tutorial packages at Columbia Junior College. *The uses of modules in college biology teaching.* Publication no. 31, Commission on Undergraduate Education in the Biological Sciences, March 1971, p. 45-52.

MCINTOSH, N. E. Research for a new institution: the Open University. *Innovation in higher education* (Edited by Colin F. Page and Harriet Greenaway). London, Society for Research into Higher Education, 1972.

MEDLEY, D. M.; MITZEL, H. E. Measuring classroom behavior by systematic observation. *Handbook of research on teaching* (Edited by N. L. Gage), Chicago, Rand McNally, 1963, p. 247-328.

MEIER, John H. Long distance microteaching. *Educational television,* November 1968, p. 20.

MEIER, J. H.; BRUDENELL, G. Remote training of early childhood educators, Report of a Title XI Institute of the National Defense Education Act. Greeley, Colorado State College, July 1968.

MEIERHENRY, W. C.; POSTLETHWAIT, S. N. New and creative uses of media in teacher education. *Frontiers in teacher education,* AACTE 19th Yearbook, Washington, D.C. 1966, p. 314-321.

MILLER, G.; GALANTER, E.; PRIBRAM, K. H. *Plans and the structure of behavior.* New York, Holt, 1960.

MOWRER, O.H. *Learning theory and the symbolic process.* New York, Wiley, 1960.

NIELSON, G. *Studies in self confrontation.* Copenhagen, 1972.

NOVAK, J. D. Relevant research on audio-tutorial methods. *School science and mathematics,* vol. 70, no. 9, p. 777-784, December 1970.

NUTHALL, G. *A comparison of the use of microteaching with two types of pupils—10-year-old pupils and peers acting as pupils.* Paper presented at the International Microteaching Symposium, University of Tübingen, April 1972.

OECD (Organization for Economic Co-operation and Development). Center for Educational Research and Development. *Transfer of curriculum development projects and learning systems (Programme Area II—Project 2).* Report of the Conference of the International Transfer of Microteaching Materials, University of Stirling, May 1972.

ONDER J. J. Personal change through self-confrontation. *Education broadcasting review,* vol. 4, p. 2332, August 1970.

OWENS, L.; HATTON, N. Telling it like it is. *Education news,* vol. 13, no. 5, October 1971.

PARAKH, J. S. *A study of relationships among teacher behavior, pupil behavior, and pupil characteristics in high school biology classes.* Western Washington State College, Bellingham, 1967.

PARAKH, J. S. *To develop a system for analyzing the reaction of teachers and students in biology classes.* Cornell University, Ithaca, 1965.

PERLBERG, A.; TINKHAM, R.; NELSON R., *The use of videotape recorders and microteaching techniques to improve instruc-*

tion in vocational-technical programs in Illinois, Part II: In-service study (Mimeo), College of Education, University of Illinois, 1968.

PERLBERG, Arye, *et. al.* The use of the Technion Diagnostic System (T.D.S.) and microteaching techniques in modifying teacher behavior. Paper presented at the Annual Meeting of the A.E.R.A., Chicago, 1972.

PERROTT, E. A break with tradition—Stirling University's plan for the preparation of graduate teachers. *Scottish journal of educational studies,* vol. 1, June 1967.

PERROTT, E. *Course design and microteaching in the context of teacher training.* Paper presented at the International Microteaching Symposium, University of Tübingen, April 1972.

PERROTT, E.; DUTHIE, J. H. Television as a feedback device: microteaching *Educational television international,* vol. 4, no. 4; p. 258-261, December 1970.

PERROTT, E.; DUTHIE, J. H. University television in action—microteaching *Univ. television newsletter,* no. 7, 1969.

POPHAM, James, W. *Instructional objectives exchange: a project of the Center for the Study of Evaluation.* Los Angeles, The Instructional Objectives Exchange, Center for the Study of Evaluation, University of California, 1970.

POSTLETHWAIT, S. N.; MERCER, F. V. Integrated multi-media systems for science education (excluding television and radio broadcasts). Paper presented at the Unesco Conference of the Utilization of Educational Technology in the Improvement of Science Education, Paris, September 1972.

POSTLETHWAIT, S. N.; NOVAK, J.; MURRAY, H. T. *The audio tutorial approach to learning.* Minneapolis, Burgess, 1969, p. 149.

ROSENSHINE, Barak. *Teaching behaviors and students achievement.* International Association for the Evaluation of Educational Achievement, I.E.A. studies no. 1. National Foundation for Educational Research in England and Wales, 1971.

SCHUELER, Herbert; LESSER, Gerald S. *Teacher education and the new media.* Washington, D.C., The American Association of Colleges for Teacher Education, 1967.

SEIDMAN, Earl. *A critical look at microteaching.* Paper presented at the University of Massachusetts Microteaching Conference, 1969.

SIMON, Anita; BOYER, E. G., Editors. *Mirrors for behavior: an anthology of classroom observation instruments.* Volumes I-XVI. Philadelphia, Research for Better Schools, Inc., 1970.

SMITH, B. O.; COHEN, S. B.; PEARL, A. *Teachers for the real world.* American Association of Colleges for Teacher Education, 1969.

STAINES, Graham L. A comparison of approaches to therapeutic communications. *Journal of counseling psychology,* vol. 16, no. 5, p. 405-414, 1969.

TEATHER, D. C. B. *Skills analysis as an heuristic device.* Paper presented at the International Microteaching Symposium, Tübingen, April 1972.

TICKTON, Sidney G., Editor. *To improve learning: an evaluation of instructional technology.* New York, Bowker, vol. I, 1970; vol. II, 1971.

TURNEY, C. Microteaching—a promising innovation in teacher education. *Australian journal of education,* vol. 14, no. 2, p. 125-141, June 1970.

TWELKER, P. A. *Prompting as an instructional variable in classroom simulation.* Paper presented at the A.E.R.A. Annual Convention, Chicago, 1966.

URBACH, Floyd, University of Nebraska. Personal correspondence with Frank W. Broadbent, cited in *Simulation in preparing school personnel.* Cruickshank and Broadbent (Editors).

UNITED STATES DEPARTMENT OF HEALTH, EDUCATION AND WELFARE. *Creative developments in the training of educational personnel.* U.S. Dept. of Health, Education and Welfare, Office of Education, Government Printing Office, 1969, p. 52.

VLCEK, C. W. Classroom simulation in teacher education. *Audiovisual instruction,* vol. 11, no. 2, 1966.

VLECK, C. W. Assessing the effect and transfer value of a classroom simulator technique. (Unpublished doctoral dissertation.) Michigan State University, 1965.

WAGNER, Angelika C. *Is practice really necessary? An experimental study on the role of practising vs. cognitive discrimination learning in behavioral change.* Paper presented at the International Microteaching Symposium, April 1972.

WEBB, C.; BAIRD, H.; BELT, D.; HOLDER, F. Description of a large-scale micro-teaching program. *Video processes in teacher programs,* Multi-State Teacher Education Project, Baltimore, Md., September 1968, p. 7-13.

WITTROCK, M. C.; WILEY, David E. *The evolution of instruction.* New York, Holt, Rinehart and Winston, 1970.

WOOD, R. L. Construction of science carrels by elementary education students. *School science and mathematics,* vol. 69, no. 9, p. 791-798, December, 1969.

WRIGHT, E. Muriel J. Teacher-pupil interaction in the mathematics classroom: A sub-project report of the secondary mathematics evaluation project. Technical Report No. 67-5. Minnesota National Laboratory, Minnesota State Department of Education, St. Paul, 1967.

WRIGHT, E. Muriel J., et al. Development of an instrument for studying verbal behaviors in a secondary school mathematics classroom. *Journal of experimental education,* vol. 28, p. 103-21, December 1959.

YOUNG, David B. The modification of teacher behavior using audio video-taped models in a micro-teaching sequence. *Educational leadership,* vol. 26, p. 394-95, 397, 399, 401, 403, January 1969.

ZIFREUND, Walther. *The use of microteaching and interaction analysis in the Tübingen Teacher Education Programme.* Paper presented at the International Microteaching Symposium, April 1972.

ZIMBARDO, P. G. *The cognitive control of motivation.* Scott, Foresman and Company, 1969.

Educational technology applied to the learning of science in developing countries

by Isaias Raw

Education Research Center, Massachusetts Institute of Technology,
Cambridge, Mass. 02138, U.S.A.

Summary

In analyzing the use of educational technology in science education, a distinction is made between education and passive acquisition of information.

The role sought for educational technology in science education is that of providing active learning and individualized instruction, so as to cope with different learning styles, learning rates and varieties of interest that could best be satisfied by tailor-made curricula. This is a goal barely touched upon by the major stream of educational technology which is often conceived merely as a large-scale replacement for the lecture.

As most schools in the developing countries are devoid of even the simplest slide projectors, priority there should be given not to teacher-aiding tools, but to low-cost laboratory equipment and other materials for student use.

The developing countries could put more effort into designing sets of multi-media units combining printed matter, short films, slides, and low-cost laboratory kits.

In trying to apply educational technology to the field of science education, we are seeking ways to better educate future generations, believing that education, and especially science education, can provide a major contribution to development. For instance, we are interested in the education of some intellectual élites: the future scientists and those capable of using science and technology to answer society's need. We also have another target, another élite, generally with no acceptable understanding of science: the future ruling classes of our societies. We assume that all men and women must have a scientific background in order to understand life as they experience it. If the political system of their country recognizes their right to influence the shape of the society in which they live (as it is generally stated), they will have to participate in decisions on issues that have strong scientific implications.

We must make a distinction between science education and the transmission of information. Science education can and should encompass both information and its application to the life of the students, and so change their attitudes in the face of a number of daily problems. It is clear that the students' own problems are the best motivation for learning science. But we cannot accept as science education the simple transmission of information, in the form of new dogmas to replace some of the superstitions of the past.

I do not deny the possibility of training by transmitting terminal behaviour. If we want to train people to perform some limited task (e.g., teaching housewives to change fuses or take the child's temperature), we can do so by a number of teaching methods: individual tutoring, school teaching, or teaching by mass media. A child can thus learn to manipulate the many controls of a television set, factory workers can be trained, or a dogma can be transmitted (for instance some health and genetical precepts contained in certain religions). But this is quite different from the learning of science.

Educational technology has a large capability to transmit information; but this transmission of information does not necessarily ensure its utilization to change behaviour. It may make an important contribution, but it cannot replace actual manipulation and performance of the operations for which training is intended. Just looking at visual aids, films and television presentations of surgical operations will not train someone to perform them.

The goal of education is learning, not teaching. The professionals in this field call themselves 'teachers', and they frequently do not realize that their role is not 'to teach' but 'to facilitate learning'. They have in common with some 'holy men' the belief that they are a sort of living source of knowledge which students must accept and absorb unquestioningly. Even today, the majority do not fully acknowledge the fundamental impact of the most established educational technological tool: Gutenberg's invention.

We recognize that a teacher, lecturing to classes of forty to several hundreds, cannot provide good conditions for students to learn science. So we look for situations and tools that could make this possible. Good educators are concerned with the need to provide education of high quality to all. Whereas we are looking for good individualized science education, the first answer of the educational technology industry has been an attempt to replace the teacher as a lecturer. The whole of the gadgetry has amounted to little more than a 'loud-speaker' with the lecturer on film or tape, and transmitted by television, cable, computer or satellite. If the role of the school were to provide lectures, and the students' role to polish chairs as they passively absorbed the lectures, it is possible that some of the gadgetry now available would be as efficient as some of today's teachers.

It is easy to understand the interest of less developed countries in educational technology. There is a natural desire to jump stages in development, and there is the glamour of beautiful and complex machinery. Also, there is a real concern with the limited number of teachers available as compared with an increasing number of students. Educational planners are also concerned about the cost of teachers. Buildings are long-range investments, based frequently on long-term loans. Teachers are week-by-week civil servants who must be paid, and who account for over ninety per cent of the education budget. In promoting educational technology, the message that is passed on to the Ministries of Education is that it will replace teachers, being more efficient and costing less. However, advanced countries using such technology, know that the ratio of teachers to students does not decrease. [1]

In the same way, as educational technology does not replace the teacher, teacher and technology together do not replace the direct exposure of students to science in the laboratory. This goes for every level of science education, from primary school to university. The needs change, but at all levels it is possible to learn by discovery, using simple low-cost materials.

For many years I have been trying to convey the idea that low-cost laboratory equipment is a fundamental and essential part of educational technology. It is much more sophisticated than a film of a lecture with a few drawings or classical demonstrations. It is a complex operation; so complex that, even when the development is available, wide-scale utilization still poses a problem—as has been the case in some places with the very ingenious PSSC equipment. It requires low-cost, large-scale production, which is the economic basis for making it possible for the student to experiment. It implies an entrepreneurship that will pay attention to the proper selection of materials, tooling, mass-production, building-up of a market, training teachers, persuading educational authorities, efficient distribution, etc. Still, most of the attempts are no more than an amateurish effort to make a few dozen sets of equipment, or inefficient dreams leading to bureaucratic structures.

The genuine concerns of some earnest activists about introducing educational technology always find an echo in the sensitivity of administrators to fancy gadgetry and novelties which bring large publicity, inaugurations and, not infrequently, personal benefits. To those activists I would recommend caution, and a very critical attitude. Let me illustrate my meaning with two personal anecdotes. In the early sixties, I was invited to participate in a conference to discuss the future developments of college education in the United States. The host institution was a leading university, and the building we used was full of teaching machines of all shapes and models, from little plastic gadgets to talking typewriters—none of them was more than one year old, and, as a critical educator would forecast, all were completely obsolete. Not so serious for a wealthy university in a rich country; but imagine a small developing country that embarked on a wide 'progressive' change of its whole school system! On my way back, I was asked to discuss a 'brilliant' proposal with a company that operates in the field of educational technology: for a few hundred thousand dollars, two trailers would take a model equipped classroom to the remote corners of the continent, in order to train teachers in the use of the new technology. I had never imagined that one would need to take a classroom around to visit other classrooms ... and, besides, did they know that there were no roads that permitted this large trailer to travel ...?

We educators dealing with the developing world cannot afford to make mistakes with our scarce funds, or with the present generation of students who represent the future. We have to establish our priorities, and I do not believe that they include 'to provide (teachers) with time-saving ways to drill students' [1]. Given the present trend towards rapid obsoleteness and fast innovations, common sense dictates that we cannot agree to investing our limited resources in hardware before the software is available and proved to be efficient.

There are a number of economic implications. It is common to jump into long-range loans. This is only justified if we expect a large multiplying effect and a significant impact by the time the loan has to be paid. Frequently, this is forgotten, and a long-range loan becomes 'bread and circus now ... your son will pay later'. Furthermore, we must realize that the initial investment is small compared with the running costs. Several regional educational instruction centres have been proposed for the United States; each will represent

---

1. The experience thus far with the new technology (applied to instruction), however, as compared with the hopes of its early supporters, indicates that it is (a) coming along more slowly, (b) costing more money, and (c) adding to rather than replacing older approaches—as the teacher once added to what the family could offer, as writing then added to oral instruction, as the book later added to the handwritten manuscript. Carnegie Commission on Higher Education. *The fourth revolution: Instructional technology in higher education.* McGraw-Hill Book Co., 1972.

an investment of thirty-five million dollars, and will cost one hundred and fifty million dollars a year to maintain in operation.

We are looking for technology to help us in coping with more and many different types of students. We would like to satisfy their varied interests, and to provide almost tailor-made courses for each one, on the basis of what he is looking for and what would be most suitable for him. We want to be able to avoid our rigid curricular structure and scheduling, as we do not intend to keep on boring the fast-learning students or failing the slow-paced ones. We are not running a speed contest.

Our curriculum, aims, materials and styles must therefore be very variable, even for the same subject. In looking at the various technologies and means available for education, no single one is the best or the most suitable. Each one serves specific purposes, each one will have a limited life-time, just as subject matter and curricula have, there are no final answers, methods and hardware. We are constantly on the move, hopefully for the better.

Even the traditional printed matter—the book—is still a powerful tool. It can transmit ideas, symbols and images. The images do not have the glamour of the slides, but neither are they flashed by the teachers at such a pace that students cannot cope with them or take data from them. The book has its advantages—it is of low cost and portable; students use it at any pace they choose, they can to return to any point at any time, they can stop at any point, take it home or to the park and even use it during their travel to school.

But in spite of the fact that books have been used for so many centuries, we are still discovering ways to improve them. Very often, they are not properly written as a self-teaching device. With the explosion of knowledge, they expanded to a degree that made them unmanageable, given the time which students had for reading. Furthermore, the student is told to acquire a book, and then receives the clear but unspoken message from the teacher: It is what I say that counts ... so forget Gutenberg and take notes ....

More recently, writers and publishers have moved away from large treatises which are difficult to up-date, to buy, and even to hold, towards small paperbacks. Curriculum innovators have been planning for some time now to move towards small modules that can be assembled in a variety of ways, and be written with many parallel alternatives in mind. The marketing of these will be a new innovation in the printing business. The Open University is an important attempt in this direction. Although many think that the Open University has television as its main teaching device, it really provides very modern correspondence courses.

One fault is that the usual printed material tends to be passive, but this is not necessarily so and books can be written so as to demand active participation by the student. An important attempt to find new ways of using printed materials for self-instruction, with feedback and reinforcement, was programmed instruction. The orthodox Skinnerian form of programmed instruction, dividing concepts in an infinite number of small steps, demanding that the great majority, if not the totality of the students achieve the right answer, resulted in sterile, boring, unimaginative bulky books. This type of programmed instruction is dead, but it made some important contributions. One was the complete analysis of the educational objectives of the teaching unit. New forms of programmed instruction, with reasonable steps, new forms of presentation, branching, combination with laboratory work, are now arising. Meanwhile, the principles of programmed instruction are providing new models for carefully written, self-paced, self-teaching materials, accompanied by proper guidance to students and self-evaluation tests.

Computers are being more and more used in many different experimental programmes. In many cases they are just a more complex form of programmed instruction texts, with higher levels of branching. Other, more imaginative uses, like simulations, have been designed. But, in many circumstances, computer education remains a myth of debatable educational value, even apart from any consideration of cost. This does not change the importance of the computer as a tool in science, mathematics and administration. If universities in developing countries have staff for computer science, it would be recommendable, in my opinion, that they keep abreast of computer instruction, and conduct limited experiments. In the meantime, the hardware and its usefulness will progress to the point where it will deserve closer and more intensive examination by the less developed countries.

The spoken word is the most traditional teaching tool, and a perfect weapon for encouraging student passivity. Frequently, in the traditional cultures, lectures are not supposed to be interrupted by questions. To provide such lectures in recorded form often merely means eliminating the main actor while retaining the inherent defects. Still, in some multimedia packages, such as autolectures, the recorded instruction is a component of a complex learning situation that may lead to active learning.

New sorts of library items are being created by recording important lectures and even discussions among leading scientists. They are important teaching tools, bringing in the motivation arising from hearing, first hand, men who have made important contributions. Unfortunately, those materials completely lose their 'flavour' when translated.

Sound tapes may find their role in some special applications. There are some tapes containing data of physical experiments between Earth and Moon that can be analysed by students. A lecture on sound could be produced on a tape, and analysed both by listening and by observing an osciloscope. Movies can be as expendable, passive and boring as any other teaching device, if not worse. Their transmission by television does not change this. Artistic interludes and background music have never enabled students to endure them for more than half an hour at a time. To crowd lots of material in a single movie, as is frequently done is to make it completely indigestible.

But movies do not have to be passive. They may present a problem—for discussion; or an experiment that is difficult to perform or to observe—for interpretation. The BSCS films represent such an effort. Furthermore, short films, on 8 mm, can be projected on small acrylic screens or even on a paper or a wall. Eight-millimetre projectors and films are of low cost, and are useful for creating small group discussions or for stimulating individualized instruction. A five-minute film can have the core of what normally is presented in a half-hour movie, and will cost correspondingly less. Sound is frequently unnecessary, and translations are therefore not required. There are very few films that demand active student participation. This could well be an area where developing countries may make valuable contributions.

Another new and important development is the creation of films with computers. It is possible to show important dynamic models in the fields of mathematics and physics, as represented by the films prepared at the MIT Education Research Center. Computer programmes exist today which create on film rotating structures of complex molecules (proteins, nucleic acids, polymers) that can be examined by students.

In this review, I have strongly stressed the limitations of some of the available instructional technology for science education. If I have over-emphasized, it was to try to impress the need for caution before adopting what is at present available.

The development of multi-media forms of individualized instruction, by confining the use of printed materials, short films, slides and low-cost laboratory kits is, in my opinion, an area of choice for less developed countries trying to provide better education for more students, in a variety of contents and styles. I recently tried to persuade some United States teachers to use the low-cost Brazilian science kits. The answer I got was that the schools in the United States could not afford the cost of individualized materials, as the normal budget for laboratory equipment was of the order of two dollars per high-school student. If this is a problem for the schools in the United States, it is easy to imagine what it is in poorer countries. Creative leaders in science education in the developing countries will have to face it, and design and produce learning materials that are of high scientific and educational quality but also cheap enough so that all students and pupils can work with them.

Appendix

Participants: Meeting of experts on the utilization of educational technology in the improvement of science education. Unesco, Paris, 13-16 September 1972

Invited experts:

John C. H. Ball	Schools Broadcasting, Voice of Kenya, P.O. Box 30456, Nairobi, Kenya.	Gunnar Markesjö	Department of Applied Electronics, Royal Institute of Technology, KTH, Fack, 10044, Stockholm 70, Sweden.
Arthur I. Berman	Institute for Studies in Higher Education University of Copenhagen, Fiolstraede 24, 1171 Copenhagen K, Denmark.	M. J. Pentz	Faculty of Science, Open University, Walton, Bletchley, Bucks., United Kingdom.
Michel Y. Bernard	Conservatoire National des Arts et Métiers, 229 Avenue Victor-Hugo, 92 Clamart (Hauts-de-Seine), France.	Arye Perlberg	Department of Teacher Training, Technion—Israel Institute of Technology Haifa, Israel.
Donald L. Bitzer	Computer-Based Education Research Laboratory, University of Illinois at Urbana-Champaign, Urbana, Illinois 61801, U.S.A.	Elizabeth Perrott	OECD International Microteaching Project, Lonsdale College, University of Lancaster, Bailrigg, Lancaster, United Kingdom.
Klaus Haefner	Project CUU, Universität Freiburg, Schänzlestrasse 9-11, 78 Freiburg i.Br., Federal Republic of Germany.	S. N. Postlethwait	Purdue University, Department of Biological Sciences, Lafayette, Indiana 47907, U.S.A.
André Jones	Centre Interfacultaire IMAGO, Université Catholique de Louvain, Celestijnenlaan 200-C, 3030 Héverlé-Louvain, Belgium.	R. A. Sutton	Physics Interface Project, Department of Physics, University College, Cardiff, P.O. Box 78, Cardiff CF1 1 XL, United Kingdom.
A. R. Kaye	The Open University, Institute of Educational Technology, Walton Hall, Walton, Bletchley, Bucks., United Kingdom.	Nina Talyzina	Department of Psychology, Lomonosov State University, 18 Marx Avenue, Moscow, U.S.S.R.
Yves Le Corre	Université-Paris VII, 2 Place Jussieu, 75005 Paris, France.	J. Valérien	Département des Actions Educatives, OFRATEME, 31 Rue de la Vanne, 92 Montrouge, France.
G. O. M. Leith	Department of Research and Development in Education, Rijksuniversiteit te Utrecht, Maliebaan 5, Utrecht, Netherlands.	J. R. Zacharias	Physics Department, Massachusetts Institute of Technology, Cambridge, Mass. 02139, U.S.A.

New trends in the utilization of educational technology for science education

ICSU Committee on Science Teaching:

M. Matyas	Chairman of the Committee, Czechoslovak Academy of Science, Institute of Solid State Physics, Cukrovarnicka 10, Prague 6, Czechoslovakia.
Dennis G. Chisman	Secretary of the Committee, Centre for Educational Development Overseas (CEDO), Tavistock House South, Tavistock Square, London WC1H 9LL, United Kingdom.
Albert V. Baez	La Ranchería, Carmel Valley, California 93924, U.S.A.
Isaias Raw	Massachusetts Institute of Technology Education Research Center, Room 20-C-004, Cambridge, Mass. 02139, U.S.A.

Unesco

S. O. Awokoya	Director, Department of Science Teaching and of Technological Education and Research.
Harold A. Foecke	Director, Division of Science Teaching.
E. Brunswic	Division of Methods, Materials and Techniques.
N. Joel	Division of Science Teaching.
D. Behrman	Press Division, Office of Public Information.

[A.35] ED. 73/D72/A